Topological Foundations of Electromagnetism

Second Edition

Topological Foundations of Electromagnetism

Second Edition

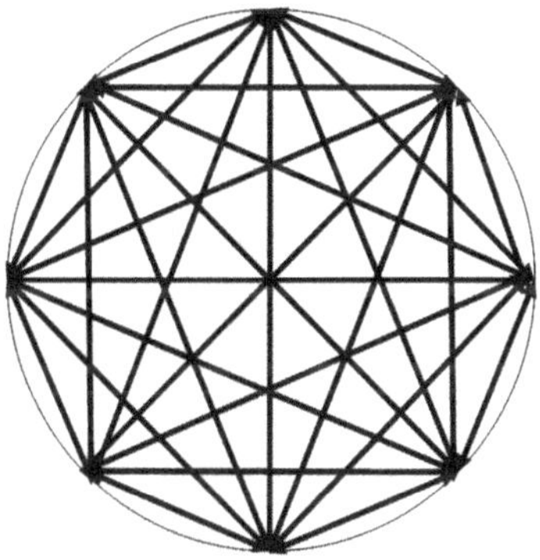

Terence William Barrett

BSEI, USA

World Scientific

NEW JERSEY • LONDON • SINGAPORE • BEIJING • SHANGHAI • HONG KONG • TAIPEI • CHENNAI • TOKYO

Published by

World Scientific Publishing Co. Pte. Ltd.
5 Toh Tuck Link, Singapore 596224
USA office: 27 Warren Street, Suite 401-402, Hackensack, NJ 07601
UK office: 57 Shelton Street, Covent Garden, London WC2H 9HE

British Library Cataloguing-in-Publication Data
A catalogue record for this book is available from the British Library.

First published 2022 (hardcover)
Reprinted 2025 (in paperback edition)
ISBN 9789819819669 (pbk)

TOPOLOGICAL FOUNDATIONS OF ELECTROMAGNETISM
Second Edition

ISBN 978-981-125-329-4 (hardcover)
ISBN 978-981-125-330-0 (ebook for institutions)
ISBN 978-981-125-331-7 (ebook for individuals)

For any available supplementary material, please visit
https://www.worldscientific.com/worldscibooks/10.1142/12755#t=suppl

Desk Editor: Nur Syarfeena Binte Mohd Fauzi

Typeset by Stallion Press
Email: enquiries@stallionpress.com

Preface

Maxwell's equations are foundational to electromagnetic theory. They are the cornerstone of a myriad of technologies and are basic to the understanding of innumerable effects. Yet there are some effects or phenomena, that cannot be explained by conventional Maxwell theory. This book examines these anomalous effects and shows that they can be interpreted by a Maxwell theory that is subsumed under gauge theory. Moreover in the case of these anomalous effects, and when Maxwell theory finds its place in gauge theory, the conventional Maxwell theory must be extended, or generalized, to a non-Abelian form.

The tried-and-tested conventional Maxwell theory is of Abelian form. It is correctly and appropriately applied to, and explains, the great majority of cases in electromagnetism. What, then, distinguishes these great majority of cases from the anomalous phenomena, aforementioned? It is the thesis of this book that it is the topology, the theory of continuous transforms, of the spatio-temporal situation that both distinguishes the two classes of effects or phenomena, and the topology is also the final arbiter of the correct choice of group algebra: Abelian or non-Abelian, to use in describing anomalous effects.

Therefore, the most fundamental, or basic, explanation of electromagnetic phenomena and their physical models lies not in differential calculus or group theory, useful as they are, but rather in the topological description, i.e., the theory of continuous transforms, of the (spatio-temporal) situation. Thus, this book shows that only after the topological description is provided can our understanding move to an appropriate application of differential calculus or group theory.

But the journey toward topology with a movement toward group description should not end there. There are further interdependencies that permit trades and translations across normally considered separate disciplinary descriptions and pictures. In particular, a group symmetry description addressing energy conservation also includes the dynamic of symmetry breaking and thereby of energy release; and connections and interdependence can be drawn between topology and (a) the theory of groups and group symmetry, (b) the thermodynamic concepts of energy conservation and release by symmetry breaking, (c) entropy, permitting a connection to (d) information (negentropy) and (e) communication systems. Such connections and translations permit *multidisciplinary* approaches in which dynamic changes and trades in one disciplinary description effect simultaneous changes in another, as opposed to *interdisciplinary* translations. Therefore, this book also addresses all of the above, which include effects of radio frequency transmission modulation and light beam modulation.

Furthermore, the application and influence of physical laws are not confined to their discipline of provenance, hence the need of a multidisciplinary approach to seemingly single discipline problems. It is thus important to realize that changes in one field of study's description always imply changes in another, and we have attempted to depict some linkages below in pictorial form. This figure in no way is intended to be canonical, but as a reminder of the multidisciplinary and cross-disciplinary nature of the topics addressed and that the approach taken here addresses multidisciplinary dynamics, rather than single disciplinary reductions. The view presented here is that no discipline is unconnected to other disciplines and their physical laws, with topology as the concert master, or the one discipline that rules them all. This book attempts to demonstrate this primacy.

The appropriate algebra for this approach is not that of Josiah Willard Gibbs, but rather that of William Kingdon Clifford, who in 1873 invented an algebra unifying Cartesian coordinates with complex numbers, now known as Clifford algebra or geometric algebra. There are then links to quaternion algebra and Pauli matrices permitting a fuller description of modulated electromagnetic fields.

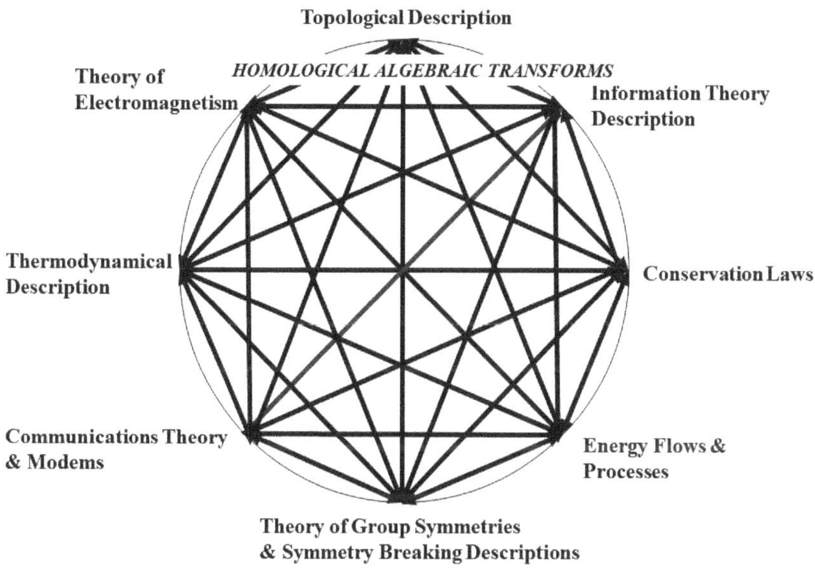

Homological topology: the one discipline that rules them all.

Contents

Chapter 1

Electromagnetic Phenomena Not Explained by Maxwell's Equations[1]

Overview

Conventional Maxwell's theory is a classical linear theory in which the scalar and vector potentials appear to be arbitrary and defined by boundary conditions and choice of gauge. Conventional wisdom in engineering is that potentials have only mathematical, not physical, significance. However, besides the case of quantum theory, in which it is well known that the potentials are physical constructs, there are a number of physical phenomena — *both classical and quantum mechanical* — which indicate that the $\mathbf{A}_\mu$ fields, $\mu = 0, 1, 2, 3$, *do* possess physical significance as many-to-one operators or gauge fields, in precisely constrained topologies.

[1]Barrett, T.W. Maxwell's theory extended. Part I. Empirical reasons for questioning the completeness of Maxwell's theory — effects demonstrating the physical significance of the A potentials, *Ann. Fondation Louis de Broglie*, 15, 143–183, 1990.

———, Maxwell's theory extended. Part II. Theoretical and pragmatic reasons for questioning the completeness of Maxwell's theory, *Ann. Fondation Louis de Broglie*, 12, 253–283, 1990.

———, Electromagnetic phenomena not explained by Maxwell's equations in Lakhtakia, A. (Ed.) *Essays on the Formal Aspects of Maxwell's Theory*, World Scientific, Singapore, pp. 6–86, 1993.

———, Sagnac effect, in Barrett, T.W. and Grimes, D.M., (Eds) *Advanced Electromagnetism: Foundations, Theory, Applications*, World Scientific, Singapore, pp. 278–313, 1995.

For some time, the present writer has been engaged in showing that the spacetime topology defines electromagnetic field equations[1] — whether the fields be of force or of phase. The unifying concept is the classification of spaces and mappings into discrete topological spaces (Dyson, 1987). That is to say, the premise of this enterprise is that a set of field equations involving $\mathbf{A}_\mu$ are only valid physical constructs with respect to set-defined topological descriptions of the physical spaces. In particular, the $\mathbf{A}_\mu$ potentials are not just mathematical conveniences, but — *in certain well-defined spaces* — are measurable, i.e., physical. In electromagnetism, those situations, in which the $\mathbf{A}_\mu$ potentials *are measurable and physical*, are defined by a topology, the transformation rules of which are describable by the SU(2) group or higher-order groups; and those situations, in which the $\mathbf{A}_\mu$ potentials *are not measurable or physical*, are defined by a topology the transformation rules of which are describable by the U(1) group (Belavin *et al.* 1975).

Maxwell's original theory, reinterpreted by others into the linear and conventional form, is of U(1) symmetry with Abelian commutation relations. It can be extended to include physically meaningful $\mathbf{A}_\mu$ effects by its reformulation in complex spatial configurations or SU(2) and higher symmetry forms. The commutation relations of conventional classical Maxwell theory are Abelian. When extended to SU(2) or higher symmetry forms, Maxwell's theory possesses non-Abelian commutation relations, and addresses many-to-one mappings, e.g., $2\rightarrow1$, with the potentials acting as many-to-one operators.

An adapted Yang–Mills interpretation of *low-energy fields* is applied in the following — an adaptation previously applied only to *high-energy fields.* This adaptation is permitted by precise definition of the topological boundary and spatial conditions of those low energy electromagnetic fields. The Wu–Yang interpretation of Maxwell's theory implicates the existence of magnetic monopoles and magnetic charge. As the classical fields considered here are low-energy fields, these theoretical constructs are *pseudoparticles, or instantons, low-energy monopoles and charges*, rather than high-energy monopoles and magnetic charges (cf. Balavin *et al.* 1975; Atiyah, 1988).

Instantons and the related solitons are particle-like solutions of classical field equations of SU(2) symmetry form as is the classical Yang–Mills equation. Unlike solitons, instantons are structures in time. They are solutions to the equations of motion on a Euclidean spacetime and can be considered as critical points of the action.

Although the term classical Maxwell's theory has a conventional meaning, this meaning actually refers to the interpretation of Maxwell's original writings by Heaviside, Fitzgerald, Lodge and Hertz. These *later interpretations of Maxwell* actually depart in a number of significant ways from Maxwell's original formulation. In this original formulation, *Faraday's electrotonic state, the A-field, was central*, making this prior-to-interpretation, Maxwell formulation compatible with Yang–Mills theory, and naturally extendable.

This reinterpreted classical Maxwell's theory is, as stated, a linear theory of U(1) gauge symmetry. The mathematical dynamic entities called solitons can be either classical or quantum mechanical, *linear or non-linear* (cf. Barut, 1991; Sharaarawi *et al.* 1990) and describe electromagnetic waves propagating through media. However, solitons are of SU(2) symmetry form (Palais, 1997). In order for the conventional reinterpreted classical Maxwell's theory of U(1) symmetry to describe such entities, the theory must be extended to SU(2) form.

This recent extension of soliton theory to linear equations of motion together with the recent demonstration that the non-linear Schrödinger equation and the Korteweg–de-Vries equation — equations of motion with soliton solutions — are reductions of the self-dual Yang–Mills equation (SDYM) and pivotal in understanding the extension of Maxwell's U(1) theory to higher-order symmetry forms such as SU(2) (Mason *et al.* 1989). Instantons are solutions of SDYM equations which have minimum action. The use of Penrose's SDYM twistor correspondence for universal integrable systems (e.g., Ward and Wells, 1990) means that instantons, twistor forms, magnetic monopole constructs and soliton forms, all have a pseudoparticle SU(2) correspondence.

Differential equations or a set of differential equations describe a *system* and its evolution. Group symmetry principles summarize both invariances and the laws of nature independent of a system's specific dynamics. It is necessary that the symmetry transformations be continuous or specified by a set of parameters which can be varied continuously; and the symmetry of continuous transformations leads to conservation laws.

There are a variety of special methods used to solve ordinary differential equations. It was Sophus Lie (1842–1899) in the 19th century who showed that all the methods are special cases of integration procedures which are based on the invariance of a differential equation under a continuous group of symmetries. These groups became known as Lie groups[2]. A symmetry group of a system of differential equations is a group which transforms solutions of the system to other solutions. In other words, there is an invariance of a differential equation under a transformation of independent and dependent variables. This invariance results in a diffeomorphism[2] on the space of independent and dependent variables, permitting the mapping of solutions to solutions.

This relationship was made more explicit by Emmy (Amalie) Noether (1882–1935) in theorems now known as *Noether's theorems* (Noether, 1918) which related symmetry groups of a variational integral to properties of its associated Euler–Lagrange equations. The most important consequences of this relationship are that (i) conservation of energy arises from invariance under a group of time translations; (ii) conservation of linear momentum arises from invariance under (spatial) translational groups; (iii) conservation of angular momentum arises from invariance under (spatial) rotational groups; and (iv) conservation of charge arises from invariance under change of phase of the wave function of charged particles. Conservation and group symmetry laws have been vastly extended to other systems of equations, e.g., the standard model of modern high energy physics, and also, of importance to the present interest: soliton equations. For example, the *Korteweg–de Vries* "soliton" equation (Korteweg and de Vries, 1895) yields a symmetry algebra spanned

[2]Diffeomorphism: an invertible function that maps one differentiable manifold to another such that both the function and its inverse are smooth.

by the four vector fields of (i) space translation; (ii) time translation; (iii) Galilean translation; and (iv) scaling.

Prolegomenon A: Physical Effects Challenging a Conventional Maxwell Interpretation

A number of physical effects strongly suggest that the Maxwell field theory of electromagnetism is incomplete. Representing the independent variable as x, the dependent variable as y, and the influence as $x \rightarrow y$, these effects address: field(s)→free electron (F→FE), field(s)→conducting electron (F→CE), field(s)→particle (F→P), wave-guide→field (WG→F), conducting electron→field(s) (CE→F) and rotating frame→field(s) (RF→F) interactions. A non-exhaustive list of these experimentally observed effects, all of which involve the $\mathbf{A}_\mu$ four potentials (vector and scalar potentials) in a physically effective role, includes:

1. The *Aharonov–Bohm* and *Altshuler–Aronov–Spivak effects* (F→FE & F→CE): Ehrenberg and Siday, Aharonov and Bohm, and Altshuler, Aronov and Spivak predicted experimental results by attributing physical effects to the $\mathbf{A}_\mu$ potentials. Most commentaries in classical field theory still show these potentials as mathematical conveniences without gauge invariance and with no physical significance.
2. The *Topological phase effects* of Berry, Aharonov, Anandan, Pancharatnam, Chiao and Wu (WG→F) and (F→P): in the WG→F version, the polarization of light is changed by changing the spatial trajectory adiabatically. The Berry–Aharonov–Anandan phase has also been demonstrated at the quantum, as well as the classical level. This phase effect in parameter (momentum) space is the correlate of the Aharonov–Bohm effect in metric (ordinary) space, both involving adiabatic transport.
3. The *Josephson effect* (CE→F): both at the quantum and macrophysical level, the free energy of the barrier is defined with respect to an $\mathbf{A}_\mu$ potential variable (phase).
4. The *quantum Hall effect* (F→CE): gauge invariance of the $\mathbf{A}_\mu$ vector potential, being an exact symmetry, forces the addition of

a flux quantum to result in an excitation without dependence on the electron density.

5. The *De Haas–Van Alphen effect* (F→CE): the periodicity of oscillations in this effect is determined by $\mathbf{A}_\mu$ potential dependency and gauge invariance.
6. The *Sagnac effect* (RF→F): exhibited in the well-known and well-used ring laser gyro, this effect demonstrates that the Maxwell theory, as presently formulated, does not make explicit the constitutive relations of free space, and does not have a built-in Lorentz invariance as its field equations are independent of metric.

The $\mathbf{A}_\mu$ potentials have been demonstrated to be physically meaningful constructs at the quantum level (Effects 1–5), at the classical level (Effects 2, 3 and 6), and at relatively long range in the case of Effect 2. In the F→CE and CE→F cases (Effects 1, 3–5), the effect is limited by the temperature-dependent electron coherence length with respect to the device/antenna length.

The Wu–Yang theory attempted the completion of Maxwell's theory of electromagnetism by the introduction of a non-integrable (path-dependent) phase factor (NIP) as a physically meaningful quantity. The introduction of this construct permitted the demonstration of $\mathbf{A}_\mu$ potential gauge invariance and gave an explanation of the Aharonov–Bohm effect. The NIP is implied by the magnetic monopole and magnetic charge theoretical constructs viewed as *pseudoparticles* or *instantons* (Shaarawi *et al.* 1990).

The recently formulated Harmuth Ansatz also addresses the incompleteness of Maxwell theory: an amended version of Maxwell's equations can be used to calculate e.m. signal velocities provided that (a) a magnetic current density, and (b) a magnetic monopole theoretical constructs, are assumed.

Formerly, treatment of the $\mathbf{A}_\mu$ potentials as anything more than mathematical conveniences was prevented by their obvious lack of gauge invariance (Panofsky and Phillips, 1962; Jackson, 1975) However, gauge invariance for the $\mathbf{A}_\mu$ potentials is obtained from situations in which fields, firstly, have a history of separate spatiotemporal conditioning and then, secondly, are mapped in a many-to-one fashion (in holonomy[3]). Such conditions are satisfied by $\mathbf{A}_\mu$ potentials with boundary conditions, i.e., the usual empirically encountered situation. Thus, with the correct geometry and topology (e.g., defined with appropriate boundary conditions) the $\mathbf{A}_\mu$ potentials always have physical meaning. This indicates that Maxwell's theory can be extended by the appropriate use of topological and gauge symmetrical concepts.

The $\mathbf{A}_\mu$ potentials are local operators mapping many-to-one spatiotemporal conditions onto the local e.m. fields. The effect of this operation is measurable as a phase change, if there is a second comparative mapping of differentially conditioned fields in a many-to-one summation. With coherent fields the possibility of measurement (detection) after the second mapping is maximized.

The conventional Maxwell theory is incomplete due to the neglect of (1) a definition of the $\mathbf{A}_\mu$ potentials as operators on the local intensity fields dependent on gauge, topology, geometry and global-boundary-conditions; and (2) a definition of the constitutive relations between medium-independent fields and the topology of the medium.[4] Addressing these issues extends the conventional Maxwell theory to cover physical phenomena which cannot be presently explained by that theory.

---

[3]Holonomy: Given a smooth closed curve C on a surface M, and picking any point P on that curve, the holonomy of C in M is the angle by which some vector turns as it is parallel transported along the curve C from point P all the way around and back to point P. Or: a global manifestation of curvature which comes from parallel transport.

[4]The paper by Konopinski (1978) provides a notable exception to the general lack of appreciation of the central role of the **A** potentials to electromagnetism. Konopinski shows that the equations, from which the Lorentz potentials $\mathbf{A}_\nu$ $(\mathbf{A},\phi)$ arising from the sources j_n $(\boldsymbol{j},\rho)$ are derived, are: $\partial_\mu^2 \mathbf{A}_\nu = -4\pi j_\nu/c, \partial_\mu \mathbf{A}_\mu = 0$, and can displace the Maxwell equations as the basis of electromagenetic theory. The Maxwell equations follow from these equations whenever the antisymmetric field tensor $F_{\mu\nu}(\boldsymbol{E},\boldsymbol{B}) = \partial_\mu \mathbf{A}_\nu - \partial_\nu \mathbf{A}_\mu$ is defined.

Prolegomenon B: Interpretation of Maxwell's Original Formulation

B.1 *The Faraday and Maxwell formulation*

Central to the Maxwell formulation of electromagnetism[5] was the Faraday concept of *electrotonic state* (from new Latin *tonicus*, of tension or tone, from Greek *tonicus*, from *tonos*, a stretching). Maxwell defined this state as the "fundamental quantity in the theory of electromagnetism" the *changes in which* (and not on its absolute magnitude) the induction current depends (Maxwell, 1954, vol. 2, p. 540). Faraday had clearly indicated the fundamental role of this state in his two circuit experiments (Faraday, 1965, vol. 1, series I, 60); and Maxwell endorsed its importance: "The scientific value of Faraday's conception of an electrotonic state consists in its directing the mind to lay hold of a certain quantity, on the *changes* of which the actual phenomena depend" (Maxwell, 1954, vol. 2, p. 541).

---

[5]Maxwell, J.C., "On Faraday's lines of force", *Trans. Camb. Phil. Soc.*, 10, (Part I), 27–83, 1856.
Maxwell, J.C., "A dynamical theory of the electromagnetic field," *Phil. Trans. Roy. Soc.*, 155, 459–512, 1865.
Maxwell, J.C., *A Treatise on Electricity and Magnetism*, 2 Vols, 1st edition 1873; 2nd edition 1881; 3rd edition, 1891, Oxford, Clarendon Press, republished in two volumes, Dover, New York, 1954.
Bork, A.M., Maxwell and the vector potential, *Isis*, 58, 210–222, 1967.
Bork, A.M., Maxwell and the electromagnetic wave equation, *Am. J. Phys.*, 35, 844–849, 1967.
Everitt, C.W.F., *James Clerk Maxwell*, Scribner's, New York, 1975.
Hunt, B.J., *The Maxwellians*, a dissertation submitted to the Johns Hopkins University with the requirements for the degree of Doctor of Philosophy, Baltimore, MD, 1984.
Buchwald, J.Z., *From Maxwell to Microphysics*, University of Chicago Press, 1985.
Hendry, J., *James Clerk Maxwell and the Theory of the Electromagnetic Field*, Adam Hilger Ltd, Bristol, 1986.
O'Hara, J.G. and Pricha, W., *Hertz and the Maxwellians: A Study and Documentation of the Discovery of Electromagnetic Wave Radiation, 1873–1894*, Peter Peregrinus Ltd, London, 1987.
Nahin, P.J., *Oliver Heaviside: sage in solitude*, IEEE Press, New York, 1988.

The continental European views of the time concerning propagation (e.g., those of Weber, Biot, Savart and Neumann), which Maxwell opposed, were based on the concept of action-at-a-distance. In place of action-at-a-distance, Maxwell offered *a medium characterized by polarization and strain* through which radiation propagated from one local region to another local region. Furthermore, instead of force residing in the medium, Maxwell adopted another Faraday concept: force fields, or *magnetic lines of force independent of matter or magnet.* In contrast, Weber's position was that force is dependent on relative velocity and acceleration. However, Maxwell's response was not made in rebuttal of Weber's. Rather, Maxwell believed that there are two ways of looking at the subject: his own and Weber's (Hendry, 1986). Certainly, Helmholtz achieved a form of synthesis of the two pictures (Nahin, 1988).

For Maxwell the distinction between *quantity* and *intensity* was also central. For example, magnetic intensity was represented by a line integral and referred to the magnetic polarization of the medium. Magnetic quantity was represented by a surface integral and referred to the magnetic induction in the medium. In all cases a medium was required and *the medium was the seat of electromagnetic phenomena.* The electromagnetic and *luminiferous medium* were identified. Faraday's lines of force were said to indicate the direction of minimum pressure at every point in the medium and constituted the field concept. Thus for Maxwell, the electromagnetic field did not exist, *sui generis*, but as a state of the medium and the mechanical cause of differences in pressure in the medium was accounted for by the hypothesis of vortices, i.e., polarization vectors. The medium was also restricted to be one in which there was only a displacement current (with no conduction currents) and in which there were no electrical or magnetic sources. Furthermore, rather than electricity *producing* a disturbance in the medium (which was W. Thomson's view), Maxwell's field theory described the presence of electricity as a disturbance, i.e., the electricity *was* the disturbance.

Maxwell conceived of electric current as a moving system with forces communicating the motion from one part of the system to another. The nature of the forces was undefined, and did not need to be defined, as for Maxwell the forces can be eliminated from the equations of motion by Lagrangian methods for any connected system. That is, *the equations of motion were defined only locally.* Thus, Maxwell dispensed with the dynamic forces permitting propagation through the medium, only the beginning and end of the propagating process was examined and that only locally. This view is reminiscent of a communications system, linked by a channel, in which only the contents of the transmitter (TX) and receiver (RX) ae measured. Therefore, only the local state of the medium, or electrotonic intensity or state, was primary, and that corresponds to Maxwell's "F" or "α_o", or in in modern symbols: the **A** field. This is where matters stood up to 1873 with the vector potential playing a pivotal *physical* role in Maxwell's theory.

B.2 *The British Maxwellians and the Maxwell–Heaviside formulation*

Subsequently, the **A** field was *banished from playing the central role in Maxwell's theory and relegated to being a mathematical (but not physical) auxiliary.* This banishment took place during the interpretation of Maxwell's theory by the Maxwellians (Hunt, 1984), i.e., chiefly, Heaviside, Fitzgerald and Lodge The "Maxwell theory" and "Maxwell's equations" we know today are really the interpretation of Maxwell by these Maxwellians (Nahin, 1988). It was Heaviside who "murdered the **A** field" (Heaviside's description) and whose work influenced the crucial discussion which took place at the 1888 Bath meeting of the British Association (although Heaviside was not present). The "Maxwell's equations" of today are due to Heaviside's "redressing" of Maxwell's work, and should, more accurately, be known as the "Maxwell–Heaviside equations". Essentially, Heaviside took the twenty equations[6] of Maxwell and reduced them to the four, now known as "Maxwell's equations".

[6]Actually, there are eight original equations as Maxwell wrote six equations in Cartesian coordinate (x, y, z) form separately, which is unnecessary.

The British Maxwellians, Heaviside, Fitzgerald and Lodge,[7] may have banished scalar and vector potentials from the propagation equations, but the center of concern for them remained the *dynamic state of the medium or ether.* The banishment of the potentials can today be justified in the light of the discussion to follow in that the Maxwell theory focused on local phenomena, and the **A** field, as we shall see, lives in global connectivity and spatial volumes. Therefore in order for the theory to progress, it was perhaps appropriate that the **A** field was put aside, or at least assigned an auxiliary role, *at that time.* Heaviside's comment that the electrostatic potential was a "physical inanity" was probably correct for the 19th century, but as will be shown below, the potential regained its sanity in the 20th century — starting with the work of Hermann Weyl.

But it should be emphasized that the British Maxwellians retained the focus of theory *on the medium.* Both Heaviside and Poynting[8] agreed that the function of a wire is as a sink into which energy passes from the medium (ether) and is convected into heat. For them, wires conduct electricity with the Poynting vector pointing at right angles to the conducting wire (Feynman *et al.* 1964, Sec. 27-5). The modern conventional view on conduction in wires is similar, but modern theory is not straightforward about where this energy goes, yet still retains Poynting's theorem. The energy flows not through a current carrying wire itself, but through the medium (ether) around it — or rather through whatever energy storing substance a modern theorist imagines exists in the absence of the ether. Nonetheless, Heaviside was probably correct to banish scalar and vector potentials from propagation equations due to the fact that the notion of gauge invariance (*Mass-stabinvarianz* — see below) was not yet conceived and thus not known to the Maxwellians. However, these **A** fields still remained as a repository of energy in the electrotonic state of the medium and the redressed and interpreted

[7]Oliver Heaviside (1850–1925; George Francis Fitzgerald (1851–1901); Oliver Lodge (1851–1940).
[8]John Henry Poynting (1852–1914).

Maxwell theory of the British Maxwellians remained a *true dynamic* theory of electromagnetism.

B.3 *The Hertzian and current classical formulation*

But all dynamics were banished by Hertz. Hertz[9] banished even the stresses and strains of the medium (ether) and was vigorously opposed in this by the British Maxwellians (Hunt, 1984). Hertz even went far beyond his mentor, Helmholtz, in his austere operational formulation. Nonetheless, the Hertz orientation finally prevailed, and modern "Maxwell's theory" is today a system of equations describing electrodynamics *which has lost its dynamical basis.*

Another significant reinterpretation of Maxwell took place in which Heaviside was involved. The 19th century battle between Heaviside and Tait[10] concerning the use of quaternions and culminating in the victory of Heaviside and vector analysis, may also be reassessed in the light of modern developments. *Without* the concepts of gauge, global (as opposed to local) fields, non-integrable phase factors (see below) and topological connections, the use of quaternions was getting in the way of progress. That is not to say that either quaternionic algebra or the potentials were, or are, unphysical or unimportant. It is to say, rather, that the potentials could not be understood *then* with the limited theory and mathematical tools available *then.* Certainly it is *now* realized that the algebraic formulation of electromagnetism is more complicated than to be described completely even by quaternionic algebra, and certainly more complicated than to be described by simple vector analysis (cf. Chisholm and Common, 1986).

But to return to Maxwell's original formulation: Maxwell *did* place the **A** field at center stage and *did* use quaternionic algebra to dress his theory. We know now that quaternionic algebra is described by the SU(2) group of transformations, and vector algebra by the U(1) group of transformations. As such modern propagation

---

[9]Heinrich Hertz (1857–1894).
[10]Peter Tait (1831–1901).

phenomena as solitons are of SU(2) form, we might even view the original Maxwell formulation as more comprehensive than that offered by the British Maxwellian interpretation, and certainly more of a dynamic theory than the physically unintuitive local theory finally adopted by Hertz. That said, when it is stated, below, that the "Maxwell equations" need extension, it is really the modern Heaviside–Hertz interpretation that is meant. The *original* Maxwell's theory could have been easily extended into Yang–Mills form.

Between the time of Hertz's interpretation of Maxwell's theory and the appearance of the gauge field concepts of Hermann Weyl, there appeared Whittaker's notable *mathematical* statement (Whittaker, 1903, 1904) that (i) the force potential can be defined in terms of both standing waves and propagating waves and (ii) any electromagnetic field — e.g., dielectric displacement, magnetic force, etc. — can be expressed in terms of the derivatives of two scalar potential functions, and also be related to an inverse square law of attraction.

Whittaker commenced his statement with Laplace's equation:

$$\frac{\partial^2 V}{dx^2} + \frac{\partial^2 V}{dy^2} + \frac{\partial^2 V}{dz^2} = 0, \tag{1.1}$$

which is satisfied by the potential of any distribution of matter which attracts according to Newton's law. The potential at any point (x, y, z) of any distribution of matter of mass m, situated at the point (a, b, c) which attracts according to this law is:

$$\int_0^{2\pi} f(z + ix\cos u + iy\sin u, u)du, \tag{1.2}$$

where u is a periodic argument. The most general solution of Laplace's equation using this expression is

$$V = \int_0^{2\pi} f(z + ix\cos u + iy\sin u, u)du, \tag{1.3}$$

where f is an arbitrary function of the two arguments:

$$z + ix\cos u + iy\sin u, \quad \text{and} \quad u \tag{1.4}$$

In order to express this solution as a series of harmonic terms, Whittaker (1903) showed that it is only necessary to expand the function f as a *Taylor series* with respect to the first argument $z + ix\cos\, u + iy\sin\, u$, and as a *Fourier series* with respect to the second argument u.

Whittaker also showed that the general solution of the partial differential wave equation:

$$\frac{\partial^2 V}{dx^2} + \frac{\partial^2 V}{dy^2} + \frac{\partial^2 V}{dz^2} = \frac{k^2 \partial^2 V}{dt^2} \tag{1.5}$$

is:

$$V = \int_0^{2\pi} \int_0^{\pi} f\left(x\sin\, u\cos\, v + y\sin\, u\sin\, v + z\cos\, u + \frac{t}{k}, u, v\right) dudv, \tag{1.6}$$

where f is now an arbitrary function of the three arguments:

$$x\sin\, u\cos\, v + y\sin\, u\sin\, v + z\cos\, u + \frac{t}{k}, u \textit{ and } v, \tag{1.7}$$

and can be analyzed into a simple plane wave solution. Therefore, for any force varying as the inverse square of the distance, the potential of such a force satisfies *both* Laplace's equation *and* also the wave equation, and can be analyzed into simple plane waves propagating with constant velocity. The *sum of these waves*, however, does not vary with time, i.e., *they are standing waves.* Therefore, the force potential can be defined in terms of *both* standing waves, i.e., by a global, or *non-local* solution, *and* by propagating waves, i.e., by a *local* solution changing in time.

Furthermore, Whittaker demonstrated that any electromagnetic field, e.g., dielectric displacement, magnetic force etc., can be expressed in terms of the derivatives of *two* scalar potential functions, F and G, satisfying:

$$\begin{aligned} &\partial^2 F/dx^2 + \partial^2 F/dy^2 + \partial^2 F/dz^2 - [1/c^2]\partial^2 F/dt^2 = 0, \\ &\partial^2 G/dx^2 + \partial^2 G/dy^2 + \partial^2 G/dz^2 - [1/c^2]\partial^2 G/dt^2 = 0. \end{aligned} \tag{1.8}$$

Thus, Whittaker's mathematical statement related the inverse square law of force to the force potential defined in terms of both standing wave (i.e., global) and propagating wave (i.e., local) solutions. The analysis also showed that the electromagnetic force fields could be defined in terms of the derivatives of two scalar potentials. This was the state of affairs prior to Weyl's introduction of gauge fields. The landmark work of Weyl[11] and Yang and Mills[12] has been matched by the conception of pseudoparticle or minimum action solutions of the Yang–Mills equations,[13] i.e., instantons. Such phenomena and the appearance of gauge structure are found in simple dynamical, or classical, systems,[14] and the concept of instanton has been the focus of intense activity in recent years.[15]

The demonstration that the non-linear Schrödinger equation and the Korteweg–de-Vries equation — equations with soliton solutions — are reductions of the self-dual Yang-Mills equations (Mason and Sparling, 1989) with correspondences to twistor formulations (Barrett, 1989) has provided additional evidence concerning the direction that Maxwell's theory must take. These reductions of self-dual Yang–Mills equations are known to apply to various classical systems depending on the choice of Lie algebra associated with the self-dual fields (Chakravarty *et al.* 1990).

---

[11]Weyl, H., Gravitation und Electrizität, *Sitz. Ber. Preuss. Ak. Wiss.*, 465–480, 1918a.

Weyl, H., Reine Infinitesimalgeometrie, *Math. Zeit.*, 2, 384–411, 1918b.

Weyl, H., Gravitation and the Electron, *Proc. Natl. Acad. Sci. USA*, 15, 323–334, 1929.

Weyl, H., *The Classical Groups, Their Invariants and Representations*, Princeton University Press, 1939 and 1946.

Weyl, H., *Gesammelte Abhandlungen*, in 4 volumes, K. Chandrasekharan (ed), Springer-Verlag, Heidelberg, 1968.

[12]Yang, C.N. and Mills, R., Isotopic spin conservation and generalized gauge invariance. *Phys. Rev.*, 95, M7, 631, 1954a.

Yang, C.N. and Mills, R.N., Conservation of isotopic spin and isotopic gauge invariance. *Phys. Rev.*, 96, 191–195, 1954b.

[13]Belavin, A.A., Polyakov, A.M., Schwartz, A.S. and Tyupkin, Y.S., Pseudoparticle solutions of the Yang–Mills equations, *Phys. Lett.*, 59B, 85–87, 1975.

[14]Wilczck, F. and Zee, A., Appearance of gauge structure in simple dynamical systems, *Phys. Rev. Lett.*, vol. 52 no. 24, 2111–2114, 1984.

[15]Atiyah, (1984, 1988); Atiyah *et al.* (1977, 1978); Atiyah *et al.* (1977); Atiyah and Jones (1978).

It is also relevant that the soliton mathematical concept need not result only from non-linear equations. Barut (1991) and Shaarawi *et al.* (1990) demonstrated that soliton solutions are possible in the case of *linear* de Broglie-like wave equations. Localized oscillating finite energy solutions of the massless wave equation are derived which move like massive relativistic particles with energy $\mathbf{E} = \lambda\omega$ and momentum $\mathbf{p} = \lambda\mathbf{k}$ ($\lambda = \text{const}$). Such soliton solutions reduced to linear wave equations do not spread and have a finite energy field.

1.1. Introduction

There are a number of reasons for questioning the completeness of the conventionally interpreted Maxwell's theory of electromagnetism. It is well known that there is an arbitrariness in the definition of the **A** vector and scalar potentials, which, nevertheless, have been found very useful when used in calculations with boundary conditions known (Konopinski, 1978) The reasons for questioning completeness are due to experimental evidence (Sec. 1.3), theoretical (Sec. 1.4), and pragmatic (Sec. 1.5).

An examination of the Maxwell theory may begin with the well-known Maxwell equations[16]:

Coulomb's Law:

$$\nabla \cdot \mathbf{D} = 4\pi\rho. \tag{1.9}$$

Maxwell's generalization of Ampere's Law:

$$\nabla \times \mathbf{H} = (4\pi/\mathbf{c})\mathbf{J} + (1/\mathbf{c})\ \partial\mathbf{D}/\partial\mathbf{t}; \tag{1.10}$$

the absence of free *local* magnetic poles postulate or the differential form of Gauss' law:

$$\nabla \cdot \mathbf{B} = \mathbf{0}; \tag{1.11}$$

and Faraday's law:

$$\nabla \times \mathbf{E} + (1/\mathbf{c})\ \partial\mathbf{B}/\partial\mathbf{t} = \mathbf{0}. \tag{1.12}$$

[16]The equations are in Gaussian or cgs units: centimeter, gram and second. The Systeme International (SI) or mks units are: meter, kilogram and second.

The constitutive relations of the medium-independent fields to matter are well-known to be:

$$\mathbf{D} = \varepsilon \mathbf{E}, \tag{1.13}$$

$$\mathbf{J} = \sigma \mathbf{E}, \tag{1.14}$$

$$\mathbf{B} = \mu \mathbf{H}. \tag{1.15}$$

Because of the postulate of an absence of free local magnetic monopoles (Eq. (1.11)), the following relation is permitted:

$$\mathbf{B} = \boldsymbol{\nabla} \times \mathbf{A}, \tag{1.16}$$

but the vector potential **A** is thus always arbitrarily defined, because the gradient of some scalar function **Λ** can be added leaving **B** unchanged, i.e., **B** is unchanged by the gauge transformations:

$$\mathbf{A} \to \mathbf{A}' = \mathbf{A} + \boldsymbol{\nabla\Lambda}; \quad \boldsymbol{\Phi} \to \boldsymbol{\Phi}' = \boldsymbol{\Phi} - (\mathbf{1/c})\boldsymbol{\partial\Lambda/\partial}\mathbf{t}. \tag{1.17}$$

This arbitrary definition of the potentials means that *any* gauge chosen is arbitrary, or, an appeal must be made to boundary conditions for any choice.

Now Eq. (1.16) permits a redefinition of Eq. (1.12):

$$\boldsymbol{\nabla} \times (\mathbf{E} + (\mathbf{1}/c)\boldsymbol{\partial}\mathbf{A}/\boldsymbol{\partial}\mathbf{t}) = \mathbf{0} \quad \text{(Faraday's law rewritten)}, \tag{1.18}$$

which means that the quantity in brackets is the gradient of a scalar function, Φ, so:

$$\mathbf{E} + (\mathbf{1}/c)\boldsymbol{\partial}\mathbf{A}/\boldsymbol{\partial}\mathbf{t} = -\boldsymbol{\nabla\Phi}, \quad \text{or} \quad \mathbf{E} = -\boldsymbol{\partial\Phi} - (1/\mathrm{c})\boldsymbol{\partial}\mathbf{A}/\boldsymbol{\partial}\mathbf{t} \tag{1.19}$$

and the Maxwell equations (1.11) and (1.12) can be redefined by use of Eqs. (1.16) and (1.19).

Maxwell equations (1.9) and (1.10) can also be written as

$$\boldsymbol{\partial}^{\mathbf{2}}\boldsymbol{\Phi} + (\mathbf{1/c})\boldsymbol{\partial}(\boldsymbol{\nabla}\cdot\mathbf{A})/\boldsymbol{\partial}\mathbf{t} = -\mathbf{4}\boldsymbol{\pi\rho}, \tag{1.20}$$

$$\begin{aligned}\boldsymbol{\nabla}^{\mathbf{2}}\mathbf{A} - (\mathbf{1}/c^{\mathbf{2}})(\boldsymbol{\partial}^{\mathbf{2}}\mathbf{A}/\boldsymbol{\partial}\mathbf{t}^{\mathbf{2}}) - \boldsymbol{\nabla}(\boldsymbol{\nabla}\cdot\mathbf{A}) + (\mathbf{1/c})\boldsymbol{\partial\Phi}/\boldsymbol{\partial}\mathbf{t})\\ = -(\mathbf{4}\boldsymbol{\pi}/c)\mathbf{J}.\end{aligned} \tag{1.21}$$

Since the gauge conditions (1.17) are arbitrary, a set of potentials ($\mathbf{A}$,Φ) can be chosen so that:

$$\boldsymbol{\nabla}\cdot \mathbf{A} + ((\mathbf{1}/\mathbf{c})\partial\boldsymbol{\Phi}/\partial\mathbf{t}) = \mathbf{0}. \tag{1.22}$$

This choice is called the Lorenz[17] condition or the Lorenz gauge equations (1.20) and (1.21) can then be decoupled to obtain:

$$\boldsymbol{\nabla}^{\mathbf{2}}\boldsymbol{\Phi} + (\mathbf{1}/\mathbf{c}^{\mathbf{2}})\partial^{\mathbf{2}}\boldsymbol{\Phi}/\partial\mathbf{t}^{\mathbf{2}} = -\mathbf{4}\pi\rho, \tag{1.23}$$

$$\boldsymbol{\nabla}^{\mathbf{2}}\mathbf{A} - (\mathbf{1}/\mathbf{c}^{\mathbf{2}})\partial^{\mathbf{2}}\mathbf{A}/\partial\mathbf{t}^{\mathbf{2}} = -(\mathbf{4}\pi/\mathbf{c})\mathbf{J}, \tag{1.24}$$

which is useful because the Maxwell equations are then independent of the coordinate system chosen. Nonetheless, as $\mathbf{A}$ and $\boldsymbol{\Phi}$ are not gauge invariant, the original choice of the Lorenz gauge is arbitrary — a choice which is not an inevitable consequence of the Maxwell theory — and the resultants from that choice, namely Eqs. (1.23) and (1.24) are equally arbitrary.[18]

Then again, the arbitrariness of Eqs. (1.17) is useful because the choice is permitted:

$$\nabla \cdot \mathbf{A} = 0. \tag{1.25}$$

Equation (1.20), which is Maxwell equation (1.9), then permits:

$$\boldsymbol{\nabla}^{\mathbf{2}}\boldsymbol{\Phi} = -\mathbf{4}\pi\rho, \tag{1.26}$$

which is the instantaneous Coulomb potential, and hence condition (1.25) is called the Coulomb or transverse gauge because the wave equation for $\mathbf{A}$ can be expressed in terms of the transverse current:

$$\boldsymbol{\nabla}^{\mathbf{2}}\mathbf{A} - (\mathbf{1}/\mathbf{c}^{\mathbf{2}})\partial^{\mathbf{2}}\mathbf{A}/\partial\mathbf{t}^{\mathbf{2}} = -(\mathbf{4}\pi/c)\mathbf{J}_{\mathrm{t}}, \tag{1.27}$$

where $\mathbf{J}_{\mathrm{t}} = \mathbf{J} - \mathbf{J}_{\mathrm{l}}$ and $\mathbf{J}_{\mathrm{l}}$ is the longitudinal current. This is a useful thing to do when no sources are present, but, again, as $\mathbf{A}$ and $\boldsymbol{\Phi}$

[17]Danish physicist Ludvig Lorenz (1829–1891).

[18]The above account applies to the Hertzian potential and Hertz vector which are related to $\mathbf{A}$ and $\boldsymbol{\Phi}$. However, the Hertzian vector obeys an inhomogeneous wave equation with the polarization vector as source, whereas $\mathbf{A}$ and $\boldsymbol{\Phi}$ obey their respective wave equations with electric current and charge as source. Furthermore, the Hertzian potential is a three component potential, whereas $\mathbf{A}$ and $\boldsymbol{\Phi}$ amount to a four potential (cf. Panofsky and Phillips, 1962, p. 254).

are not gauge invariant, i.e., considered to have no physical meaning, the original choice of the Coulomb gauge is arbitrary, and so is the resultant from that choice, namely Eq. (1.27).

For all that, the absence of gauge invariance (physical meaning) of the $\mathbf{A}$ vector potential and the Φ scalar potential may seem a fortunate circumstance to those using the Maxwell theory to calculate predictions. These potentials have long been considered a fortunate mathematical convenience, but *just* a mathematical convenience, with no physical meaning. These constructs lack gauge invariance, a defining characteristic of physical, rather than merely mathematical, constructs. What then is meant by a gauge and gauge invariance?

1.2. What is a Gauge?

In 1918 Weyl (see also Yang, 1986) treated Einstein's general theory of relativity as if the Lorentz[19] symmetry were an example of *global* symmetry but with only *local* coordinates defineable, i.e., the general theory was considered as a *local* theory. A consequence of Weyl's theory is that the absolute magnitude or norm of a physical vector is not treated as an absolute quantity but depends on its location in spacetime. This notion was called scale (*Mass-stab*) or gauge invariance.

This concept can be understood as follows. Consider a vector at position x with norm given by $f(x)$. If the coordinates are transformed, so that the vector is now at $x + dx$, the norm is $f(x + dx)$. Using the abbreviation ∂/∂^{μ}, $\mu = 0, 1, 2, 3$, expanding to first order, and using Einstein's summation convention:

$$f(x + dx) = f(x) + \partial_{\mu} f dx^{\mu}. \tag{1.28}$$

If a gauge change is introduced by a multiplicative scaling factor, $\mathbf{S}$ (x), which equals unity at x, then

$$\mathbf{S}(x + dx) = 1 + \partial_{\mu}\mathbf{S}dx^{\mu}. \tag{1.29}$$

[19]Dutch physicist Hendrik Antoon Lorentz (1853–1928).

If a vector is to be constant under change of location then:

$$\mathbf{S}\, f = f + [\partial_\mu \mathbf{S}] f\, dx^\mu + [\partial_\mu\, f] dx^\mu \tag{1.30}$$

and, on moving, the norm changes by an amount

$$[\partial_\mu + \partial_\mu\, \mathbf{S}]\, f\, dx^\mu. \tag{1.31}$$

Weyl identified $\partial_\mu \mathbf{S}$ with the electromagnetic potential $\mathbf{A}_\mu$.

However, this suggestion was rejected (by Einstein) because the natural scale for matter is the Compton wavelength, λ, and as the wave description of matter is $\lambda = \hbar/mc$ (where $\hbar$ is Planck's constant and c is the speed of light), then if, as is always assumed, the wavelength is determined by the particle's mass, m, and with $\hbar$ and c constant (according to the special theory of relativity), λ cannot depend on position without violating the special theory. When made aware of this reasoning, Weyl abandoned his proposal. So the term: gauge change, originally meant "change in length", and was withdrawn from consideration for this particular metric connotation shortly after its introduction.

But the term did not die. "Gauge invariance" managed to survive in classical mechanics because, with the potentials arbitrary, Maxwell's equations for the **E**, **B**, **H** and **D** fields have a built-in symmetry and such arbitrary potentials became a useful mathematical device for simplifying many calculations in electrodynamics, as we have seen. Nevertheless, the gauge invariance in electromagnetism for the **E**, **B**, **H** and **D** fields was regarded as only an "accidental" symmetry, and the lack of gauge invariance of the electromagnetic vector and scalar potentials was interpreted as an example of the well-known arbitrariness of the concept of the potential in classical mechanics.

But this arbitrariness in the concept of the potential did, and does not, exist in quantum mechanics. The electromagnetic vector and scalar potentials were viewed in quantum mechanics in yet another way. Upon the development of quantum mechanics, Weyl and others realized that the original gauge theory could be given a new meaning. They realized that the phase of a wavefunction could be a new *local* variable. Instead of a change of scale or metric, for which it was originally introduced, a gauge transformation was reinterpreted as a change in the phase of the wavefunction:

$$\Psi \to \Psi \exp[-ie^{\lambda}] \tag{1.32}$$

and the gauge transformation for the potential $\mathbf{A}_\mu$ became:

$$\mathbf{A}_\mu \to \mathbf{A}_\mu - \partial\lambda/\partial x_\mu. \tag{1.33}$$

Equations (1.32) and (1.33) together ensure that the Schrödinger formulation for a single-charged particle in an electromagnetic field remains invariant to phase changes because they self-cancel.[20] Thus any change in location, for that *single* charged particle, which produces a change in the phase (Eq. (1.32)) is compensated by a corresponding change in the potential (Eq. (1.33)). Therefore, Weyl's original idea, reinterpreted, was accepted, and the potential in quantum mechanics was viewed as a *connection* which relates phases at different locations. Nevertheless, this use and interpretation did not carry over into classical mechanics and a schizoid attitude exists to this day regarding the physical meaning of the potentials in classical and quantum mechanics. In classical mechanics the potentials were, up until recently, viewed as having only an arbitrary mathematical, not physical, meaning, as they seemed to lack gauge invariance. In quantum mechanics, however, they *are* viewed as gauge invariant and do possess a physical meaning. It is an aim of this book to show that in classical mechanics the potentials can also be taken to have, under special circumstances, a physical meaning, i.e., possess the required gauge invariance.

A major impetus to rethink the physical meaning of the potentials in classical mechanics came about from the experiments examined in the next section.

---

[20]Wignall (1989) has shown that no phase change occurs for de Broglie waves under the low velocity limit of the Lorentz transformation. However, the phase of de Broglie waves is not invariant under a change of frame described by a Galilean transformation.

1.3. Empirical Reasons for Questioning the Completeness of Maxwell's Theory

1.3.1. *Aharonov–Bohm (AB) and Altshuler–Aronov–Spivak (AAS) effects*

Beginning in 1959 Aharonov and Bohm challenged the view that the classical vector potential produces no observable physical effects by proposing two experiments. The one which is most discussed is shown in schematic form in Fig. 1.1. A beam of monoenergetic electrons exits from the source at X and is diffracted into two beams, I and II, by the two slits in the wall at Y1 and Y2. The two beams produce an interference pattern at III which is measured. Behind the wall is a solenoid, the **B** field of which points out of the paper. The absence of a free local magnetic monopole postulate (Maxwell equation (1.11) above) predicts the magnetic field outside the solenoid to be zero. Before the current is turned on in the solenoid, there should be the usually expected interference patterns seen at III. Aharonov and Bohm predicted that if the current is turned on and due to the differently directed **A** fields in paths I and II

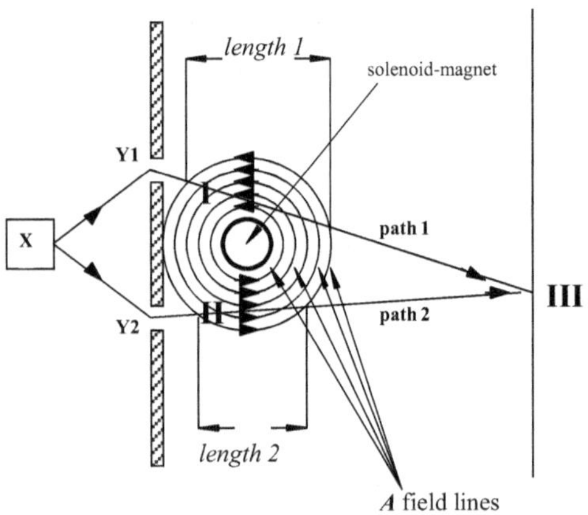

Fig. 1.1. Two-slit diffraction experiment of the Aharonov–Bohm effect. Electrons are produced by a source at X, diffracted by the slits at Y1 and Y2 and their diffraction pattern is detected at III. The solenoid is between the slits and directed out of the page. The different orientations of the **A** field at the points of interaction with the two paths are indicated by the arrows $>$ and $<$ following the right-hand rule.

indicated by the arrows in Fig. 1.1, additional phase shifts should be discernible at III. This prediction was confirmed experimentally[21] and the evidence has been extensively reviewed.[22] Aharonov and Casher (1984) have extended the theoretical treatment of the AB effect to neutral particles with a magnetic moment; and Botelho and de Mello (1985) have analyzed a non-Abelian AB effect in the framework of pseudoclassical mechanics.

One explanation of the effect is as follows. Let ψ_0 be the wavefunction when there is no current in the solenoid. After the current is turned on the Hamiltonian is

$$H = \frac{1}{2m}(-i\hbar\nabla - e\mathbf{A})^2, \tag{1.34}$$

and the new wavefunction is

$$\psi = \psi_0 \exp\left[\frac{-ie\mathbf{S}}{\hbar}\right], \tag{1.35}$$

where $\mathbf{S}$ the flux, is defined as

$$\mathbf{S} = \oint \mathbf{A} \cdot d\mathbf{x}, \tag{1.36}$$

which is the quantum analog of the classical action evaluated along the paths I and II. At point III the wavefunctions of the two electron beams are:

$$\psi_1 = \psi_0 \exp\left[\frac{-eS_1}{\hbar}\right], \tag{1.37a}$$

$$\psi_2 = \psi_0 \exp\left[\frac{-eS_2}{\hbar}\right], \tag{1.37b}$$

and the phase difference is

$$\left(\frac{e}{\hbar}\right)(\mathbf{S}_1 - \mathbf{S}_2) = \left(\frac{e}{\hbar}\right)\left(\int_1 \mathbf{A} \cdot d\mathbf{x} - \int_2 \mathbf{A} \cdot d\mathbf{x}\right) = 2\pi\left(\frac{e}{\hbar}\right)\mathbf{\Phi}. \tag{1.38}$$

---

[21]Chambers (1960); Boersch *et al.* (1960); Möllenstedt and Bayh (1962); Matteucci and Pozzi (1985); Tonomura *et al.* (1982, 1983, 1986); Tonomura and Callen, (1987).

[22]Berry (1980); Peshkin (1981); Olariu and Popescu (1985); Horvathy (1986); Peshkin and Tonomura (1989).

By Stokes' theorem, this is directly proportional to the magnetic flux, $\mathbf{\Phi} = \oint \mathbf{A} \cdot d\mathbf{x}$, in the solenoid.

However, the phase difference given by Eq. (1.38) is not single-valued. Therefore, the value of the phase change will only be determined to within an arbitrary multiple, n, of $2\pi e\mathbf{\Phi}/\hbar$, where n is the number of times the measured charge circulated the solenoid. The topological feature of the background space of the Aharonov–Bohm effect is its *multiple-connectedness* (Schulman, 1981). Therefore, the mathematical object to be computed in this framework is a propagator expressed as a path integral in the covering space of the background physical space (cf. Benido and Inomata, 1981). This means that for a *simply-connected* space, all paths between two points are in the same homotopy class,[23] and the effect of the potential, $\mathbf{A}_\mu$, is as a multiplier of the free-particle propagator with a single gauge phase factor. *In this case, the potential has no physically discernible effect.* However, for a *multiply-connected* manifold, the potential *can have a physically discernible effect* because the gauge factors can be different for different homotopy classes (Aharonov, 1983; Sundrum and Tassie, 1986).

The AB effect was confirmed experimentally in the originally proposed field→free electron (F→FE) situation (cf. Chambers, 1960). That is, the effect predicted by Aharonov and Bohm refers to the influence of the $\mathbf{A}$ vector potential on electrons confined to a *multiply connected region*, but within which the magnetic field is zero. As a consequence of the gauge invariance, the energy levels of the electrons have a period $\hbar/e$ of the enclosed flux. More recent experiments address the appearance of the effect in the field→conduction electron (F→CE) situation (cf. Aronov and Sharvin, 1987). This situation is also not strictly the same as in the originally proposed AB experiment in another respect — the magnetic flux is produced by a large

---

[23]Homotopy groups record information about the basic shape, or holes, of a topological space. To define the nth homotopy group, the base-point-preserving maps from an n-dimensional sphere (with base point) into a given space (with base point) are collected into equivalence classes, called Homotopy classes. In the case of two continuous functions, if one can be "continuously deformed" into the other, such a deformation is called a homotopy between the two functions.

solenoid surrounding the influenced condensed matter, usually a loop or a cylinder — so that the **B** field is not set to zero within the material. However, the preponderance of the **B** field is always in the hole encompassed by that cylinder or ring, and the magnetic field causes only secondary effects in the material. In this situation, periodic oscillations in the conductance of the ring appear as a function of the applied magnetic field, **B**. The periodicity of the oscillations is

$$\nabla \mathbf{B} = \hbar / e\mathbb{A}, \tag{1.39}$$

where $\mathbb{A}$ is the area enclosed by the ring.

Under these conditions, the AB effect is seen in normal metal[24] bulk Mg[25]; semiconductors[26] and on doubly connected geometries on GaAs/AlGaAs heterostructures.[27] The effect has also been seen in structures such as: cylindrical Mg films[28] and Li films,[29] wire arrays,[30] arrays of Ag loops,[31] small metal loops,[32] and MBE-grown double quantum wells.[33]

Bandyopadhyay *et al.* (1986) and Datta and Bandyopadhyay (1987) have also discussed a novel concept for a transistor based on the electrostatic AB effect in MBE-grown quantum wells, where the current is modulated by quantum interference of electrons in two contiguous channels of a gate voltage. They predict that transistors based on this effect will have power-delay products orders of magnitude better than those of existing devices such as MODFETs and Josephson junctions. The transconductance will also be much higher

---

[24]Tonomura *et al.* (1982, 1983, 1986); Webb *et al.* (1985, 1987), Benoit *et al.* (1986); Washburn *et al.* (1985, 1987); Chandrasekhar *et al.* (1985); Datta *et al.* (1985); Cavalloni and Joss (1987).
[25]Sandesara and Stark (1984).
[26]Datta *et al.* (1986); Datta and Bandyopadhyay (1987).
[27]Timp *et al.* (1987).
[28]Sharvib and Sharvin (1981); Gijs *et al.* (1984).
[29]Altshuler *et al.* (1982).
[30]Pannetier *et al.* (1984); Bishop *et al.* (1985).
[31]Umbach *et al.* (1986).
[32]Webb *et al.* (1987); Chandrasekhar *et al.* (1985).
[33]Datta *et al.* (1985).

than that of MODFETS. Unlike previous experimental treatments which assumed diffusive transport with negligible inelastic scattering, Datta and Bandyopadhyay (1987) assume ballistic transport and perfect symmetry in the arms of the interferometer and in the voltage along the interferometer or two channel structure.

Now, the AB (F→CE) effect is temperature-dependent as coherent transport is required. The effect has only been seen at very low temperature. Measurements were made on parallel GaAs quantum wells at 4.2 K and below[34]; on 860 nm-i.d. Au loops at 0.003 K[35] and $0.05\,\mathrm{K} < T < 0.7\,\mathrm{K}$,[36] on 75 nm-o.d. Sb loops at $0.01 < T < 1\,\mathrm{K}$[37] and at 0.04 K[38] and on Ag loop arrays at 4.2 K.[39] Measurements on 1.5–2.0 micron diameter Mg cylinders of length 1 cm were made at 1.12 K.[40] The Thouless scaling parameter, V, or the sensitivity of energy levels to a change in the phase of the wavefunctions at the boundaries[41] implies that the necessary energy correlation range for small rings is accessible in the temperature range 0.0001–10 K.[42]

What is remarkable is that these experiments on the (F→CE) AB effect demonstrate that the effect *can* occur in *disordered* electrical conductors if the temperature is low enough. The effect in metals is a small magnetoresistance oscillation superimposed on the ohmic resistance in multiply-connected conductors at low temperatures.[43] *This means that the conducting electrons must possess a high degree of phase coherence (internal correlation) over distances larger than the atomic spacing or the free path length.* It was initially thought that the effects of finite temperature and the scattering from, and collision with, impurities, would cause incoherence and prevent the

[34] Datta *et al.* (1985).
[35] Webb *et al.* (1987).
[36] Webb *et al.* (1985); Washburn *et al.* (1985).
[37] Milliken *et al.* (1987).
[38] Washburn *et al.* (1987).
[39] Umbach *et al.* (1986).
[40] Sharvin and Sharvin (1981).
[41] Edwards and Thouless (1972); Lee *et al.* (1987).
[42] Stone and Imry (1986).
[43] Sharvin and Sharvin (1981); Altshuler *et al.* (1982); Stone and Imry (1986).

observation of the Aharonov–Bohm effect in bulk samples.[44] The metal loops used measure, e.g., less than a micron in diameter and less than 0.1 microns in line thickness. Therefore, the electron is thought to be represented by a pair of waves — one traveling around the ring in the clockwise direction, and the other in the opposite direction, but following the time-reversed path of the first wave. Thus, although each wave has been scattered many times, *each wave collides with the same impurities*, i.e., acquires the same phase shifts, resulting in constructive interference at the origin. *The total path length of both waves is twice the circumference of the ring, meeting the requirement that the phase coherence of the electrons be larger than the circumference of the ring*, or, the transport through the metals arms considered as disordered systems is determined by the eigenvalues of a large random matrix (Imry, 1986).

Thus, the conductance, G, of a one-dimensional ring in the presence of elastic scattering is (Landauer, 1970):

$$G = \left(\frac{2e^2}{\hbar}\right)\left(\frac{t}{1-t}\right), \tag{1.39}$$

and an AB flux applied to the ring results in periodic oscillations of G, *provided that the phase coherence length of the ring is longer than the size of the system.* (The AB oscillations can be suppressed by magnetic fields, vanishing near resistance minima associated with plateaus in the Hall effect (Timp *et al.* 1989). Dupuis and Montambaux (1991) have shown that in the case of the AB effect in metallic rings, the statistics of levels shows a transition from the Gaussian Orthogonal Ensemble (GOE), in which the statistical ensemble shows time reversal, to the Gaussian Unitary Ensemble (GUE), in which time-reversal symmetry is broken.

A related effect is the Altshuler–Aronov–Spivak (AAS) effect (Altshuler *et al.* 1981). These authors considered an ultrathin normal metal cylindrical shell of moderate length but very small transverse dimensions at low temperature and how the magnetoresistance would depend on the intensity of magnetic flux axially threading the

[44] Imry (1986).

cylinder. They concluded that it would be an oscillating function of the total flux with a period of $\hbar/2e$, i.e., the same as the flux of the superconductive state. The analogous "flux quantum" of the AB effect is $\hbar/e$ (Webb *et al.* 1985, 1987) and differs from the AAS situation which involves coherent "backscattering". The AAS effect has been observed in a 1000 Å thick magnesium layer on a quartz fiber several millimeters long (Sharvin and Sharvin, 1981). Other tests of the AAS effect (Stone and Imry, 1986; Büttiker *et al.* 1983, 1985) are based on the quantum mechanical transmission (t) coefficients of electrons and, unlike the original AAS treatment, find an $\hbar/e$ periodic component as well as the $\hbar/2e$ harmonic. Raising the temperature above a crossover, T_c, changes the flux periodicity of magnetic resistance oscillations from $\hbar/e$ to $\hbar/2e$, where T_c is determined by the energy correlation range $\hbar D/L^2$, D is the elastic diffusion constant, L is the length of sample and the quantity $\hbar D/L^2$ is the Thouless scaling parameter V for a metal.

The AAS effect arises because of a special set of trajectories — time-reversed pairs which form a closed loop — which have a fixed relative phase for any material impurity configuration. These trajectories do not average to zero and contribute to the reflection coefficients which oscillate with period $\hbar/2e$. The $\hbar/e$ oscillations of the AB effect, on the other hand, arise from oscillations in the transmission coefficients and can at higher temperature average to zero. Below T_c both contributions are of order $e^2/\hbar$ (Stone and Imry, 1986).

Xie and DasSarma (1987) studied the AB and AAS effects in the transport regime of a strongly disordered system in which electron transport is via a hopping process, specifically, via variable-range-hopping transport. Their numerical results indicate that only the $\hbar/2e$ (AAS) flux-periodic oscillations survive at finite temperatures in the presence of any finite disorder.

The results of the metal loop experiments demonstrated that elastic scattering does not destroy the phase memory of the electron wave functions (Webb *et al.* 1987; Washburn and Webb, 1986). The flux periodicity in a condensed matter system due to the

Aharonov–Bohm effect would not be surprising in a superconductor. However, the same periodicities in *finite* conductors is remarkable (Xie and DasSarma, 1987). Numerical simulation of variable-range-hopping conduction only finds AB oscillations ($\varphi_0 = \hbar/e$) in hopping conductance when a metal ring is small and at low temperature. At the large ring limit and higher temperature AAS ($\varphi_0 = \hbar/2e$) oscillations survive — a finding consistent with the experimental findings of Polyarkov *et al.* (1986). A suggested reason for the retention of long range phase coherence is that the phase memory is only destroyed exponentially as $\exp[-L/L_i]$, where L is a "typical inelastic scattering length" and the destruction depends on the energy changes in the hopping process dependent on long wavelength, low-energy acoustic phonons. Search for an explanation for both AB and AAS effects has resulted in consideration of systems as *neither precisely quantum mechanical nor classical, but inbetween, i.e., "mesoscopic"*.

"Mesoscopic" systems have been studied by Stone (1985) in which the energy and spacing is only a few orders of magnitude smaller than kT at low temperatures. The prediction was made that large AB oscillations should be seen in the transport coefficients of such systems. Such systems have a sample length which is much longer than the elastic mean free path, but shorter than the localization length. The magnetic field through a loop connected to leads changes the relative phase of the contribution from each arm of the loop by $2\pi\Phi/\Phi_0$, where $\Phi_0 = \hbar c/e$ is the one electron flux quantum and Φ is the flux through the hole in the loop — but only if the phase dependent terms do not average to zero. In the mesoscopic range, if inelastic scattering is absent, these phase-dependent contributions do not self-average to zero.

Washburn *et al.* (1985) and Stone and Imry (1986) demonstrated experimentally that the amplitude of aperiodic and periodic conductance fluctuations decrease for the F→CE AB effect with increasing temperature. There is a characteristic correlation energy:

$$E_c = \frac{\pi\hbar D}{2L^2}, \tag{1.40}$$

where D is the diffusion constant of the electrons, L is the minimum length of the sample length. If thermal energy $k_B T > E_C$, the conductance fluctuations decrease as $(E_C/k_B T)^{1/2}$. The conductance fluctuations also decrease when L_ϕ, the phase coherence length is shorter than the length, L or the distance between voltage probes, the decrease being described by a factor $\exp(-L/L_\phi)$ (Milliken *et al.* 1987). This gives a conductance fluctuation:

$$\Delta Gn = (L_\varphi/L)^{3/2}. \tag{1.41}$$

In condensed matter, therefore, the AB effect appears as the modulation of the electron wave functions by the $\mathbf{A}_\mu$ potential. The phase of the wave function can also be changed by the application of an electric field (Washburn *et al.* 1987). The electric field contributes to the fourth term in the four-vector product $\mathbf{A}_\mu(dx)^\mu$ which contains the scalar potential Φ associated with transverse electric fields and time. The phase shift in the wavefunction is

$$\Delta\varphi = \int e\Phi dt/\hbar. \tag{1.42}$$

Experiments on Sb metal loop devices on silicon substrate have demonstrated that the voltage on capacitative probes can be used to tune the position (phase) of $\hbar/e$ oscillations in the loop. Thus, there appears two ways to modulate the phase of electrons in condensed matter: application of the $\mathbf{A}_\mu$ potential by threading magnetic flux between two paths of electrons; and also by application of a scalar potential by means of a transverse electric field. Aharonov–Bohm fluctuations in metal loops are also not symmetric about $H = 0$: four probe measurements yield resistances which depend on the lead configurations (Benoit *et al.* 1986).

More recently, it has been shown that due to transport via edge states and penetration of a strong magnetic field into the conducting region, periodic magnetoconductance oscillations can occur in a singly-connected geometry, e.g., as in a point contact or a "quantum dot" (a disc-shaped region in a two-dimensional electron gas) (Beenakker, 1991; Beenakker *et al.* 1991). As this effect is dependent on transport via edge states circulating along the boundary of a quantum dot and enclosing a well-defined amount of flux, *the geometry is effectively doubly connected.* The claim of single-connectedness is thus more apparent than real. However, there is a difference between the AB effect in a ring and in a dot is that in each period $\nabla\mathbf{B}$, the number of states below a given energy stays constant in a ring, but increases by one in a quantum dot. In the case of a dot the AB magnetoconductance oscillations are accompanied by an increase in the charge of the dot by one elementary charge per period. This can result in an increase in Coulomb repulsion which can block the AB magnetoconductance oscillations. This effect, occurring in quantum dots, has been called the Coulomb blockade of the Aharonov–Bohm effect.

Finally, Boulware and Deser (1989) explain the AB effect in terms of a vector potential coupling minimally to matter, i.e., a vector potential not considered as a gauge field. They provide an experimental bound on the range of such a potential as 10^2 km.

In summary, the AB and AAS effects, whether F→FE or F→CE, demonstrate that the phase of a *composite* particle's wavefunction is a physical degree of freedom which is dependent on differences in $\mathbf{A}_\mu$ potential influences on the spacetime position or path of a *first* particle's wave function with respect to that of another *second* particle's wavefunction. But the *connection*, or *mapping*, between spatiotemporally different fields or particles which originated at, or passed through, spatiotemporally separated points or paths with differential $\mathbf{A}_\mu$ potential influences, is only measurable by many-to-one mapping of those different fields or particles. By interpreting the phase of a wavefunction as a *local* variable instead of the norm of a vector, electromagnetism can be interpreted as a *local* gauge (phase) theory, if not exactly, then very close to the way Weyl originally envisioned it to be.

Below, the interaction of the $\mathbf{A}_\mu$ field (x), whether vector or potential, as independent variable with dependent variable constructs, will be referred to in a x→y notation. For example, field→free electron, field→conducting electron, field→wave guide, field→neutral particle and field→rotating frame interactions will be referred to as (F→FE), (F→CE), (F→WG), (F→P) and (F→RF) interactions.

Although the $\mathbf{A}_\mu$ field is a classical field, the AB and AAS effects are either F→FE and F→CE effects and might be considered "special" in that they involve quantum mechanical particles, i.e., electrons. In the next section, however, we examine a phase rotation which can only be considered classical, as *both* independent and dependent variables are classical. Nonetheless, the same result: the $\mathbf{A}_\mu$ potentials demonstrating physical effects, applies.

1.3.2. *Topological phases: Berry, Aharonov–Anandan, Pancharatnam and Chiao–Wu phase rotation effects*

When addressing the AB effect, Wu and Yang (1975) argued that the wavefunction of a system will be multiplied by a non-integrable (path-dependent) phase factor after its transport around a closed curve in the presence of an $\mathbf{A}_\mu$ potential *in ordinary space.* In the case of the Berry–Aharonov–Anandan–Pancharatnam (BAAP) phase, another non-integrable phase factor, arises from the adiabatic transport of a system around a closed path in *parameter* (*momentum*) *space*, i.e., this topological phase is the AB effect in parameter space.[45] The WG→F version of this effect has been experimentally verified (Tomita and Chiao, 1986) and the phase effect in general interpreted as due to parallel transport in the presence of a gauge field (Simon, 1983). The effect exists at both the classical and quantum levels (cf. Thomas, 1988; Cai *et al.* 1990).

There has been, however, an evolution of understanding concerning the origins of topological phase effects. Berry (1984a) originally proposed a geometrical (beside the usual dynamical) phase acquisition for a non-degenerate quantum state which varies adiabatically through a circuit in parameter space. Later, the constraint of an adiabatic approximation was removed (Berry, 1987) and also the constraint of degenerate states (Wilczek and Zee, 1984). Then Aharonov and Anandan (Aharonov and Anandan, 1987) showed that the effect can be defined for any cyclic evolution of a quantum system. Bhandari and Samuel (1988) have also pointed out that Berry's

[45]Berry (1984a,b; 1985; 1987a,b); Wilkinson (1984a,b); Chiao and Wu (1986); Chiao and Tomita (1987); Haldane (1987).

phase is closely connected with a phase discovered by Pancharatnam (1956, 1975). These authors also demonstrated that unitary time evolution of a system is not essential for the appearance of the phase by the measurement of the phase change in one beam of a laser interferometer as the polarization state of light is taken along a closed circuit on the Poincaré sphere. Thus current thought is that the history of "windings" of a particle is "remembered", or registered and indicated, by changes in phase in both either a quantum mechanical particle's state, or in a classical wave's polarization. The topological phase effect appears to arise from the nontrivial topology of the complex projective Hilbert space — whether classical or quantum mechanical (Kiritis, 1987) — and to be equivalent to a gauge potential in the parameter space of the system — again, whether classical or quantum mechanical.

Jiao *et al.* (1989) have also indicated at least three variations of topological phases: (i) the phase which arises from a cycling in the direction of a beam of light so that the tip of the spin vector of a photon in this beam traces out a closed curve on the sphere of spin directions — which is that originally studied by Chiao and Wu (1986); (ii) Pancharatnam's phase arising from a cycling in the polarization states of the light while keeping the direction of the beam of light fixed, so that the Stokes' vector traces out a closed curve on the Poincaré sphere, i.e., the phase change is due to a polarization change; (iii) the phase change due to a cycle of changes in squeezed states of light.

Topological phase change effects, in its Field→Photon version, have been observed in NMR interferometry experiments (Suter *et al.* 1987, 1988) and using ultracold neutrons (Richardson *et al.* 1988); in coherent states (Giavarini *et al.* 1989a,b); optical resonance (Ellinas *et al.* 1989) and the degenerate parametric amplifier (Gerry, 1989). However, topological phase change effects are more commonly studied in a classical waveguide→classical field (WG→F) version, in which the parameter space is the momentum **k**-space.[46]

[46] Chiao and Wu (1986); Jiao *et al.* (1989); Bialynicki-Birula and Bialynicki-Birula (1987).

For example, the helicity or polarization state, σ, is (Chiao and Wu, 1986):

$$\sigma = \mathbf{s} \cdot \mathbf{k}, \tag{1.43}$$

where $\mathbf{s}$ is a spin or helicity operator and $\mathbf{k}$ is the direction of propagation (k_x, k_y, k_z). If τ is the optical path length, then $|\mathbf{k}(\tau), \sigma\rangle$ is the spin or polarization state. Interpreted classically, the constraint of keeping $\mathbf{k}$ parallel to the axis of a waveguide is due to the linear momentum being in that direction. This means that a waveguide can act as a polarization rotator. Furthermore, as helicity (polarization), σ, is adiabatically conserved, $\mathbf{s}$ is also constrained to remain parallel to the local axis of the waveguide. Therefore, the topology of a waveguide, e.g., a helix-shape, will constrain $\mathbf{k}$ and also $\mathbf{s}$ to perform a trajectory $\mathbf{C}$ on the surface of a sphere in the parameter space (k_x, k_y, k_z) which prescribes the linear momentum. Thus the topology of the constrained trajectory of radiation progressing between two *local* positions has a *global* effect indicated by a polarization (spin) change. If $\gamma\mathbf{C}$ is the topological phase, and $\beta = \exp(i\gamma\mathbf{C})$ is a phase factor, the final polarization state after progression along a constrained trajectory, i.e., "momentum conditioning", is

$$\sigma_2 = \beta \cdot \mathbf{s} \cdot \mathbf{k}, \tag{1.44}$$

where the subscript indicates a second location on the trajectory.

As a monopole is theoretically required at $\mathbf{k} = 0$, due to the radial symmetry of the parameter space and resulting singularity, a solid angle $\Omega(C)$ can be defined on a parameter space sphere with respect to the origin $\mathbf{k} = 0$. Thus, $\Omega(C)$ can be said to define the "excited states" of the monopole at $\mathbf{k} = 0$. Therefore,

$$\sigma_2 - \sigma_1 = \beta \cdot \mathbf{s} \cdot \mathbf{k} - \sigma_1 = \sigma_1\Omega(\mathbf{C}) - \sigma_1 = \gamma(\mathbf{C}). \tag{1.45}$$

The question can be asked: what conservation law underlies the topological phase? A clue is provided by Kitano *et al.* (1987), who point out that the phase change can also be seen in discrete optical systems which contain no waveguides, e.g., in a configuration of (ideal or infinitely conducting) mirrors. Now, mirrors do not conserve helicity; they reverse it and the local tangent vector, $\mathbf{t}$, must be replaced by $-\mathbf{t}$ on alternate segments of the light path. Mirror configurations of this type have been used in a laser gyro (Chow *et al.* 1985). This suggests that changes of acceleration, whether along a waveguide, or in mirror reflection, under equivalence principle conditions is the compensatory change which matches changes in the topological phase, giving the conservation equation:

$$\gamma(C) + \oint \mathbf{A} \cdot dl = 0. \tag{1.46}$$

That the phase effect change can occur in classical mechanical form is witnessed by changes in polarization rotation resulting from changes in the topological path of a light beam. Tomita and Chiao (1986) demonstrated effective optical activity of a helically wound single mode optical fiber in confirmation of Berry's prediction. The angle of rotation of linearly polarized light in the fiber gives a direct measure of the topological phase *at the classical level.* (Hannay (1985) has also discussed the classical limit of the topological phase in the case of a symmetric top.) The effect arises from the overall geometry of the path taken by the light and is thus a *global topological effect* independent of the material properties of the fiber. The optical rotation is *independent of geometry* and therefore may be said to quantify the *"topological charge"* of the system, i.e., the helicity of the photon, which is a relativistic quantum number.

Referring to Fig. 1.2, the fiber length is

$$\mathrm{s} = [p^2 + (2\pi r)^2]^{1/2}, \tag{1.47}$$

and the solid angle in momentum space $\Omega(C)$ spanned by the fiber's closed path C, a circle in the case considered, is

$$\Omega(C) = 2\pi(1 - \cos\theta). \tag{1.48}$$

The topological phase is

$$\gamma(C) = -2\pi\sigma(1 - p/\mathrm{s}), \tag{1.49}$$

where $\sigma = \pm 1$ is the helicity quantum number of the photon.

By wrapping a piece of paper with a computer generated curve on a cylinder to which the fiber is fitted, and then unwrapping the

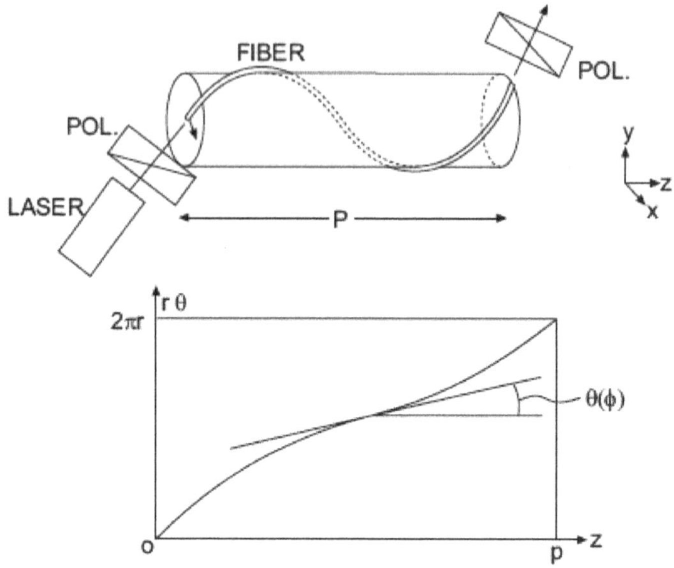

Fig. 1.2. (a) Experimental setup; (b) geometry used to calculate the solid angle in momentum space of a non-uniformly wound fiber on a cylinder. After Tomita and Chiao (1986).

paper, the local pitch angle, or tangent to the curve followed by the fiber can be estimated to be (Fig. 1.2(b)):

$$\theta(\varphi) = \tan^{-1}(rd\varphi/dz), \tag{1.50}$$

which is the angle between the local waveguide and the helix axes. In momentum space, $\theta(\varphi + \pi/2)$ traces out a closed curve C, the fiber path on the surface of a sphere. The solid angle subtended by C to the center of the sphere is

$$\Omega(C) = \int_0^{2\pi} [1 - \cos\theta(\varphi)]d\varphi. \tag{1.51}$$

The topological phase is then, more correctly:

$$\begin{aligned} \gamma(C) &= -\sigma\Omega(C) \text{ or} \\ \gamma(C) &= \nu\Omega(C) \end{aligned} \tag{1.52}$$

(where $\nu = 1/2$ in the case of polarization charges, i.e., Pancharatnam's phase). There is thus a linear relation between the angle of rotation of linearly polarized light, and the solid angle $\Omega(C)$ subtended by C at the origin of the momentum space of the photon (Tomita and Chiao, 1986).

Regarding the C in Eq. (1.52), Chiao *et al.* (1988) demonstrated a topological phase shift in a Mach–Zehnder interferometer in which light travels along nonplanar paths in two arms. They interpret their results in terms of the Aharonov–Anandan phase and changes in projective Hilbert space, i.e., the sphere of spin directions of the photon, rather than parameter (momentum) space. The hypothesis tested was that the evolution of the state of a system is cyclic, i.e., that it returns to its starting point adiabatically or not. Thus the C in Eq. (1.52) is to be interpreted as a closed circuit on the sphere of spin directions.

There are varying interpretations of rotation effects. Chiao and Wu (1986) considered topological phase rotation effects to be "topological features of the Maxwell theory which originate at the quantum level, but which survive the correspondence principle limit ($h \to 0$) into the classical level." However, this opinion is contested and the effect is viewed as classical by other authors.[47] For example, the evolution of the polarization vector can be viewed as determined by a connection on the tangent bundle of the two-dimensional sphere (Segert, 1987a,b) The effect is then viewed as non-Abelian. The situation can then be described with a family of Hamiltonian operators, $H_0 + \mathbf{k} \cdot \mathbf{V}$, where H_0 is rotationally invariant, $\mathbf{V}$ is a vector operator and $\mathbf{k}$ varies over the unit vectors in $\mathbb{R}^3$.

The Chiao–Wu phase and the Pancharatnam phase are additive. This is because the two topological phase effects arise in different parameter spaces: the former in $\mathbf{k}$-space, and the latter in polarization vector (Poincaré sphere) space. To see this, Maxwell's equations can be recast into six-component spinor form (Jiao *et al.* 1989):

$$\nabla \times (\mathbf{E} \pm \mathbf{B}) = \pm\partial(\mathbf{E} \pm \mathbf{iB})/\partial\mathbf{t}, \tag{1.53}$$

$$\mathbf{\Psi} = \mathbf{col}(\mathbf{E} + \mathbf{B}, \mathbf{E} - \mathbf{B}) = \mathbf{col}((\mathbf{\Psi}^+), (\mathbf{\Psi}^-)), \tag{1.54}$$

[47] Berry (1987a,b); Segert (1987a,b).

where **col** denotes a column vector, to obtain a Schrödinger-like equation:

$$i\partial\mathbf{\Psi}/\partial t = H\mathbf{\Psi}, \tag{1.55}$$

and where the Hamiltonian, H, is given by

$$H = \begin{bmatrix} (\nabla x) & 0 \\ 0 & (-\nabla x) \end{bmatrix} \tag{1.56}$$

(where ∇x represents the curl). This spinor representation of Maxwell's equations has a natural correspondence with natural optical activity in the frequency domain (Bohron, 1975).

The conventional dynamical phase becomes:

$$\delta(H) = -\int_0^T (\mathbf{\Psi}, H\mathbf{\Psi})dt, \tag{1.57}$$

and the geometrical (topological) phase is, as before:

$$\gamma(C) = \oint \mathbf{A} \cdot dl, \tag{1.58}$$

but where the vector potential, $\mathbf{A}$, is explicitly defined as

$$\mathbf{A} = -(\mathbf{\Psi}, \mathbf{\nabla\Psi}), \tag{1.59}$$

i.e., a connection defined on the state space.

By means of Stokes' theorem:

$$\gamma(C) = \oint \mathbf{A} \cdot dl = \int_S \nabla \times AdS, \tag{1.60}$$

$$\gamma(C) = \nu\Omega(C), \tag{1.61}$$

as before.

Figure 1.3 shows the different manifestations and representations of the topological phase effect. Top (i) is a sphere of *spin directions* for representing the (Chiao–Wu) phase arising from the spin vector of a photon tracing out a closed curve on the sphere (Jiao *et al.* 1989). The topological phase is equal to the angle, Ω, shown. Middle (ii) is a Poincaré sphere of polarization states, or *helicity*, of a photon for representing the (Pancharatnam) phase effect arising from cycling in

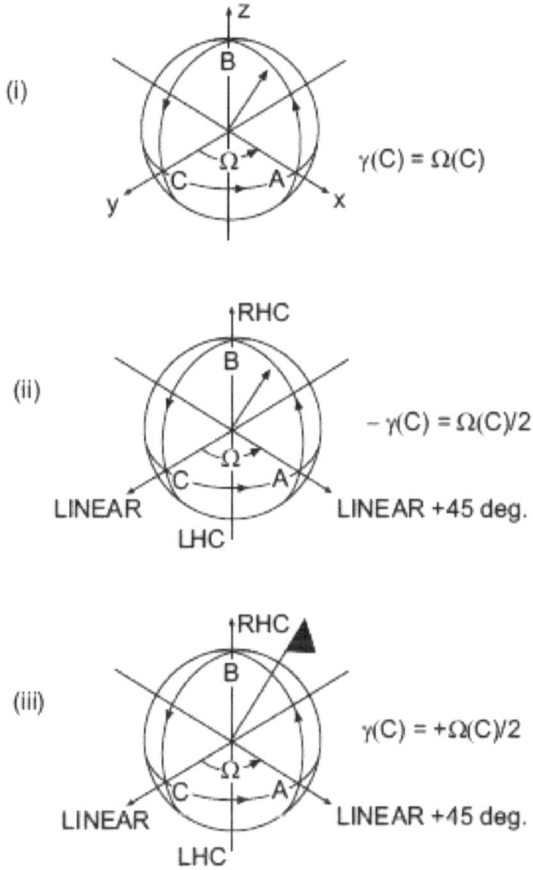

Fig. 1.3. (i) A sphere of *spin directions* for representing the (Chiao–Wu) phase arising from the spin vector of a photon tracing out a closed curve on the sphere. After Jiao *et al.* (1989). The topological phase is equal to the angle, Ω, shown. (ii), A Poincaré sphere of polarization states, or *helicity*, of a photon for representing the (Pancharatnam) phase effect arising from cycling in the polarization states of the photon while keeping the direction of the beam fixed.[48] The topological phase is equal to the *negative* of one half the angle, Ω, shown. (iii) A *generalized* Poincaré sphere for representing the angular momentum of light (with space-fixed axis),[49] or *null flag or twistor* representation[50] for representing the Pancharatnam topological phase but with the phase equal to the *positive* value of one half the angle, Ω, shown. The topological phase effects represented in (i) and (iii) are additive.

[48] After Bhandari and Samuel (1988).
[49] After Jiao *et al.* (1989).
[50] After Penrose and Rindler (1984, 1986); Barrett (1989).

the polarization states of the photon while keeping the direction of the beam fixed (Bhandari and Samuel, 1988). The topological phase is equal to the *negative* of one half the angle, Ω, shown. Bottom (iii) is a *generalized* Poincaré sphere for representing the angular momentum of light (with space-fixed axis), or *null flag or twistor* representation (Penrose and Rindler, 1984, 1986; Barrett, 1989) for representing the Pancharatnam topological phase but with the phase equal to the *positive* value of one half the angle, Ω, shown. The topological phase effects represented in (i) and (iii) are additive.

A more explicit relation between the topological phase effect and Maxwell theory is obtained within the formulation of Maxwell's theory by Biakynicki-Birula & Bialynicka-Birula (1975). Within this formulation, the intrinsic properties of an electromagnetic wave are its wave vector, $\mathbf{k}$, and its polarization, $\mathbf{e}(\mathbf{k})$. As Maxwell's theory can be formulated as a representation of Poincaré symmetry, all wave vectors form a vector space. Thus, implicit in Maxwell's theory is the topology of the surface of a sphere (i.e., the submanifold, $\mathbf{S}^2$). The generators of the Poincaré group involve a covariant derivative in momentum space, whose curvature is given by a magnetic monopole field. However, the Maxwell equations can only determine the polarization tensor, $\mathbf{e}(\mathbf{k})$, *up to an arbitrary phase factor* as Maxwell's theory corresponds to the structure group U(1), i.e., Maxwell's theory *cannot* determine the phase of polarization in momentum space. This arbitrariness permits additional topological phase effects.

On the other hand, the topological phase is *precisely* obtained from a set of angles associated with a group element and there is just one such angle corresponding to a holonomy transformation of a vector bundle around a closed curve on a sphere (Anandan and Stodolsky, 1987). The parameter space is the based manifold and each fiber is isomorphic to an N-dimensional Hilbert space. In particular, for the SU(2) case there is a single angle from the holonomy of the the Riemannian connection on a sphere. The observation that gauge structure appears in simple dynamical systems — both quantum mechanical and classical — has been made (Wilczek and Zee, 1984). For the special case of Fermi systems, the differential geometric background for the occurrence of SU(2) topological phases is the quaternionic projective space with a time evolution corresponding to the SU(2) Yang–Mills instanton (Lévy, 1990). *Locally*, the non-Abelian phase generated can be reduced to an Abelian form. However, it is not possible to define the connection defined on the bundle space except *globally*. This reflects the truly non-Abelian nature of the topological phase. The topological phase effect can be described in a *generalized* Bloch-sphere model and an SU(2) Lie-group formulation in the spin-coherent state (Layton *et al.* 1990). Furthermore, while acknowledging that in general terms, and formally, Berry's phase is a geometrical object in projective Hilbert space (ray space), the non-adiabatic Berry's phase, physically, is related to the expectation value of spin (spin alignment), and Berry's phase quantization is related to spin-alignment quantization (Wang, 1990).

The topological phase effect even appears in quantum systems constrained by molecular geometry. For example, the topological phase effect appears in the molecular system Na_3 (Delcrétaz *et al.* 1986). Suppose a system in an eigenstate $C(r, t)$ responds to slowly varying changes in its parameters $R(t)$, such that the system remains in the *same eigenstate apart from an acquired phase*. If the parameters, $R(t)$, completed a circuit in parameter space, then that acquired phase is not simply the familiar dynamical phase, $[(i\hbar)^{-1}E(R(t)]dt$, but rather an additional geometrical phase factor $\gamma_n(c)$. The origins of this additional phase factor depend only on the geometry of the parameter space and the topology of the circuit traversed. Therefore, adiabatic excursions of molecular wave functions in the neighborhood of an electronic degeneracy results in a change of phase. That is, if the internuclear coordinates of a wave function traverse a circuit in which the state is degenerate with another, then the electronic wave function acquires an additional phase, i.e., it changes its sign. This change was predicted[51] and is a special case of the topological phase applying to a large class of molecular systems exhibiting conical intersections. Delacrétaz *et al.* (1986) reported the evidence for half-odd quantization of free molecular pseudorotation and offered the

[51]Herzberg and Longuet-Higgens (1963); Longuet-Higgens (1975); Mead and Truhlar (1979).

first experimental confirmation of the sign-change theorem and a direct measurement of the phase. The topological phase has also been observed in fast-rotating superfluid nuclei, i.e., oscillations of pair-transfer matrix elements as a function of the angular velocity (Nikam and Ring, 1987) and in neutron spin rotation (Bitter and Dubbers, 1987).

The appearance of the topological phase effect in both classical and quantum mechanical systems thus gives credence to the view that the $\mathbf{A}_\mu$ potentials register physical effects both at the classical and quantum mechanical levels. That such a role for these potentials exists at the quantum mechanical level is not new. *It is new to consider the $\mathbf{A}_\mu$ potentials for such a role at the classical level.* One may ask how the schism in viewing the $\mathbf{A}_\mu$ potentials came about; that is, why are they viewed as physical constructs in quantum mechanics, but as merely arbitrary mathematical conveniences in classical mechanics? *The answer is that whereas quantum theory is defined with respect to boundary conditions, in the formal presentation of Maxwell theory boundary conditions are undefined.* Stokes' theorem demonstrates this.

1.3.3. *Stokes theorem re-examined*

Stokes' theorem of potential theory applied to classical electromagnetism relates diverging potentials on line elements to rotating potentials on surface elements. Thus, Stokes's theorem describes a many-to-one field relationship.

If $\mathbf{A}(\mathrm{x})$ is a vector field, $\mathbf{S}$ is an open, orientable surface, $\mathbf{C}$ is the closed curve bounding $\mathbf{S}$, $\mathbf{dl}$ is a line element of $\mathbf{C}$, $\mathbf{n}$ is the normal to $\mathbf{S}$ and $\mathbf{C}$ is traversed in a right-hand screw sense (positive direction) relative to $\mathbf{n}$, then the line integral of $\mathbf{A}$ is equal to the surface integral over $\mathbf{S}$ of $(\nabla \times \mathbf{A}) \cdot \mathbf{n}$:

$$\oint \mathbf{A} \cdot \mathbf{dl} = \int_S (\nabla \times \mathbf{A}) \cdot \mathbf{n}\, da. \tag{1.62}$$

It is also necessary that $\boldsymbol{S}$ be the union of a finite number of smooth surface elements and that the first-order partial derivation of the components of $\mathbf{A}$ be continuous on $\mathbf{S}$. Thus Stokes' theorem, as described, takes no account of: (i) spacetime overlap in a region with fields derived from different sources; and (ii) the exact form of the boundary conditions.

This neglect of the exact form of the boundary conditions in the Stokes' theorem of classical mechanics can be contrasted with the situation in quantum mechanics. In quantum mechanics the wave function satisfies a partial differential equation coupled to boundary conditions because the Schrödinger equation describes a minimum path solution to a trajectory between two points. The boundary condition in the doubly connected (overlap) region outside of the shielded volume in an Aharonov–Bohm experiment is the reason for the single valuedness of the wavefunction, and also the reason for quantization. The situation is also different with spatial symmetries other than the usual, Abelian, spatial symmetry.

A non-Abelian Stokes' theorem is (Goddard and Olive, 1978):

$$h^{-1}(dh/ds) \approx ie \int_0^1 g^{-1}\mathbf{G}_{ij}g\left(\partial r^i/\partial t\right)\left(\partial r^j/\partial s\right)dt \tag{1.63}$$

where $h(s)$ is a path-dependent phase factor associated with a closed loop and defines a closed loop $r(s,t)$, $0 \leq t \leq 1$, s fixed, in the U(1)-symmetry space, $\mathbf{H}$ (equivalent to $\mathbf{A}_\mu$); $\mathbf{G}$ is a gauge field tensor for the SU(2) non-Abelian group; and g is magnetic charge. Here, the boundary conditions, i.e., the path dependencies, are made explicit, and we have a local field (with U(1) symmetry) to overlapped field (with SU(2) symmetry) connection.

In classical electromagnetism, therefore, Stokes' theorem appears merely as a useful mathematical relation between a vector field and its curl. *In gauge theory, on the other hand, an amended Stokes' theorem would provide the value for the net comparative phase change in the internal direction of a particle traversing a closed path, i.e., a many-to-one connection.*

Lest it be thought that the $\mathbf{A}_\mu$ field which functions as the independent variable in the Aharonov–Bohm experiment is only a quantum effect with no relevance to classical behavior, the relation of the $\mathbf{A}_\mu$ potential to the properties of bulk condensed matter is examined in the following section. A more complete definition of Stokes' theorem is also given in Sec. 1.3.4 (Eq. (1.63)).

Use of Stokes' theorem has a price: that of the (covert) adoption of a gauge for many-to-one connections. This is because Stokes' theorem applies directly to propagation issues, which are defined by many-to-one connections. Such connections are also required in propagation through matter. Thus there is a requirement for Stokes' theorem in any realistic definition of macroscopic properties of matter, and in the next section we see that the physical effects of the $\mathbf{A}_\mu$ potentials exist not merely in fields traversing through various connecting topologies, but in radiation-matter interactions.

1.3.4. *Properties of bulk condensed matter — Ehrenberg & Siday's observation*

In the Aharonov–Bohm F→FE situation, when the size of the solenoid is much larger than the de Broglie wavelength of the incident electrons, the scattering amplitude is essentially dominated by simple classical trajectories. But the classical manifestation of quantum influences is not peculiar to the Aharonov–Bohm effect. For example, macroscopic quantum tunneling is observable in Josephson tunnel junctions in which the phase difference of the junction can be regarded as a macroscopic degree of freedom, i.e., a classical variable (Martinis *et al.* 1987; Clarke *et al.* 1988).

Even without known quantum influences or quantum mechanical explanation, there is a classical justification for the $\mathbf{A}_\mu$ potential as a physical effect. For example, on the basis of optical arguments, the $\mathbf{A}_\mu$ potential must be chosen so as to satisfy Stokes's theorem *thereby removing the arbitrariness with respect to gauge*. Furthermore, an argument, originating with Ehrenberg and Siday (1949) shows that a gauge-invariant $\mathbf{A}_\mu$ potential is *presupposed* in any definition of the refractive index.

This argument is a derivation of the refractive index based on *Fermat's Principle*: in any optical medium, a scalar quantity, e.g., the refractive index, finite everywhere in space, can be defined so that the line integral in the three-dimensional space taken between any fixed points must be an extremum which passes through these points. The optical path along a given line connecting a point 1 and 2 is

$$\int_1^2 mds = \int_1^2 [mv + (\mathbf{A}_\mu \cdot \mathbf{n})]\, ds, \tag{1.64}$$

where $\mathbf{n}$ is the unit vector in the direction of the line, v is the velocity of the electron, and m is its mass. Defined in this way, an unambiguous definition of the refractive index indicates the *necessity* of a unique (gauge invariant) definition of the $\mathbf{A}_\mu$ potential. Stated differently: an unambiguous definition of the refractive index implies defining the boundary conditions through which test radiation moves. These boundary conditions define a definite gauge and thereby definite $\mathbf{A}_\mu$ potentials.

This an example of physical $\mathbf{A}_\mu$-dependent effects (the refractive properties of matter) seen when radiation propagates through matter — from one point to another. In the next section, $\mathbf{A}_\mu$ effects are described when two fields are in close proximity. This is the Josephson effect, and again, the potential functions as a many-to-one operator.

1.3.5. *Josephson effect*

Josephson (1962–1974) predicted that a d.c. voltage, V, across the partitioning barrier of a superconductor gives rise to an alternating current of frequency:

$$\omega = 2eV/\hbar. \tag{1.65}$$

The equivalent induced voltage is (Bloch, 1968):

$$V = (1/c)d\mathbf{\Phi}/dt, \tag{1.66}$$

where $\mathbf{\Phi}$ is the magnetic flux through a superconducting ring containing a barrier. The circulating current, I, exhibits a periodic dependence upon $\mathbf{\Phi}$:

$$I(\alpha) = \sum\nolimits_n a_n \sin 2\pi\alpha, \tag{1.67}$$

where

$$\alpha = \mathbf{\Phi}/(\hbar c/e). \tag{1.68}$$

The validity of Eq. (1.68) depends upon the substitution of

$$\mathbf{p} - \mathbf{eA}/\mathbf{c} \tag{1.69}$$

for the momentum, **p**, of any particle with charge and with a required gauge invariance for the **A** potential.

The phase factor existing in the junction gap of a Josephson junction is an exponential of the integral of the **A** potential. The fluxon, or the decrementlessly conducting wave in the long Josephson junction and in a SQUID, is the equivalent of an **A**-wave in one-dimensional phase space. The phenomenological equations are:

$$\partial\varphi/\partial x = (2ed/\hbar c)H_y; \tag{1.70}$$

$$\partial\varphi/\partial t = (2e/\hbar)V; \tag{1.71}$$

$$J_x = j\sin\varphi + \sigma V, \tag{1.72}$$

where φ is the phase difference between two superconductors, H is the magnetic field in the barrier, V is the voltage across the barrier, $d = 2\lambda + l$, λ is the penetration depth, and l is the barrier thickness.[52]

If the barrier is regarded as having a capacitance, **C**, per unit area, then Eq. (1.70) and Maxwell's equations give:

$$\left[\partial^2/\partial x^2 - (1/\mathbf{c}^2)(\partial^2/\partial t^2) - (\beta/\mathbf{c}^2)(\partial/\partial t)\right]\varphi = (1/\lambda_0^2)\sin\varphi, \tag{1.73}$$

where $\mathbf{c}^2 = c^2/4\pi dC$ is the phase velocity in the barrier, $\lambda_0^2 = hc^2/\delta\pi edj$ is the penetration depth and $\beta = 4\pi d\mathbf{c}^2\sigma = \sigma/C$ is the damping constant. Anderson (1964) demonstrated that solutions of this equation, representing vortex lines in the barrier, are obtained as solutions of:

$$\partial\varphi^2/\partial x^2 = (1/\lambda_0^2)\sin\varphi, \tag{1.74}$$

which, except for sign, is the equation of a pendulum.

[52]Lenstra *et al.* (1986) have shown an analogy between Josephson-like oscillations and the Sagnac effect.

The Josephson effect is remarkable in the present context for three reasons: (i) with well-defined boundary conditions (the barrier), the phase, ϕ, is a well-defined gauge-invariant variable; and (ii) an equation of motion can be defined in terms of the well-studied pendulum (Barrett, 1987) relating a phase variable to potential energy; (iii) the "free" energy in the barrier is (Lebwohl and Stephen, 1967):

$$F = (\hbar j/2e) \int dx[(1 - \cos\varphi) + 1/2\lambda_0^2(\partial\varphi/\partial x)^2 + 1/2(\lambda_0/\mathbf{c})^2(\partial\varphi/\partial t)^2], \tag{1.75}$$

an equation which provides a (free) energy measure in terms of the differential of a phase variable. The Josephson effect, like the AB effect, demonstrates the registration of physical influences by means of phase changes. The Josephson phase, also like the AB phase, registers field influences.

Jaklevic *et al.* (1965) studied multiply-connected superconductors utilizing Josephson junction tunneling and modulated the supercurrent with an applied magnetic field. The interference "fringes" obtained were found to occur even when the magnetic flux is confined to a region not accessible to the superconductor, i.e., there occurs vector potential modulation of superconducting electron drift velocity. As always, the superconductive state had overlapped field phase coherence, indicating that the modulation effect studied was a local ($\mathbf{A}_\mu$) influence on global phase effects (i.e., the phase order parameter in the barrier).

In the case of the next effect examined, the quantized Hall effect, the effect is crucially dependent upon the gauge invariance of the $\mathbf{A}_\mu$ potential. The result of such gauge invariance is remarkably significant: *an independence of the quantization condition on the density of mobile electrons in a test sample.*

This independence was seen above while examining the remarkable independence in preservation of phase coherence in electrons over distances larger than the atomic spacing or the free path length in the F→CE Aharonov–Bohm effect. In both cases, the primacy and importance of macroscopic, and "mesoscopic", effects are indicated.

1.3.6. *Quantized Hall effect*

The quantized Hall effect[53] has the following attributes:

(1) there is the presence of a Hall conductance $\sigma_{\rm xx}$ in a two-dimensional gas within a narrow potential well at a semiconductor-heterostructure interface e.g., in MOS, quantum well and MOSFET;
(2) the temperature is low enough that the electrons are all in the ground state of the potential well and with the Fermi level being between the Landau levels;
(3) the conductance is quantized with a plateau having $\sigma_{xy} = n\hbar/e^2\sigma_{xy}$ (n an integer) for finite ranges of the gate voltage in which the regular conductance is severely reduced;
(4) together with the well-known Hall effect (1879) condition, (a magnetic field perpendicular to the plane and an electric field in the plane and the electrons drifting in the direction $\mathbf{E} \times \mathbf{B}$), the energy associated with the cyclotron motion of each electron takes on quantized values $(n+1/2)\hbar\omega_c$, where ω_c is the cyclotron frequency at the imposed magnetic field and n is the quantum number corresponding to the Landau level.

The Aharonov–Bohm flux, or **A**-wave, can be generated in such two-dimensional systems and be increased by one flux quantum by changing the phase of the ground state wave function around the system. The quantized Hall effect is thus a macroscopic quantum Hall phenomenon related to the fundamental role of the phase and the $\mathbf{A}_\mu$ potential in quantum mechanics.

An important feature of the quantized Hall effect is the lack of dependence of quantization (integral multiples of $e^2/\hbar$*) on the density of the mobile electrons in the sample tested (but rather on the symmetry of the charge density wave (Tsui et al.* 1982*). Underlying this lack of dependence is a required gauge invariance of the* A_μ *potential.* For example, the current around a metallic loop is equal to the derivative of the total electronic energy, U, of the system with

[53] Von Klitzing *et al.* (1980); Stormer and Tsui (1983).

respect to the magnetic flux through the loop, i.e., with respect to the $\mathbf{A}_\mu$ potential pointing around the loop (Laughlin, 1981):

$$I = (c/L)\partial U/\partial \mathbf{A}. \tag{1.76}$$

As this derivative is non-zero only with phase coherence around the loop, i.e., with an extended state, Eq. (1.76) is valid only if:

$$\mathbf{A} = \mathbf{n}\hbar\mathbf{c}/(\mathbf{eL}), \tag{1.78}$$

i.e., only with a gauge invariance for **A**.

With a gauge invariance defined for **A**, and with the Fermi level in a mobility gap, a vector potential increment changes the total energy, U, by forcing the filled states toward one edge of the total density of states spectrum and the wave functions are affected by a vector potential increment only through the location of their centers. Therefore, *gauge invariance of the* **A** *potential, being an exact symmetry, forces the addition of a flux quantum to result in only an excitation or de-excitation of the total system* (Laughlin, 1981). Furthermore, the energy gap exists in field overlap between the electrons and holes affected by such a perturbation in the way described, rather than in specific local density of states. Thus, the Fermi level lies *in the overlapped field* in a gap in an extended state spectrum and there is no dependence of Hall conductivity on the density of mobile electrons.

Post (1982, 1983) has also implicated the vector potential in the conversion of a voltage/current ratio of the quantized Hall effect into a ratio of period integrals. If **V** is the Hall voltage observed transversely from the Hall current **I**, the relation is

$$\mathbf{V}/\mathbf{I} = \oint \mathbf{A} \Big/ \oint \mathbf{G} = \mathbf{Z}_H = \text{ quantized Hall impedance,} \tag{1.79}$$

where **G** defines the displacement field **D** and the magnetic field **H**. The implication is that:

$$\mathbf{V}/\mathbf{I} = \int_0^T \mathbf{V}dt \Big/ \int_0^T \mathbf{I}dt, \tag{1.80}$$

where

$$\int_0^T \mathbf{V}dt = \oint \mathbf{A} \text{— the quantization of magnetic flux,} \quad (1.81)$$

$$\int_0^T \mathbf{I}dt = \oint \mathbf{G} \text{— the quantization of electric flux,} \quad (1.82)$$

and T is the cyclotron period.

Aoki and Ando (1986) also attribute the universal nature of the quantum Hall effect, i.e., the quantization in units of $e^2/\hbar$ at $T = 0$ for every energy level in a finite system, to a topological invariant in a mapping from the gauge field to the complex wave function. These authors assume that in the presence of external Aharonov–Bohm magnetic fluxes, the vector potential $\mathbf{A}_0$, is replaced by $\mathbf{A}_0 + \mathbf{A}$, where $\mathbf{A} = (A_x, A_y)$. In cylindrical geometry, a magnetic flux penetrates the opening of the cylinder and the vector potential is thought of as two magnetic fluxes, $(\mathbf{\Phi}_x, \mathbf{\Phi}_y) = (A_xL, A_yL)$ penetrating inside and through the opening of a torus when periodic boundary conditions are imposed in both x and y directions for a system of size L. According to the Byers–Yang theorem (Byers and Yang, 1961), the physical system assumes its original state when A_x or A_y increases by $\mathbf{\Phi}_0/L$, where $\mathbf{\Phi}_0 = \hbar c/e$, the magnetic flux quantum.

The next effect examined, the De Haas–Van Alphen effect, also pivots on $\mathbf{A}_\mu$ potential gauge invariance.

1.3.7. *De Haas–Van Alphen effect*

In 1930 de Haas and van Alphen observed what turned out to be susceptibility oscillations with a changing magnetic field which were periodic with the reciprocal field. Landau showed in the same year that for a system of free electrons in a magnetic field, the motion of the electrons *parallel* to the field is *classical*, while the motion of the electrons *perpendicular* to the field is *quantized*; and Peierls showed in 1933 that this holds for free electrons in a metal (with spherical Fermi surface). Therefore, the free energy of the system and thus the magnetic moment ($\mathbf{M} = \partial\mathbf{F}/\partial\mathbf{H}$) oscillates with the magnetic field **H**. This oscillation is the major cause of the de Haas–van Alphen effect.

In 1952 Onsager showed that the frequencies of oscillations are directly proportional to the extremal cross-sections of the Fermi surface perpendicular to the magnetic field. If **p** is the electronic momentum and

$$[\mathbf{p} - (e/c)\mathbf{A}] \tag{1.83}$$

is the canonical momentum (cf. Eq. (1.69), section 1.3.5 Josephson effect), then

$$\oint [\mathbf{p} - (e\mathbf{A}/c) \cdot dl] = (n + \gamma)h, \tag{1.84}$$

where n is an integer and γ is a phase factor. The relation of the **A** vector potential and the real space orbit is

$$\oint \mathbf{A} \cdot dl = \int \nabla \times \mathbf{A} \cdot d\mathbf{S} = \mathbf{HS}, \tag{1.85}$$

where **S** is the area of the orbit in real space. Furthermore, electron paths in momentum space have the same shape as those in real space but changed in scale and turned through 90°, due to the Lorentz force relation: $d\mathbf{p}/dt - (e/c)(\nabla \times \mathbf{H})$.

Therefore, as (i) the area of the orbit in momentum space is $\mathbf{S} = (n + \gamma)(e\hbar\mathbf{H}/c)$, and (ii) the susceptibility is $-(1/\mathbf{H})(\partial F/\partial \mathbf{H})$, which is periodic in $(1/H)$ with period $\Delta(1/H) = 2\pi e/c\hbar S$, there is a direct influence of the **A** vector potential on the de Haas–van Alphen effect due to the phase factor dependence (Eq. (1.84)). Thus the validation of Eq.s (1.83) and (1.84) requires $\mathbf{A}_\mu$ potential gauge invariance. (The relation between the Aharonov-Bohm effect and the quantized Hall effect has been observed by Timp *et al.* 1987).

Two effects have now been examined pivoting on $\mathbf{A}_\mu$ potential gauge invariance. *This gauge invariance implies flux conservation, i.e., an overlapped field conservation law.* The next effect examined, the Sagnac effect, makes explicit the consequences of this overlapped field conservation.

1.3.8. *Sagnac effect*

In 1913 Sagnac demonstrated a fringe shift by rotating an interferometer (with a polygonal interference loop traversed in opposite senses) at high speed[54] (Fig. 1.4). Einstein's general theory of relativity predicts a phase shift proportional to the angular velocity and to the area enclosed by the light path — not because the velocity of the two beams is different, but because they each have their own time. However, *the AB, AAS and the topological phase effects deny Lorentz invariance to the electromagnetic field as any field's natural and inevitable implication, i.e., Lorentz invariance is not "built-in" to the Maxwell theory — it is a gauge implied by special* $\mathbf{A}_\mu$ *potential conditions, i.e., special boundary conditions imposed on the electromagnetic field.* Therefore, the Einstein interpretation pivots

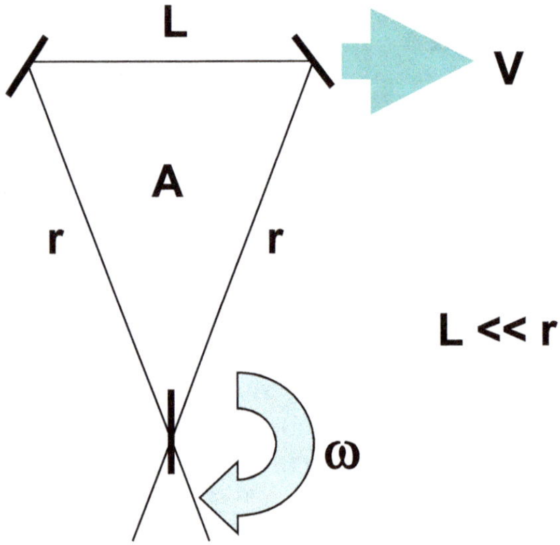

Fig. 1.4. The Sagnac interferometer in which the center of rotation coincides with the beam splitter location. The Sagnac phase shift is independent of the location of the center of rotation and the shape of the area. The phase shift along L is independent of r. After Silvertooth (1986).

[54]Sagnac (1913a,b; 1914).

on unproven boundary conditions and the effect is open to other, competing, explanations (cf. Forder, 1984).

A different explanation is offered by the Michelson *et al.* experiments[55] of 1924–1925. These investigators predicted a phase shift more simply on the basis of a difference in the velocity of the counter propagating beams and the earth rotating in a stationary ether without entrainment. (It should be noted that the beam path in the well-known Michelson–Morley 1886 interferometer does not enclose a finite surface area. Therefore this experiment cannot be compared with the experiments and effects examined in the present review, and, in fact, according to these more recent experiments, no fringe shift can be expected as an outcome of a Michelson–Morley experiment, i.e, the experiment was not a test for the presence of an ether.)

Post (1967) argued that the Sagnac effect demonstrates that the spacetime formulations of the Maxwell equations do not make explicit the constitutive properties of free space. The identification: $\mathbf{E} = \mathbf{D}, \mathbf{H} = \mathbf{B}$, in the absence of material polarization mechanisms in free space is the so called *Gaussian field identification* (Post, 1978). This identification is equivalent to an unjustified adoption of Lorentz invariance. However, the Sagnac effect and the well-used ring laser gyro on which it is based indicate that in a rotating frame the Gaussian identity does not apply. This requirement of metric independence was proposed by Van Dantzig (1934). In order to define the constitutive relations between the fields $\mathbf{E}$ and $\mathbf{B}$ constituting a covariant six-vector $\mathbf{F}_{\lambda\nu}$, and the fields $\mathbf{D}$ and $\mathbf{H}$, constituting a contravariant six-vector, $\mathbf{G}^{\lambda\nu}$, the algebraic relation (Post, 1978):

$$\mathbf{G}^{\lambda\nu} = 1/2\chi^{\lambda\nu\sigma\kappa}\mathbf{F}_{\lambda\nu} \tag{1.86}$$

was proposed, where $\chi^{\lambda\nu\sigma\kappa}$ is the constitutive tensor and Eq. (1.86) is the constitutive map. The generally invariant vector d'Alembertian (wave equation) is

$$\partial_\nu\chi^{\lambda\nu\sigma\kappa}\partial_\sigma\mathbf{A}_\kappa = 0, \tag{1.87}$$

indicating the vector potential dependence.

[55]Michaelson (1924, 1925); Michaelson *et al.* (1925).

The pivotal role of the vector potential is due to the flux conservation, which is a *global* conservation law (Post, 1974, 1982). The *local* conservation law of flux:

$$d\mathbf{F} = 0 \tag{1.88}$$

excludes a role for the **A** potential (**F** is *inexact*). However, only if:

$$\oint \mathbf{F} = 0 \tag{1.89}$$

is it possible to state $d\mathbf{A} = \mathbf{F}$ (**F** is *exact*). In other words, $d\mathbf{F} = 0$ implies $\oint \mathbf{F} = 0$ *only if the manifold over which* **F** *is defined is compact and simply connected,*[56] contrasting with e.g., 1-connectedness (contractable circles), 2-connectedness (contractable spheres), and 3-connectedness (contractable 3-spheres).

Post (1962; 1972a,b) argued that the constitutive relations of the medium-free fields **E** and **H** to the medium left out treatment of free space as a "medium". If **C** is the differential 3-form of charge and current density, then the *local* conservation of charge is expressed by

$$d\mathbf{C} = 0, \tag{1.90}$$

and the *overlapped field* definition is:

$$\mathbf{C} = d\mathbf{G}. \tag{1.91}$$

The Post relation is in accord with the symmetry of spacetime and momentum–energy required by the reciprocity theory of Born (1949) and, more recently, that of Ali (Ali, 1985; Ali and Prugovecki, 1986). Placing these issues in a larger context, Hayden (1991a) has argued that the classical interpretation of the Sagnac effect indicates that the speed of light is not constant. Furthermore, Hayden (1991b), emphasizing the distinction between kinematics (which concerns space and time) and mechanics (which concerns mass, energy, force and momentum), argues that time dilation (a kinematic issue) is

[56] For example, simply-connected or 1-connectedness is associated (with contractable circles), 2-connectedness (with contractable spheres), and 3-connectedness (with contractable 3-spheres).

difficult to distinguish from changes in mass (a mechanics issue) due to the way time is measured. The demonstration of an analogy between Josephson-like oscillations and the Sagnac effect (Lenstra *et al.* 1986) supports this viewpoint.

1.3.9. *Summary*

In summary the following effects have been examined:

(i) The Aharonov–Bohm and Altshuler–Aronov–Spival effects in which changes in the $\mathbf{A}_\mu$ potential at a third location indicates differences in the $\mathbf{A}_\mu$ field along *two* trajectories at *two* other locations.

(ii) Topological phase effects in which changes spin direction or polarization defined by the $\mathbf{A}_\mu$ potential at one location, a, is different from that at another location, b, due to topological winding of the trajectory between the *two* locations a and b.

(iii) Stokes' theorem which requires precise boundary conditions for *two* fields — the local and overlapped fields — for exact definition in terms of the $\mathbf{A}_\mu$ potential.

(iv) Ehrenberg and Siday's derivation of the refractive index which describes propagation between *two* points in a medium and which requires gauge-invariance of the $\mathbf{A}_\mu$ potential.

(v) The Josephson effect which implicates the $\mathbf{A}_\mu$ potential as a many-to-one operator connecting *two* fields.

(vi) The quantized Hall effect which requires gauge invariance of the $\mathbf{A}_\mu$ potential in the presence of *two* fields.

(vii) The de Haas–van Alphen effect which requires gauge invariance of the $\mathbf{A}_\mu$ potential in the presence of *two* fields.

(viii) The Sagnac effect which requires flux conservation, i.e., gauge invariance of the $\mathbf{A}_\mu$ potential in comparing *two* fields before and after movement.

All these effects pivot on a physical definition of $\mathbf{A}_\mu$ potentials. There are other effects that also exemplify field many-to-one mapping. But here we wish only to make a case, not a comprehensive survey. In the next section, the theoretical reasons for questioning the completeness of Maxwell's theory are examined as well as the reasons for the physical effectiveness of the $\mathbf{A}_\mu$ potentials in the presence of two fields. The $\mathbf{A}_\mu$ potentials have an ontology or physical meaning as *local* operators mapping *local* e.m. fields into *overlapped field* spatiotemporal conditions. This operation is measurable if there is a second comparative mapping of the local conditioned fields in a many-to-one fashion (multiple connection).

1.4. Theoretical Reasons for Questioning the Completeness of Maxwell's Theory

Yang[57] interpreted the electromagnetic field in terms of a non-integrable (i.e., path dependent) phase factor by an examination of Dirac's monopole field.[58] According to this interpretation, the Aharonov–Bohm effect is due to the existence of this phase factor whose origin is due to the topology of connections on a fiber bundle.

The *phase factor*,

$$\exp\left[(ie/\hbar c)\oint \mathbf{A}_\mu dx^\mu\right], \tag{1.92}$$

according to this view, is physically meaningful, but not the *phase*,

$$\left(ie/\hbar c\oint \mathbf{A}_\mu dx^\mu\right), \tag{1.93}$$

which is ambiguous because different phases in a region may describe the same physical situation. The phase factor, on the other hand, can distinguish different physical situations having the same field strength but different action. Referring to Fig. 1.5, the phase factor for any path from, say, P to Q is

$$\mathbf{\Phi}_{QP} = \exp\left[(ie/\hbar c)\int_Q^P \mathbf{A}_\mu dx^\mu\right]. \tag{1.94}$$

For a *static magnetic monopole* at an origin defined by the spherical coordinate, $r = 0$, θ with azimuthal angle φ, and considering

[57] Yang and Mills (1954); Wu and Yang, (1975); Yang (1970, 1974).
[58] Dirac (1931, 1948).

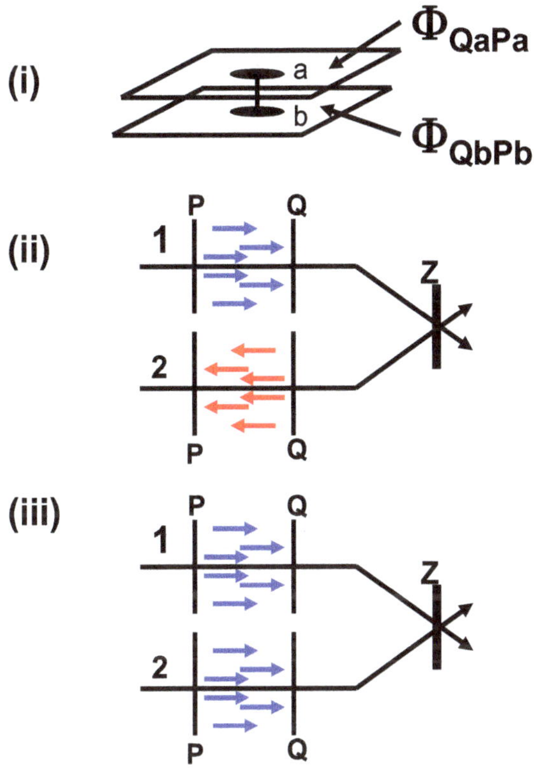

Fig. 1.5. (i) The overlap area (Z in (ii) and (iii)) showing a mapping from location a to b. The phase factor Φ_{QaPa} is associated with the e.m. field which arrived at Z through path 1 in (ii) and (iii) and Φ_{QbPb} with the e.m. field which arrived at Z through path 2 in (ii) and (iii). (ii) In paths 1 and 2 the e.m. fields are conditioned by an **A** field between P and Q oriented in the direction indicated by the arrows. *Note the reversal in direction of the* **A** *field in paths* 1 *and* 2, hence $S(\mathrm{P}) \neq S(\mathrm{Q})$ and $\Phi_{\mathrm{QaPa}} \neq \Phi_{\mathrm{QbPb}}$. (iii) Here the conditioning **A** fields *are oriented in the same direction*, hence there is no noticeable gauge transformation and no difference noticeable in the phase factors $S(\mathrm{P}) = S(\mathrm{Q})$ and $\Phi_{\mathrm{QaPa}} = \Phi_{\mathrm{QbPb}}$. After Wu and Yang (1978).

the region R of all spacetime *other than this origin*, the gauge transformation in the overlap of two regions, a and b, is

$$S_{ab} = \exp(-i\alpha) = \exp[(2ige/\hbar c)\varphi], \quad 0 \leq \varphi \leq 2\pi, \tag{1.95}$$

where g is the monopole strength.

This is an allowed gauge transformation if, and only if

$$2ge/\hbar c = \text{an integer} = D \tag{1.96}$$

which is Dirac's quantization. Therefore

$$S_{ab} = \exp(iD\varphi). \tag{1.97}$$

In the overlapping region there are two possible phase factors $\mathbf{\Phi}_{\mathrm{QaPa}}$ and $\mathbf{\Phi}_{\mathrm{QbPb}}$ and

$$\mathbf{\Phi}_{\mathrm{QaPa}} S(\mathrm{P}) = S(\mathrm{Q}) \mathbf{\Phi}_{\mathrm{QbPb}}, \tag{1.98}$$

a relation which states that $(\mathbf{A}_\mu)_{\mathrm{a}}$ and $(\mathbf{A}_\mu)_{\mathrm{b}}$ are related by a gauge transformation factor.

The general implication is that for a gauge with any field defined on it, the total magnetic flux through a sphere around the origin $r = 0$ is independent of the gauge field and only depends on the gauge (phase):

$$\varphi_{\mu\nu} dx^\mu dx^\nu = (-i\hbar c/e) \int \partial/\partial x^\mu (\ln S_{ab}) dx^\mu, \tag{1.99}$$

where the integral is taken around any loop around the origin $r = 0$ in the overlap between the R_{a} and R_{b}, as, for example, in an equation for a sphere $r = 1$. As S_{ab} is single valued, this integral must be equal to an integral multiple of a constant (in this case $2\pi\mathrm{i}$).

Another implication is that if the $\mathbf{A}_\mu$ potentials originating from, or passing through, two or more different local positions are gauge invariant when compared at another, again different, local position, then the referent providing the basis or metric for the comparison of the phase differences at this local position is a unit magnetic monopole. The unit monopole, defined at $r = 0$, is unique in not having any internal degrees of freedom (Weinberg, 1980). Furthermore, both the monopole and charges are topologically conserved, *but whereas electric charge is topologically conserved in* U(1) *symmetry, magnetic charge is only conserved in* SU(2) *symmetry.*

Usually there is no need to invoke the monopole concept as the $\mathbf{A}_\mu$ field is, as emphasized here, treated as a mathematical, not physical, construct in contemporary classical physics. However, in quantum physics the wave function satisfies a partial differential equation coupled to boundary conditions. The boundary conditions in the doubly-connected region outside of the solenoid volume in an Aharonov–Bohm experiment results in the single valuedness of the wavefunction, which is the reason for quantization. Usually, e.g., in textbooks explaining the theory of electromagnetism as noted above, Stokes' theorem is written:

$$\oint \mathbf{A} dx = \mathbf{H} \cdot ds = \int_S (\nabla \times \mathbf{A}) \cdot \mathbf{n} da, \tag{1.100}$$

and no account is taken of spacetime overlap of regions with fields derived from different sources having undergone different spatiotemporal conditioning, and no boundary conditions are taken into account. Therefore, no quantization is required.

There is no lack of competing opinions on what the theoretical basis is for the magnetic monopole implied by gauge-invariant $\mathbf{A}_\mu$ potentials[59] (cf. Goldhaber and Tower, 1990 for a guide to the literature). The Dirac magnetic monopole is an anomalously-shaped (string) magnetic dipole at a singularity. The Schwinger magnetic monopole[60] is essentially a double singularity line. However, gauge-invariant $\mathbf{A}_\mu$ potentials are the local manifestation of overlapped field constructs. This precludes the existence of *isolated* magnetic monopoles, but permits them to exist *in overlapped fields* in any situation with the requisite energy conditioning. Others[61] have described such situations.

More recently, Zeleny (1991) showed that Maxwell's equations and the Lorentz force law can be derived, not by using invariance of the action (Hamilton's principle) nor by using constants of the motion (Lagrange's equations), but by considerations of symmetry. If, in the derivation, the classical $\mathbf{A}$ field is dispensed with in favor of the electromagnetic tensor $\mathbf{F}$, classical magnetic monopoles are

[59]Goddard and Olive (1978); Bogomol'nyi (1976); Montonen and Olive (1977); Atiyah and Hitchen (1980); Craigie (1986).

[60]Schwinger (1966, 1969).

[61]Wu and Yang (1975, 1976a,b); t'Hooft (1971, 1974); Polyakov (1974); Prasad and Sommerfeld (1975).

obtained, which are without strings and can be extended particles. Such particles are accelerated by a magnetic field and bent by an electric field.

Related to mechanisms of monopole generation is the Higgs field, $\mathbf{\Phi}$, approach to the vacuum state,[62] which has recently been detected. The field, in some scenarios, breaks a higher-order symmetry field, e.g., SU(2), **G**, into **H** of U(1) form. The **H** field is then proportional to the electric charge.

There are at least five types of monopoles presently under consideration: (1) the Dirac monopole, a point singularity with a string source. The $\mathbf{A}_\mu$ field is defined everywhere except on a line joining the origin to infinity, which is occupied by an infinitely long solenoid, so that $\mathbf{B} = \nabla \cdot \mathbf{A}$ (a condition for the existence of a magnetic monopole). Dirac's approach assumes that a particle has either electric or magnetic charge but not both. (2) Schwinger's approach, on the other hand, permits the consideration of particles with both electric and magnetic charge, i.e., dyons. (3) The 't Hooft–Polyakov monopole, which has a smooth internal structure but without the need for an external source. There is, however, the requirement for a Higgs field. The 't Hooft–Polyakov model can be put in the Dirac form by a gauge transformation (Goddard and Olive, 1978). (4) The Bogomol'nyi–Prasad–Sommerfeld monopole is one in which the Higgs field is massless, long range and with a force which is always attractive. (5) The Wu–Yang monopole requires no Higgs field, has no internal structure and is located at the origin. It requires multiply-connected fields. The singular string of the Dirac monopole can be moved arbitrarily by a gauge transformation (Brandt and Primack, 1977). Therefore, the Dirac and the Wu–Yang monopoles can be made compatible. The Higgs field formalism can also be related to that of Wu–Yang in which only the exact symmetry group appears.

Goddard and Olive (1978) demonstrated that there are two conserved currents for a monopole solution: the usual Noether current whose conservation depends on the equations of motion;

[62] Higgs (1964a,b, 1968).

and a topological current whose conservation is independent of the equations of motion.

Yang (1983) showed that if spacetime is divided into two overlapping regions in both of which there is a vector potential $\mathbf{A}$ with gauge transformation between them in the overlap regions, then the proper definition of Stokes' theorem when the path integral goes from region, 1, to another, 2, is (Wu and Yang, 1976a,b):

$$\int_{\mathbf{A}}^{\mathbf{C}} \mathbf{A} dx = \int_{\mathbf{A}}^{\mathbf{B}} \mathbf{A}_1 dx = \int_{\mathbf{B}}^{\mathbf{C}} \mathbf{A}_2 dx + \beta(B). \tag{1.101}$$

The β function is defined by the observation that in the region of overlap the difference of the vector potentials $\mathbf{A}_1 - \mathbf{A}_2$ is curlless as both potentials give the same local electromagnetic field.

There are also general implications. Gates (1986) takes the position that all the fundamental forces in nature arise as an expression of gauge invariance. If a phase angle $\theta(x,t) = -(i/2)\ln[\psi/\hat{\psi}]$ is defined for quantum mechanical systems, then although the difference $\theta(x_1,t) - \theta(x_2,t)$ is a *gauge-dependent* quantity, the expression:

$$\theta(x_1,t) - \theta(x_2,t) + (e/\hbar c)\int_{x_1}^{x_2} ds\mathbf{A}(s,t) \tag{1.102}$$

is *gauge invariant.* (Note that according to the Wu–Yang interpretation, the last expression should be $\exp\left[(e_0/\hbar c)\int_{x_1}^{x_2} ds\mathbf{A}(s,t)\right]$. Therefore, any measureable quantity which is a function of such a difference in phase angles must also depend on the vector potentials shown. Setting the expression (1.102) to zero gives a general description of both the Aharonov–Bohm and Josephson effects. Substituting $\exp\left[(e_0/\hbar c)\int_{x_1}^{x_2} ds\mathbf{A}(s,t)\right]$ for the final term gives a description of the topological phase effect.

The phenomena described above are a sampling of a range of effects. There are probably many yet to be discovered, or provided with the honorific title of an "effect". A unifying theme of all of them is that the physical effect of the $\mathbf{A}_\mu$ potentials is only describable (a) when *two* or more fields undergo different spatiotemporal conditioning *and* there is also a possibility of cross-comparison (many-to-one mapping) or, equivalently, (b) in the situation of a field trajectory with a *beginning* (giving the field before the spatiotemporal conditioning) and an *end* (giving the field after the conditioning) *and* again a possibility of cross comparison. Setting boundary conditions to an e.m. field gives gauge invariance but without necessarily providing the conditions for detection of the gauge invariance. The gauge-invariant $\mathbf{A}_\mu$ potential field operates on an e.m. field state to an extent determined by overlapped field symmetries defined by spatiotemporal conditions, but the effect of this operation or conditioning is only detectable under the overlapped field conditions (a) and (b). With no interfield mapping or comparison, as in the case of the solitary electromagnetic field, the $\mathbf{A}_\mu$ fields remain ambiguous, but this situation occurs only if no boundary conditions are defined — an ambiguous situation even for the electric and magnetic fields. Therefore, the $\mathbf{A}_\mu$ potentials in all useful situations have a meaningful physical existence related to boundary condition choice — even when no situation exists for their comparative detection. *What is different between the $\mathbf{A}_\mu$ field and the electric and magnetic fields is that the ontology of the $\mathbf{A}_\mu$ potentials is related to the overlapped field spatiotemporal boundary conditions in a way in which the local electric and magnetic fields are not.* Due to this overlapped field spatiotemporal (boundary condition) dependence, the operation of the $\mathbf{A}_\mu$ potentials is a more-than-one-to-one, or local to overlapped field mapping of individual e.m. fields, the nature of which is examined in Sec. 1.5.2. The detection of such mappings is only within the context of a *second* comparative projection, but this time, overlapped field-to-local.

This section addressed theoretical reasons for questioning the completness of U(1) symmetry, or Abelian, Maxwell theory in the presence of *two* local fields separated from the overlapped condition. In the next section, a pragmatic reason is offered: propagating velocities of e.m. fields in lossy media cannot be calculated in U(1) Maxwell theory. The theoretical justification for physically defined $\mathbf{A}_\mu$ potentials lies in the application of Yang–Mills theory — not to high-energy fields, where the theory first found application — but to low-energy fields crafted to a specific group of transformation rules by boundary conditions. This is a new application of Yang–Mills theory.

1.5. Pragmatic Reasons for Questioning the Completeness of Maxwell's Theory

1.5.1. *Harmuth's Ansatz*

A satisfactory concept permitting the prediction of the propagation velocity of e.m. signals does not exist within the framework of Maxwell's theory (Harmuth, 1986a,b,c).[63] The calculated group velocity fails for two reasons: (i) it is almost always larger than the speed of light for RF transmission through the atmosphere; (ii) its derivation implies a transmission rate of information equal to zero. Maxwell's equations also do not permit the calculation of the propagation velocity of signals with a defined and limited bandwidth propagating in a lossy medium *and* all the published solutions for propagation velocities assume sinusoidal (linear) signals.

In order to remedy this state of affairs, Harmuth proposed an amendment of Maxwell's equations, which I have called: the *Harmuth Ansatz* (Barrett, 1988, 1989). The proposed amended equations (in Gaussian form) are as follows:

- Coulomb's law:

$$\nabla \cdot D = 4\pi\rho_e; \tag{1.103}$$

- Maxwell's generalization of Ampère's law:

$$\nabla \times \mathbf{H} = (4\pi/c)\,\mathbf{J}_e + (1/c)\,\partial\mathbf{D}/\partial t; \tag{1.104}$$

- Presence of free magnetic poles postulate:

$$\nabla \cdot \mathbf{B} = \rho_m; \tag{1.105}$$

- Faraday's law with magnetic monopole:

$$\nabla \times \mathbf{E} + (1/c)\partial\mathbf{B}/\partial t + (4\pi/c)\mathbf{J}_m = 0; \tag{1.106}$$

[63]Harmuth (1987a,b); Harmuth *et al.* (1987); Neatrour (1987); Wait (1987); Kuester (1987); Djordjvic and Sarkar (1987); Hussain (1987); Gray and Boules (1987); Barrett (1988, 1989).

and the constitutive relations:

$$\mathbf{D} = \varepsilon\mathbf{E}; \tag{1.107}$$

$$\mathbf{B} = \mu\mathbf{H}; \tag{1.108}$$

$$\mathbf{J}_e = \sigma\mathbf{E} \text{ (electric Ohm's law)}; \tag{1.109}$$

$$\mathbf{J}_m = s\mathbf{E} \text{ (magnetic Ohm's law)}, \tag{1.110}$$

where $\mathbf{J}_e$ is electric current density, $\mathbf{J}_m$ is magnetic current density, ρ_e is electric charge density, ρ_m is magnetic charge density, σ is electric conductivity and s is magnetic conductivity.

Setting $\rho_e = \rho_m = \nabla \cdot \mathbf{D} = \nabla \cdot \mathbf{B} = 0$, for free space propagation gives:

$$\nabla \times \mathbf{H} = \sigma\mathbf{E} + \varepsilon\partial\mathbf{E}/\partial t; \tag{1.111}$$

$$\nabla \times \mathbf{E} + \mu\partial\mathbf{H}/\partial t + s\mathbf{H} = 0; \tag{1.112}$$

$$\varepsilon\nabla \cdot \mathbf{B} = \mu\nabla \cdot \mathbf{B} = 0, \tag{1.113}$$

and the following equations of motion:

$$\partial\mathbf{E}/\partial y + \mu\partial\mathbf{H}/\partial t + s\mathbf{H} = 0, \tag{1.114}$$

$$\partial\mathbf{H}/\partial y + \varepsilon\partial\mathbf{E}/\partial t + \sigma\mathbf{E} = 0. \tag{1.115}$$

Differentiating Eqs. (1.114) and (1.115) with respect to y and t permits the elimination of the magnetic field resulting in (Harmuth, 1986a, Eq. (21)):

$$\partial^2\mathbf{E}/\partial y^2 - \mu\varepsilon\partial^2\mathbf{E}/\partial t^2 - (\mu\sigma + \varepsilon s)\partial\mathbf{E}/\partial t - s\sigma\mathbf{E} = 0, \tag{1.116}$$

which is a two-dimensional Klein–Gordon equation (without boundary conditions) in the sine-Gordon form:

$$\partial^2\mathbf{E}/\partial y^2 - (1/c^2)\partial^2\mathbf{E}/\partial t^2 - \alpha\sin(\beta\mathbf{E}(y,t)) = 0; \tag{1.117}$$

$$\alpha\sin(\beta\mathbf{E}(y,t)) \approx -[\alpha\beta\mathbf{E}/\partial t - O(\mathbf{E})]; \tag{1.118}$$

$$\alpha = \exp(+\mu\sigma); \tag{1.119}$$

$$\beta = \exp(+\varepsilon\sigma)\beta, \tag{1.120}$$

where $(\partial^2 E/\partial y^2 - (1/c^2)\partial^2 E/\partial t^2)$ is the "non-linear" term and $(\alpha \sin(\beta E(y,t)))$ is the dispersion term. This match of "non-linearity" and dispersion permits soliton solutions and the field described by Eq. (1.117) has a "mass", $m = \sqrt{\alpha\beta}$.

Equation (1.117) may be derived from the Lagrangian density:

$$\mathbf{L} = (1/2)[(\partial \mathbf{E}/\partial y)^2 - (\partial \mathbf{E}/\partial t)^2] - V(\mathbf{E}), \tag{1.121}$$

where

$$V(\mathbf{E}) = (\alpha/\beta)(1 - \cos \beta\mathbf{E}). \tag{1.122}$$

The wave equation for $\mathbf{E}$ has a solution which can be written in the form:

$$\mathbf{E}(y,t) = \mathbf{E}_E(y,t) = \mathbf{E}_0[w(y,t) + \mathbf{F}(y)], \tag{1.123}$$

where $\mathbf{F}(y)$ indicates that an electric step function is the excitation.

A wave equation for $\mathbf{F}(y)$ is

$$d^2\mathbf{F}/dy^2 - s\sigma\mathbf{F} = 0, \tag{1.124}$$

with solution:

$$\mathbf{F}(y) = A_{00}\exp[-yL] + A_{01}\exp[y/\mathbf{L}], \quad \mathbf{L} = (s\sigma)^{-1/2} \tag{1.125}$$

Boundary conditions require $A_{01} = 0$ and $A_{00} = 1$, therefore:

$$\mathbf{F}(y) = \exp[-y/\mathbf{L}]. \tag{1.126}$$

Insertion of Eq. (1.123) into Eq. (1.126) gives (Harmuth, 1986, Eq. (40)):

$$\partial^2 w/\partial y^2 - \mu\varepsilon\partial^2 w/\partial t^2 - (\mu\sigma + \varepsilon s)\partial w/\partial t - s\sigma w = 0,$$

which we can again put into sine-Gordon form:

$$\partial^2 w/\partial y^2 - (1/c^2)\partial^2 w/\partial t^2 - \alpha \sin(\beta w(y,t)) = 0. \tag{1.127}$$

Harmuth (1986) developed a solution of (1.123) by seeking a general solution of $w(y,t)$ using a separation of variables method (and setting s to zero after a solution is found). This solution works well, but we now indicate another solution. The solutions to the sine-Gordon equation (1.117) are the hyperbolic tangents:

$$\mathbf{E}(y) = \pm\left(\frac{8\sqrt{\alpha\beta}}{\beta^2}\right)\tan^{-1}\left(\exp\left[\frac{y-y_0-ct}{\sqrt{1-c^2}}\right]\right) \tag{1.128}$$

where $c = \sqrt{1/(\mu\varepsilon)}$, which describe solitons. It is also well-known that the sine-Gordon and the Thirring (1958) models are equivalent (Goddard and Olive, 1978) and that both admit two currents: one a Noether current, and the other a topological.

The following remarks may now be made: the introduction of $\mathbf{F}(y) = \exp[-y/\mathbf{L}]$, according to the Harmuth Ansatz (1986a, p. 253) provides integrability. It is well known that soliton solutions require complete integrability. According to the present view, $\mathbf{F}(y)$ also provides the problem with boundary conditions, the *necessary condition* for $\mathbf{A}_\mu$ potential invariance. Equation (1.126) is, in fact, a phase factor (Eqs. (1.92), (1.94), (1.95) and (1.96)). Furthermore, Eq. (1.123) is of the form of Eq. (1.102). Therefore the Harmuth Ansatz amounts to a definition of boundary conditions — i.e., obtains the condition of separate electromagnetic field comparison by overlapping fields — *which permits complete integrability and soliton solutions of the extended Maxwell's equations.* Furthermore, it was already seen, above, that with boundary conditions defined, the $\mathbf{A}_\mu$ potentials are gauge invariant implying a magnetic monopole and charge. It is also known that the magnetic monopole and charge constructs only exist under certain field symmetries. In the next section, methods are presented for conditioning fields into those higher order symmetries.

1.5.2. *Conditioning the electromagnetic field to altered symmetry: Stokes' interferometers and Lie algebras*

The theory of Lie algebras offers a convenient summary of the interaction of the $\mathbf{A}_\mu$ potential operators with the $\mathbf{E}$ fields (Eisenhart, 1933; Hodge, 1959). The relevant parts of the theory are as follows. A manifold, $\mathbf{L}$, is a set of elements in one-to-one correspondence with the points of a vector manifold $\mathbf{M}$. $\mathbf{M}$ is a set of vectors called: points of **M**. A Lie group, **L**, is a group which is also a manifold on which the group operations are continuous. There exists an invertible function, **T**, which maps each point x in **M** to a group element $X = \mathbf{T}(\mathrm{x})$ in **L**. The group **M** is a global parameterization of the group **L**.

If $\partial = \partial_x$ is the derivative with respect to a point on a manifold **M**, then *the Lie bracket* is

$$[a, b] = a \cdot \partial b - b \cdot \partial a = a\nabla \cdot b - b \cdot \nabla a, \qquad (1.129)$$

where a and b are arbitrary vector-valued functions. Alternatively, with $\wedge$ signifying the outer product (Hestenes and Sobczyk, 1984; Hestenes, 1987), then:

$$[a, b] = \partial \cdot (a \wedge b) - b\partial \cdot a + a\partial \cdot b, \qquad (1.130)$$

showing that the Lie bracket preserves tangency.

The fundamental theorem of Lie group theory is: the Lie bracket $[a, b]$ of differential fields on any manifold is again a vector field. A set of vector fields, $a, b, c \ldots$ on any manifold form a Lie algebra if it is closed under the Lie bracket and all fields satisfy the Jacobi identity:

$$[[a, b], c] + [[b, c], a] + [[c, a], b] = 0. \qquad (1.131)$$

If $c = 0$, then

$$[a, b] = 0. \qquad (1.132)$$

The $\mathbf{A}_\mu$ potentials effect mappings, $\mathbf{T_1}$, from the global field to the **E** local fields, considered as group elements in **L**; and there must be a second mapping, $\mathbf{T_2}$, of those separately conditioned **E** fields considered now global, onto a single local field for the $\mathbf{T_1}$ mappings to be detected (measureable). That is, in the Aharonov–Bohm situation (and substituting fields for electrons), if the **E** fields, traversing the two paths, are $\mathbf{E}_1$ and $\mathbf{E}_2$, and those fields before and after interaction with the $\mathbf{A}_\mu$ field are $\mathbf{E}_{1\mathrm{i}}$ and $\mathbf{E}_{1\mathrm{f}}$ and $\mathbf{E}_{2\mathrm{i}}$ and $\mathbf{E}_{2\mathrm{f}}$ respectively, then $\mathbf{E}_{1\mathrm{f}} + \mathbf{E}_{2\mathrm{f}} = \mathbf{T}(\mathbf{E}_{1\mathrm{i}} + \mathbf{E}_{2\mathrm{i}})$, where $x_1 = \mathbf{E}_{1\mathrm{i}}$ and $x_2 = \mathbf{E}_{2\mathrm{i}}$ are points in **M**, and $X = (\mathbf{E}_{1\mathrm{f}} + \mathbf{E}_{2\mathrm{f}})$ is considered a *group* point in **L** and $\mathbf{T} = \mathbf{T_1} + \mathbf{T_2}$, $\mathbf{T_1} = \mathbf{T_1}^{-1}$. In the same situation, although $(\mathbf{E}_{1\mathrm{f}} - \mathbf{E}_{2\mathrm{f}}) = \exp(i\hbar/e) \oint \mathbf{A}_\mu dx^\mu = \mathbf{\Phi}$ (i.e., the phase factor detected at Z in a separate second mapping, $\mathbf{T_2}$, in Fig. 1.6 can be

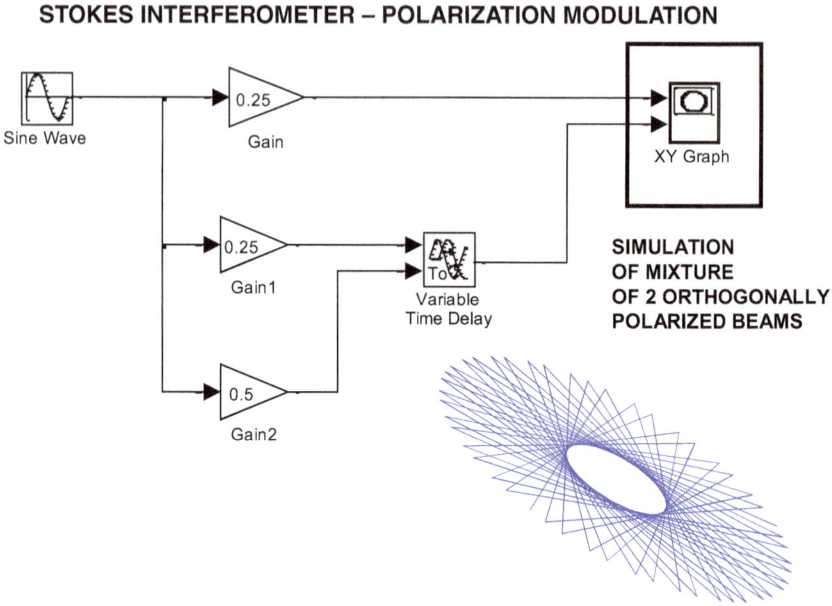

Fig. 1.6. Waveguide system paradigm for polarization modulated ($\partial\varphi/\partial t$) wave emission. This is a completely adiabatic system in which oscillating energy enters from the left and exits from the right. On entering from the left, energy is divided into two parts equally. One part, of amplitude $E/2$, is polarization rotated and used in providing phase modulation, $\partial\varphi/\partial t$ — this energy is spent (absorbed) by the system in achieving the phase modulation; the other part, of amplitude $E/2$, is divided into two parts equally, so that two oscillating waveforms of amplitude $E/4$ are formed for later superposition at the output. Due to the phase modulation of one of them with respect to the other, $0 < \phi < 360°$, and their initial orthogonal polarization, the output is of continuously varying polarization. The choice of wave division into two parts equally is arbitrary. From (Barrett, 1991a,b).

ascribed to a non-integrable (path independent) phase factor) the influence of the first, $\mathbf{T_1}$, mapping or conditioning of $\mathbf{E}_{1\mathrm{i}} + \mathbf{E}_{2\mathrm{i}}$ by the $\mathbf{A}_\mu$ operators along the separate path trajectories *preceded* that second mapping, $\mathbf{T_2}$, at Z. Therefore, the $\mathbf{A}_\mu$ potential field operators produce a mapping of the global spatiotemporal conditions onto local fields, which, in the case we are considering, are the separate $\mathbf{E}_{1\mathrm{i}} + \mathbf{E}_{2\mathrm{i}}$ fields. Thus, according to this conception, the $\mathbf{A}_\mu$ potentials are local operator fields mapping the many-to-one gauge ($\mathbf{T_1} : \mathbf{M} \to \mathbf{L}$), whose effects are detectable at a later spatiotemporal position only at an overlapping (X group) point, i.e., by a second mapping ($\mathbf{T_2} : \mathbf{L} \to \mathbf{M}$), permitting comparisons of the differently conditioned fields in a many-to-one (global-to-local) summation.

If $a = \mathbf{E}_{1i}$ and $b = \mathbf{E}_{2i}$ where $\mathbf{E}_{1i}$ and $\mathbf{E}_{2i}$ are local field intensities and $c = \mathbf{A}_\mu$, i.e., $\mathbf{A}_\mu$ is a local field mapping ($\mathbf{T_1} : \mathbf{M} \to \mathbf{L}$) according to gauge conditions specified by boundary conditions, then the field interactions of a, b and c, or $\mathbf{E}_{1i}$, $\mathbf{E}_{2i}$ and $\mathbf{A}_\mu$ are described by the Jacobi identity (Eq. (1.131)). If $c = \mathbf{A}_\mu = 0$, then $[a, b] = 0$. With the Lorentz gauge (or boundary conditions), the $\mathbf{E}_{1i}$, $\mathbf{E}_{1f}$, $\mathbf{E}_{2i}$ and $\mathbf{E}_{2f}$ field relations are described by SU(2) symmetry. With other boundary conditions and no separate $\mathbf{A}_\mu$ conditioning, the $\mathbf{E}_{1i}$ and $\mathbf{E}_{2i}$ fields (there are no $\mathbf{E}_{1f}$ and $\mathbf{E}_{2f}$ fields in this situation) are described by U(1) symmetry relations.

The $\mathbf{T_1}$,$\mathbf{T_2}$ mappings can be described by classical control theory analysis and the $\mathbf{A}_\mu$ potential conditioning can be given a physical waveguide interferometer representation (cf. Tellegen, 1948). The waveguide system considered here is completely general in that the output can be phase, frequency and amplitude modulated. It is an adiabatic system (lossless) and only three of the lines are waveguides — the input, the periodically delayed line, and the output. Other lines shown are energy-expending, phase-modulating lines. The basic design is shown in Fig. 1.6. In this figure, the input is $\mathbf{E} = \mathbf{E}\exp(i\omega t)$. The output is

$$\mathbf{E}_{\mathrm{OUT}} = (\mathbf{E}/4)\exp(i\omega t) + (\mathbf{E}/4)\exp(i[\omega + \exp(i\omega t) - 1]t), \quad (1.133)$$

and where, referring to Fig. 1.6, $\varphi = \mathbf{F}(E/2)$ and $\partial\varphi/\partial t = \mathbf{F}'(E/2)$, and it is understood that the first arm is orthogonal to the second.

The waveguide consists of two arms — the upper ($\mathbf{E}/4$) and the second ($\mathbf{E}/4$) with which the upper is combined. The lower, or third arm, merely expends energy in achieving the phase modulation of the second arm with respect to the first. This can be achieved by merely making the length of the second arm change in a sinusoidal fashion (i.e., producing a $\partial\varphi/\partial t$ with respect to the first arm), or it can be achieved electro-optically. Whichever way is used, one-half the total energy of the system ($\mathbf{E}/2$) is spent on achieving the phase modulation in the particular example shown in Fig. 1.6. *The entropy change from input to output of the waveguide is compensated by energy expenditure in achieving the phase modulation to which the entropy change is due.*

One can nest phase modulations. The next order nesting is shown in Fig. 1.7, and other, higher order nestings of order n, for the cases $\partial\varphi^n/\partial t^n, n = 2, 3, \ldots$. follow the same procedure. The input is again $\mathbf{E} = \mathbf{E}\exp(i\omega t)$. The output is

$$\mathbf{E}_{\mathrm{OUT}(n=2)} = (\mathbf{E}/4)\exp(i\omega t) + (\mathbf{E}/4)\exp(i[\varphi_1 + \exp(i\varphi_2 t) - 1]t), \tag{1.134}$$

where $\varphi_1 = \mathbf{F}_1(E/4)$; $\varphi_2 = \mathbf{F}_2(E/4)$ and $\partial\varphi^2/\partial t^2 = \mathbf{F}_1^2.\mathbf{F}_2^2$, and again, it is understood that the first arm is orthogonal to the second.

Again, the waveguide consists of two arms — the upper $(\mathbf{E}/4)$ and the second $(\mathbf{E}/4)$ with which the upper is recombined. The lower two arms, three and four, merely expend energy in achieving the phase modulation of the second arm with respect to the first. This again can be achieved by merely making the length of the second arm change in a sinusoidal fashion (i.e., producing a $\partial\varphi^2/\partial t^2$ with respect to the first arm), or it can be achieved electro-optically for visible frequencies. *Whichever way is used, one half the total energy of the system* $(\mathbf{E}/4 + \mathbf{E}/4 = \mathbf{E}/2)$ *is spent on achieving the phase modulation of the particular sample shown in Fig.* 1.6.

Both the systems shown in Figs. 1.6 and 1.7, and all higher order such systems, $\partial\varphi^n/\partial t^n$, $n = 1, 2, 3, \ldots$., are adiabatic with respect to the total field and Poynting's theorem applies to them all. However, the Poynting description, or rather limiting condition, is insufficient to describe these fields exactly, neglects the orthogonal polarization two beam picture, and a more exact analysis is provided by the control theory picture shown here.

These waveguides we shall call Stokes' interferometers. The Stokes' equation is (Eq. (1.62)):

$$\oint \mathbf{A} \cdot dl = \int_S (\nabla \times \mathbf{A}) \cdot \mathbf{n}\, da, \tag{1.135}$$

and the energy-expending lines of the two Stokes' interferometers shown are normal to the two wave guide lines. l is varied sinusoidally

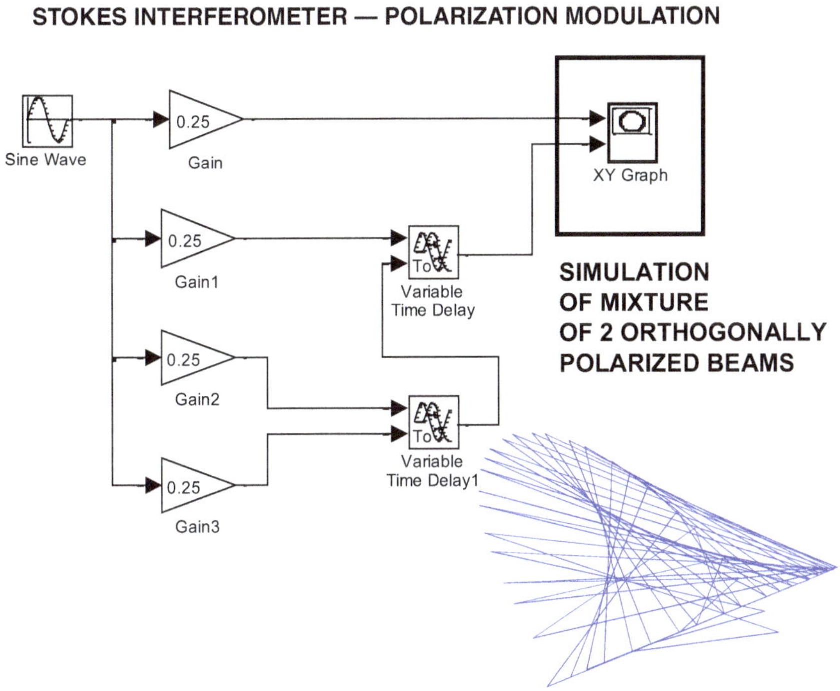

Fig. 1.7. Waveguide system paradigm for polarization modulated $(\partial\varphi^2/\partial t^2)$ wave emission. This is a completely adiabatic system in which oscillating energy enters from the left and exits from the right. On entering from the left, the energy is divided into two parts equally. One part, of amplitude $\mathbf{E}/2$, is used in providing phase modulation, $\partial\varphi^2/\partial t^2$ — this energy is spent (absorbed) by the system in obtaining the phase modulation; the other part, of amplitude $\mathbf{E}/2$, is divided into two parts equally one of which is polarization rotated, so that two oscillating waveforms of amplitude $\mathbf{E}/4$ but initially orthogonally polarized are formed for later superposition at the output. Unlike the system shown in Fig. 1.6, the energy expended on phase modulating one of these waves is divided into two parts equally, of amplitude $\mathbf{E}/4$, one of which is phase modulated, $\partial\varphi/\partial t$, with respect to the other as in Fig. 1.6. The energy of the superposition of these two waves is then expended to provide a second phase modulated $\partial\varphi^2/\partial t^2$ wave which is superposed with the non-delayed wave. Due to the phase modulation of one of them with respect to the other, $0 < \phi < 360°$, and their initial orthogonal polarization, the output is of continuously varying polarization. The choice of wave division into two parts equally is arbitrary. From Barrett (1990a,b).

so we have

$$\oint A \sin \omega t dl = \int_S (\nabla \times \mathbf{A}) \cdot \mathbf{n}\, da = E_{\mathrm{OUT}n=1} \text{ (Fig. 1.6)}, \tag{1.136}$$

$$\oint A \sin \omega t dl = \int_S (\nabla \times \mathbf{A}) \cdot \mathbf{n} da = E_{\mathrm{OUT}n=2} \text{ (Fig. 1.7)}. \tag{1.137}$$

The gauge symmetry consequences of this conditioning are shown in Figs. 1.8 and 1.9. The potential, $\mathbf{A}_\mu$, in Taylor expansion along one coordinate is

$$A = \frac{1}{4}x^4 + \frac{1}{2}ax^2 + bx + c, \tag{1.138}$$

with $b < 0$ in the case of $\mathbf{E}_{\text{in}}$ and $b > 0$ in the case of $\mathbf{E}_{\text{out}}$. A Stokes' interferometer permits the $\mathbf{E}$ field to restore a symmetry which

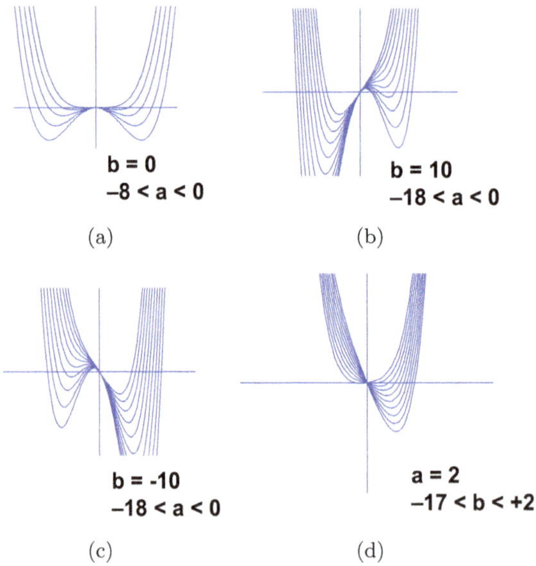

Fig. 1.8. Plots of potential $\Psi = 1/4x^4 + 1/2ax^2 + bx + c$. (a) $b = 0$ and various values of a; (b) $b = 10$ and various values of a; (c) $b = -10$ and various values of a; (d) $a = 2$ and various values of b. For positive values of a, U(1) symmetry is restored. For negative values of a as shown here, U(1) symmetry is broken and SU(2) symmetry is obtained. From Barrett (1987a).

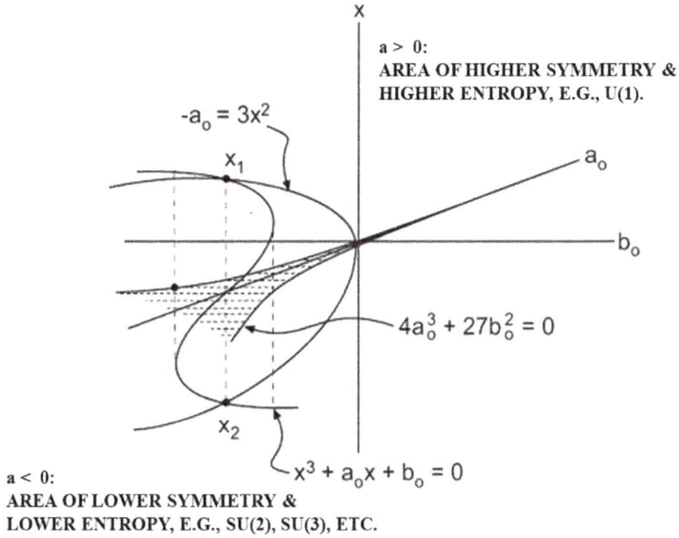

(a)

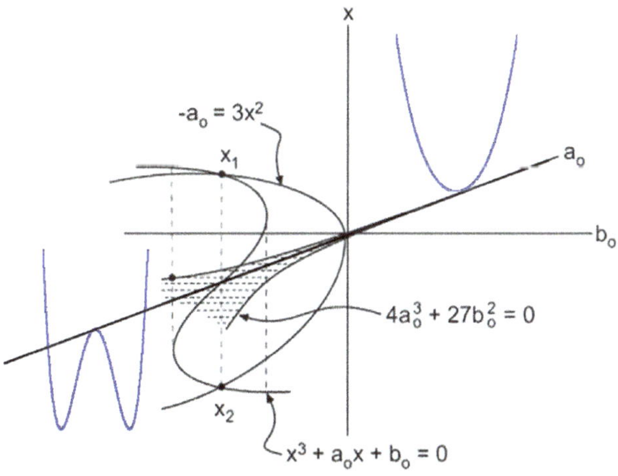

(b)

Fig. 1.9. (a) A representative system defined in the (x, a, b)-space with $x = \psi$. Other systems can be represented in the cusp area at other values of x, a and b. As in Fig. 1.8, for positive values of a, U(1) symmetry is restored. For negative values of a, symmetry is broken and SU(2) symmetry is obtained. (b) Plots of ψ–b plane. Superposed are two ψ–b depictions with $b = 0$ and a with positive and negative values. From Barrett (1987a).

was broken before this conditioning. Thus the $\mathbf{E}_{\text{in}}$ field is in U(1) symmetry and the $\mathbf{E}_{\text{out}}$ field is conditioned to be in SU(2) symmetry form. The conditioning of the **E** field to SU(2) symmetry form is the opposite of symmetry breaking. It is well known that the Maxwell theory is in U(1) symmetry form and the theoretical constructs of the magnetic monopole and charge exist in SU(2) symmetry form.[64]

Other interferometric methods beside Stokes' interferometer polarization modulators which restore symmetry are cavity waveguide interferometers. For example, The Mach–Zehnder and the Fabry–Perot are SU(2) conditioning interferometer (Yurke *et al.* 1986) (Fig. 1.10). The SU(2) group characterizes passive lossless devices with two inputs and two outputs with the boson commutation relations:

$$[E_{1*}, E_{2*}] = [E_{1*}^{\Uparrow}, E_{2*}^{\Uparrow}] = 0, \tag{1.139}$$

$$[E_{1*}, E_{2*}^{\Uparrow}] = d_{12*}, \tag{1.140}$$

where $E^{\Uparrow}$ is the Hermitian conjugate of E and * signifies both *in* (entering) and *out* (exiting) fields, i.e., before and after $\mathbf{A}_\mu$-conditioning. The Hermitian operators are as follows:

$$\mathbf{J}_x = \left(\frac{1}{2}\right)\left(\mathbf{E}_1^{\Uparrow}\mathbf{E}_{2*} + \mathbf{E}_{1*}\mathbf{E}_2^{\Uparrow}\right) = \left(\frac{1}{2}\right)(\mathbf{A}_1 \times \mathbf{B}_{1IN} + \mathbf{B}_{2IN} \times \mathbf{A}_2)$$
$$= (\mathbf{A} \times \mathbf{B} - \mathbf{B} \times \mathbf{A}), \tag{1.141a}$$

$$\mathbf{J}_y = \left(-\frac{i}{2}\right)\left(\mathbf{E}_1^{\Uparrow}\mathbf{E}_{2*} - \mathbf{E}_{1*}\mathbf{E}_2^{\Uparrow}\right) = \left(-\frac{i}{2}\right)(\mathbf{A}_1 \times \mathbf{B}_{1IN}$$
$$+ \mathbf{B}_{2IN} \times \mathbf{A}_2) = (\mathbf{A} \cdot \mathbf{B} - \mathbf{B} \cdot \mathbf{A}), \tag{1.141b}$$

$$\mathbf{J}_z = \left(\frac{1}{2}\right)\left(\mathbf{E}_1^{\Uparrow}\mathbf{E}_{1*} - \mathbf{E}_{2*}\mathbf{E}_2^{\Uparrow}\right) = \left(\frac{1}{2}\right)(\mathbf{A}_1 \times \mathbf{E}_{1\mathbf{OUT}}$$
$$- \mathbf{E}_{2\mathbf{OUT}} \times \mathbf{A}_2) = (\mathbf{A} \times \mathbf{E} - \mathbf{E} \times \mathbf{A}), \tag{1.141c}$$

$$i\mathbf{J}_z = (\mathbf{1/2})\left(\mathbf{E}_1^{\Uparrow}\mathbf{E}_{1*} + \mathbf{E}_{2*}\mathbf{E}_2^{\Uparrow}\right) = (-i/\mathbf{2})(\mathbf{A}_1 \times \mathbf{E}_{1\mathbf{OUT}}$$
$$+ \mathbf{E}_{2\mathbf{OUT}} \times \mathbf{A}_2) = (\mathbf{A} \times \mathbf{E} + \mathbf{E} \times \mathbf{A}), \tag{1.141d}$$

[64] Barrett (1987a,b; 1989a,b; 1990a,b; 1991).

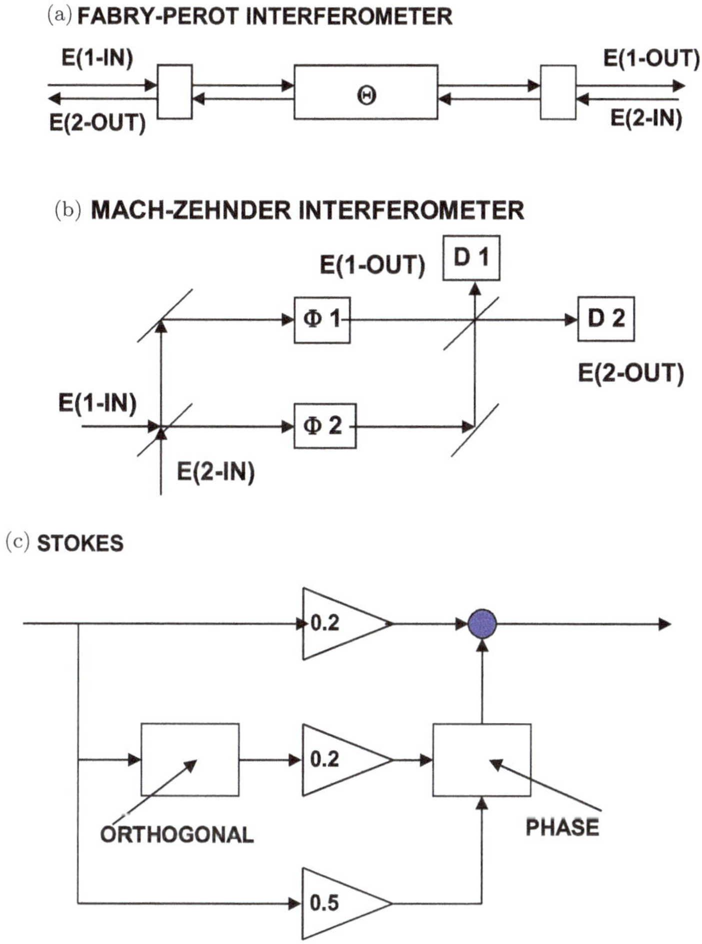

Fig. 1.10. SU(2) field conditioning interferometers: (a). Fabry-Perot; (b). Mach-Zehnder; (c) Stokes. After Ref. Barrett (1990b).

where the substitutions are as follows:

$$
\begin{aligned}
&E_{1*} = B_{2\mathrm{IN}} \text{ and } \times \mathrm{E}_{1\mathrm{OUT}},\\
&E_{2*} = \times B_{1\mathrm{IN}} \text{ and } E_{2\,\mathrm{OUT}},\\
&E_1^{\Uparrow} = A_1,\\
&E_2^{\Uparrow} = \times A_2,
\end{aligned} \tag{1.142}
$$

satisfying the Lie algebra:

$$\begin{aligned}[\mathbf{J}_x, \mathbf{J}_y] &= i\mathbf{J}_z, \\ [\mathbf{J}_y, \mathbf{J}_z] &= i\mathbf{J}_x, \\ [\mathbf{J}_z, \mathbf{J}_x] &= i\mathbf{J}_y.\end{aligned} \tag{1.143}$$

The analysis presented in this section is based on the relation of induced angular momentum to the *eduction* of gauge invariance (see also Paranjape, 1987). One gauge invariant quantity or observable in one gauge or symmetry can be covariant with another in another gauge or symmetry. The Wu–Yang condition of field overlap, permitting measurement of $\mathbf{\Phi} = i \oint \exp \mathbf{A}_\mu d^\mu$, requires coherent overlap. All other effects are observed either at low temperature where thermodynamic conditions provide coherence, or is a self-mapping, which also provides coherence. Thus the question of whether classical $\mathbf{A}_\mu$ wave effects can be observed at long range, reduces to the question of how far coherency of the two fields can be maintained.

Oh *et al.* (1988) derived the non-relativistic propagator for the generalized Aharonov–Bohm effect, which is valid for any gauge group in a general multiply connected manifold, as a gauge artifact in the universal covering space. These authors conclude that (1), if a partial propagator along a multiply connected space ($\mathbf{M}$ in the present notation) is lifted to the universal covering space ($\mathbf{L}$ in the present notation) i.e., $\mathbf{T_1} : \mathbf{M} \to \mathbf{L}$, then (2), for a gauge transformation $\mathbf{U}(x)$ of $\mathbf{A}_\mu$ on the covering space $\mathbf{L}$, an Aharonov-Bohm effect will arise if (3), $\mathbf{U}(x)$ is not projectable to be a well-defined single-valued gauge transformation on $\mathbf{M}$, but (4), $\mathbf{A}_\mu = \mathbf{U}(x)\partial_\mu \mathbf{U}(x)^{-1}$ (i.e., $\mathbf{T_1}\mathbf{T_2}^{-1}$) is nevertheless projectable, i.e., for a $\mathbf{T_2} : \mathbf{L} \to \mathbf{M}$, in agreement with the analysis presented here. We have stressed, however, that the $\mathbf{A}_\mu = \mathbf{T_1} : \mathbf{L} \to \mathbf{M}$ have a physical existence, whether the $\mathbf{T_2} : \mathbf{L} \to \mathbf{M}$ mapping is, or can be, performed or not. Naturally, if this second mapping is not performed, then no Aharonov-Bohm effect exists (i.e., no comparative mapping exists).

Although interferometric methods can condition fields into SU(2) or other symmetric form, there is, of course, no control over the spacetime metric in which those fields exist. When the conditioned field leaves the interferometer, at time $t = 0$, the field is in exact $U(1)$ form. At time $t > 0$, the field will depart from $U(1)$ form inasmuch as it is scattered or absorbed by the medium.

The gauge invariance of the phase factor requires a multiply-connected field. In the case of quantum particles, this would mean wave function overlap of two individual quanta. Classically, however, *every* polarized wave is constituted of two polarized vectorial components. Therefore, classically, every polarized wave is a multiply connected field (cf. Merzbacher, 1962) in SU(2) form. However, the extension of Maxwell's theory to SU(2) form, i.e., to non-Abelian Maxwell's theory, defines multiply-connected local fields in a global covering space, i.e., in a global simply-connected form. In the next section, the Maxwell's equations redefined in both $\mathrm{SU}(2)/Z_2$, non-Abelian, or multiply-connected form are examined.

1.6. Non-Abelian Maxwell Equations

Using Yang–Mills theory (Yang and Mills, 1954), the non-Abelian Maxwell equations, which describe SU(2) symmetry conditioned radiation, become:

- Coulomb's law:

$$\text{no existence in SU(2) symmetry,} \tag{1.144}$$

- Ampère's law:

$$\partial \mathbf{E}/\partial t - \nabla \times \mathbf{B} + iq\,[A_0, \mathbf{E}] - iq(\mathbf{A} \times \mathbf{B} - \mathbf{B} \times \mathbf{A}) = -\mathbf{J}, \tag{1.145}$$

- the presence of free "magnetic monopoles" (instantons[65]):

$$\nabla \cdot \mathbf{B} + iq(\mathbf{A} \cdot \mathbf{B} - \mathbf{B} \cdot \mathbf{A}) = 0, \tag{1.146}$$

[65]The determination of the classical field configurations in Euclidean space minimizes the action subject to appropriate asymptotic conditions in 4-space.

- Faraday's law:

$$\nabla \times \mathbf{E} + \partial \mathbf{B}/\partial t + iq\,[A_0, \mathbf{B}] + iq\,(\mathbf{A} \times \mathbf{E} - \mathbf{E} \times \mathbf{A}) = 0, \quad (1.147)$$

and the current relation:

$$\boldsymbol{\nabla} \cdot \mathbf{E} - \mathbf{J_0} + \mathbf{iq}\,(\mathbf{A} \cdot \mathbf{E} - \mathbf{E} \cdot \mathbf{A}) = \mathbf{0}. \quad (1.148)$$

The conventional Coulomb's law (Eq. (1.9)) amounts to an imposition of spherical symmetry requirements, as a single isolated source charge permits the choice of charge vector to be arbitrary at every point in spacetime. Imposition of this symmetry reduces the non-Abelian Maxwell equations to the same form as conventional electrodynamics, i.e., to Abelian form.

Harmuth's Ansatz is the addition of a magnetic current density to Maxwell's equations, an addition which may be set to zero after completion of calculations (Barrett, 1987b). With a magnetic current density, Maxwell's equations describe a spacetime field of higher order symmetry and consist of invariant physical quantities (e.g., the field $\partial_x \mathbf{F} = J$), magnetic monopole and charge. Harmuth's amended equations are (Harmuth, 1986a, Eqs. (4)–(7)):

$$\boldsymbol{\nabla} \times \mathbf{H} = \boldsymbol{\partial} \mathbf{D}/\boldsymbol{\partial} t + \mathbf{g}_e, \quad (1.149a)$$

$$-\nabla \times \mathbf{E} = \partial \mathbf{B}/\partial t + g_m, \quad (1.149b)$$

$$\boldsymbol{\nabla} \cdot \mathbf{D} = \boldsymbol{\rho}_e, \quad (1.149c)$$

$$\boldsymbol{\nabla} \cdot \mathbf{B} = \boldsymbol{\rho}_m, \quad (1.149d)$$

$$g_e = \sigma \mathbf{E}, \quad (1.149e)$$

$$g_m = s\mathbf{H}, \quad (1.149f)$$

where g_e, g_m, ρ_e, ρ_m and s are electric current density, magnetic current density, electric charge density, magnetic charge density and magnetic conductivity, respectively.

It should be noted that classical magnetic sources or instanton-like sources are not dismissed by all researchers in classical field theory. For example, Tellegren's formulation for the gyrator (Tellegen, 1948) and formulations required in descriptions of chiral media (natural optical activity) (Lakhtakia, 1991; Lakhtakia *et al.* 1989) require magnetic source terms even in the frequency domain. The ferromagnetic aerosol experiments by Mikhailov on the Ehrenhaft effect also implicate a magnetic monopole instanton or pseudoparticle interpretation,[66] the spherical symmetry of the aerosol particles providing SU(2) boundary conditions according to the present view.

Comparing the SU(2) formulation of the Maxwell equations and the Harmuth equations reveals the following identities:

$U(1)$*Symmetry*	SU(2)*Symmetry*
$\rho_e = J_0$	$\rho_e = J_0 - iq(\mathbf{A}\cdot\mathbf{E} - \mathbf{E}\cdot\mathbf{A}) = J_0 + q\mathbf{J}_z \quad (1.150a)$
$\rho_m = 0$	$\rho_m = -iq(\mathbf{A}\cdot\mathbf{B} - \mathbf{B}\cdot\mathbf{A}) = -q\mathbf{J}_y \quad (1.150b)$
$g_e = \mathbf{J}$	$\mathrm{g}_e = \mathbf{iq}[\mathbf{A_0},\mathbf{E}] - \mathbf{iq}(\mathbf{A}\times\mathbf{B} - \mathbf{B}\times\mathbf{A}) + \mathbf{J} = \mathbf{iq}[\mathbf{A_0},\mathbf{E}] - \mathbf{iq}\mathbf{J}_x + \mathbf{J} \quad (1.150c)$
$g_m = 0$	$\mathrm{g}_m = \mathbf{iq}[\mathbf{A_0},\mathbf{B}] - \mathbf{iq}(\mathbf{A}\times\mathbf{E} - \mathbf{E}\times\mathbf{A}) = \mathbf{iq}[\mathbf{A_0},\mathbf{B}] - \mathbf{iq}\mathbf{J}_z \quad (1.150d)$
$\sigma = \mathbf{J}/\mathbf{E}$	$\sigma = \dfrac{\{\mathbf{iq}[\mathbf{A_0},\mathbf{E}] - \mathbf{iq}(\mathbf{A}\times\mathbf{B} - \mathbf{B}\times\mathbf{A}) + \mathbf{J}\}}{\mathbf{E}} = \{\mathbf{iq}[\mathbf{A_0},\mathbf{E}] - \mathbf{iq}\mathbf{J}_x + \mathbf{J}\}/\mathbf{E} \quad (1.150e)$
$s = 0$	$\mathrm{s} = \dfrac{\{\mathbf{iq}[\mathbf{A_0},\mathbf{B}] - \mathbf{iq}(\mathbf{A}\times\mathbf{E} - \mathbf{E}\times\mathbf{A})\}}{\mathbf{H}} = \{\mathbf{iq}[\mathbf{A_0},\mathbf{B}] - \mathbf{iq}\mathbf{J}_z\}/\mathbf{H} \quad (1.150f)$

It is well known that only some topological charges are conserved (i.e., are gauge invariant) after symmetry breaking — electric charge is, magnetic charge is not (Weinberg, 1980). Therefore, the Harmuth Ansatz of setting magnetic conductivity (and other SU(2) symmetry constructs) to zero on conclusion of signal velocity calculations has a theoretical justification. It is also well known that some physical constructs which exist in both a lower and a higher symmetry form are more easily calculated for the higher symmetry, transforming

[66]Mikhailov (1987, 1988a,b, 1991); Mikhailov and Mikhailova (1990).

to the lower symmetry after the calculation is complete. The observables of the electromagnetic field exist in a U(1) symmetry field. Therefore, the problem is to relate invariant physical quantities to the variables employed by a particular observer. This means a mapping of spacetime vectors into space vectors, i.e., a spacetime split.

This mapping is not necessary for solving and analyzing the basic equations. As a rule, it only complicates the equations needlessly. Therefore, the appropriate time for a split is usually after the equations have been solved. It is appropriate to mention here, the interpretation of the Aharonov–Bohm effect offered by Bernido and Inomata (1981). These authors point out that a path integral can be explicitly formulated as a sum of partial propagators corresponding to homotopically different paths. In the case of the AB effect, the mathematical object to be computed in this approach is a *propagator* expressed as a path integral in the *covering space* of the background physical space. Therefore, the path-dependence of the AB phase factor is wholly of topological origin and the AB problem is reduced to showing that *the full propagator can be expressed as a sum of partial propagators belonging to all topological inequivalent paths.* The paths are partitioned into their homotopy equivalence classes; Feynman sums over paths in each class giving homotopy propagators; and the whole effect of the gauge potential being to multiply these homotopy propagators by different gauge phase factors. The relevant point, however, with respect to the Harmuth ansatz, is that the full propagator is expressed in terms of the covering space, rather than the physical space. The homotopy propagators are related to propagators in the universal covering manifold, leading to an expansion of the propagators in terms of eigenfunctions of a Hamiltonian on the covering manifold.

The approach to multiply connected spaces offered by Dowker (1972) and Sundrum and Tassie (1986) also uses the covering space concept. A multiply-connected space **M** and a universal covering space, **M***, are defined:

$$\mathbf{M}* \to \mathbf{M} = \frac{\mathbf{M}*}{\mathbf{G}}, \tag{1.151}$$

where **G** is a properly continuous, discrete group of isometries of **M***, without fixed points and **M*** is simply connected. Each group of **M** corresponds to n different points **qg** of **M***, where **g** ranges over the n elements of **G**. **M*** is then divided into subsets of a finite number of points or fibers, one fiber corresponding to one point of **M**. **M*** is a bundle or fibered space, and $\mathbf{\Gamma}$ is the group of the bundle. The major point, in the present instance, is that the propagator is given in terms of a matrix representation of the covering space **M***. Harmuth calculates the propagation in the covering space where the Hamiltonian is self-adjoint.[67] Thus, the propagation in the covering space is well-defined.

Consequently, Harmuth's ansatz can be interpreted as: (i) a mapping of Maxwell's (U(1) symmetrical) equations into a higher order symmetry field (of SU(2) symmetry) — a symmetry which permits the definition of magnetic monopoles (instantons) and magnetic charge; (ii) solving the equations for propagation velocities; and (iii) mapping the solved equations back into the U(1) symmetrical field (thereby removing the magnetic monopole and charge constructs).

1.7. Discussion

The concept of the electromagnetic field was conceived by Faraday and set in a mathematical frame by Maxwell to describe electromagnetic effects in a spacetime region. It is a concept addressing *local* effects. The Faraday–Maxwell theory which was founded on the concept of the electrotonic state *potentially* had the capacity to describe *global* effects but the manifestation of the electrotonic state, the **A** field, was abandoned in the later interpretation of Maxwell. When, in this interpretation, the theory was refounded on the field concept, and the issue of energy propagation was examined, action-at-a-distance (Newton) was replaced by contact-action (Descartes). That is, a theory (Newton's) accounting for both local and global effects was replaced by a completely local theory (Descartes'). The

[67]Self-adjointness means that non-Hermitian components are compensated (Schulman, 1971).

contemporary local theory can address global effects with the aid of the Lorentz invariance condition, or the Lorenz gauge. However, Lorentz invariance is due to a chosen gauge and chosen boundary conditions, and these are not an inevitable consequence of the (interpreted) Maxwell theory, which became a theory of local effects.

According to the conventional viewpoint, the local field strength, $\mathbf{F}_{\mu\nu}$, completely describes electromagnetism. However, due to the effects discussed here, there is reason to believe that $\mathbf{F}_{\mu\nu}$ does not describe electromagnetism completely. In particular, it does not describe *global* effects resulting in different histories of local spatiotemporal conditioning of the constituent parts of summed multiple fields.

Weyl (1918–1939) first proposed that the electromagnetic field can be formulated in terms of an Abelian gauge transformation. But the Abelian gauge only describes local effects. It was Yang and Mills (1954) who extended the idea to non-Abelian groups. The concepts of the Abelian electromagnetic field — electric charge, **E** and **H** fields, are explained within the context of the non-Abelian concepts of magnetic charge and monopole. The Yang–Mills theory *is* applicable to both local and global effects.

If the unbroken gauge group is non-Abelian, only some of the topological charges are gauge-invariant. The electric charge is, the magnetic charge is not (Weinberg, 1980). That is the reason magnetic sources are not seen in Abelian Maxwell theory which has boundary conditions which do not compactify nor reconstitute symmetry and degrees of freedom.

The $\mathbf{A}_\mu$ potentials have an ontology or physical meaning as *local* operators mapping the *local* e.m. fields onto *global* spatiotemporal conditions the *local* e.m. fields. This operation is measurable if there is a second comparative mapping of the conditioned local fields in a many-to-one fashion (multiple connection). In the case of a single local (electromagnetic) field, this second mapping is ruled out — but such an isolated local field is only imaginary, because the imposition of boundary conditions implies the existence of separate local conditions and thereby always a global condition. Therefore, practically speaking, the $\mathbf{A}_\mu$ potentials always have a gauge-invariant physical existence. The $\mathbf{A}_\mu$ potential gauge invariance implies the theoretical constructs of a magnetic monopole (instanton) and magnetic charge, but with no singularities. These latter constructs are, however, confined to SU(2) field conditioning, whereas the $\mathbf{A}_\mu$ potentials have an existence in *both* U(1) *and* SU(2) symmetries.

The physical effects of the $\mathbf{A}_\mu$ potentials are observable empirically at the quantum level (effects 1–5) and at the classical level (2, 3 and 6). The Maxwell theory of fields, restricted to a description of *local intensity* fields, and with the U(1) symmetry broken to SU(2), requires no amendment at all. If, however, the intention is to describe both local *and* global electromagnetism, then an amended Maxwell theory is required in order to include the local operator field of the $\mathbf{A}_\mu$ potentials, the integration of which describe the phase relations between local intensity fields of different spatiotemporal history after global-to-local mapping.

With only the constitutive relations of e.m. fields-to-matter defined (and not those of fields-to-vacuum), contemporary opinion is that the dynamic attribute of force resides in the medium-independent fields, i.e, they are fields of force. As the field-vacuum constitutive relations are lacking, this view can be contested, giving rise to competing accounts of where force resides, e.g., the opposing view of force not residing in the fields but in the matter (cf. Graneau, 1985).

The uninterpreted Maxwell, of course, had *two types* of constitutive relations in mind, the second one referring to the energy-medium relation:

> "...whenever energy is transmitted from one body to another in time, there must be a medium or substance in which the energy exists after it leaves one body and before it reaches the other..." (Maxwell, 1891, Vol II, p. 493)

After removal of the medium from consideration, only one constitutive relation remained and the fields have continued to exist as the classical limit of quantum mechanical exchange particles. However, that cannot be a true existence for the classical force fields because those quantum mechanical particles are in units of action, not force.

Bibliography

Aharonov, Y., Non-local phenomena and the Aharonov–Bohm effect. *Proc. Int. Symp. Foundation of Quantum Mechanics*, Tokyo, 1983, 10–19.

Aharonov, Y. and Anandan, J., Phase change during a cyclic quantum evolution. *Phys. Rev. Lett.*, 58(16), 1593–1596, 1987.

Aharonov, Y. and Bohm, D., Significance of electromagnetic potentials in the quantum theory. *Phys. Rev.*, 115(3), 485–491, 1959.

Aharonov, Y. and Bohm, D., Further considerations on electromagnetic potentials in quantum theory. *Phys. Rev.*, 123(4), 1511–1524, 1961.

Aharonov, Y. and Bohm, D., Remarks on the possibility of quantum electrodynamics without potentials. *Phys. Rev.*, 125(6), 2192–2193, 1962.

Aharonov, Y. and Bohm, D., Further discussion of the role of electromagnetic potentials in the quantum theory. *Phys. Rev.*, 130(4), 1625–1632, 1963.

Aharonov, Y. and Casher, A., Topological quantum effects for neutral particles. *Phys. Rev. Lett.*, 53(4), 319–321, 1984.

Ali, S.T., Stochastic localization, quantum mechanics on phase space and quantum spacetime. *La Rivista del Nuovo Cimento*, 8(11), 1–128, 1985.

Ali, S.T. and Prugovecki, E., Mathematical problems of stochastic quantum mechanics: harmonic analysis on phase space and quantum geometry. *Acta Appl. Mathe.*, 6, 1–18, 1986.

Altshuler, B.L., Aronov, A.G. and Spivak, B.Z., The Aharonov–Bohm effect in disordered conductors. *JETP Lett.*, 33, 94–97, 1981.

Altshuler, B.L., Aronov, A.G., Spivak, B.Z., Sharvin, Y.D. and Sharvin, Y.V., Observation of the Aaronov–Bohm effect in hollow metal cylinders. *JETP Lett.*, 35, 588–591, 1982.

Anandan, J. and Stodolsky, L., Some geometrical considerations of Berry's phase. *Phys. Rev. D*, 35, 2597–2599, 1987.

Anderson, P.W., Special effects in superconductivity, in *Lectures on the Many-Body Problem*, E.R. Caianello (ed.), Vol. 2, Academic, New York, 1964, pp. 113–135.

Aoki, H. and Ando, T., Universality of the quantum Hall effect: topological invariant and observable. *Phys. Rev. Lett.*, 57(24), 3093–3096, 1986.

Aronov, A.G. and Sharvin, Y.V., Magnetic flux effects in disordered conductors. *Rev. Mod. Phys.*, 59, 755–779, 1987.

Atiyah, M. and Hitchin, N., *The Geometry and Dynamics of Magnetic Monopoles*, Princeton University Press, 1980.

Atiyah, M.F., Instantons in two and four dimensions. *Commun. Math. Phys.*, 93, 437–451, 1984.

Atiyah, M.F., *Michael Atiyah: Collected Works, Volume 5, Gauge Theories*, Clarelndon Press, Oxford, 1988.

Atiyah, M., Hitchin, N.J. and Singer, I.M., Deformations of instantons. *Proc. Nat. Acad. Sci.*, 74, 2662–2663, 1977.

Atiyah, M., Hitchin, N.J., Drinfeld, V.G. and Manin, Y.I., Construction of instantons. *Phys. Lett. A*, 65, 185–187, 1978.

Atiyah, M. and Jones, J.D.S., Topological aspects of Yang-Mills theory. *Commun. Math. Phys.*, 61, 97–118, 1978.

Atiyah, M. and Ward, R.S., Instantons and algebraic geometry. *Commun. Math. Phys.*, 55, 117–124, 1977.

Bandyopadhyay, S., Datta, S. and Melloch, R., Aharonov–Bohm effect in semiconductor microstructures: novel device possibilities. *Superlattices Microstruct.*, 2(6), 539–542, 1986.

Barrett, T.W., *Energy Transfer Dynamics: signal* SU(2) *Symmetry Conditioning for Radiation Propagation, Discrimination and Ranging*, W.J. Schafer Associates Inc., 1987.

Barrett, T.W., Nonlinear physics of electronic and optical materials for submicron device applications. Chapter 5, pp. 24–48, in T.W. Barrett & H.A. Pohl (eds) *Energy Transfer Dynamics*, Springer-Verlag, New York, 1987.

Barrett, T.W., The transition point between classical and quantum mechanical systems deterministically defined. *Physica Scripta*, 35, 417–422, 1987.

Barrett, T.W., Comments on the Harmuth Ansatz: use of a magnetic current density in the calculation of the propagation velocity of signals by amended Maxwell theory. *IEEE Trans. Electromagn. Comput.*, 30, 419–420, 1988.

Barrett, T.W., On the distinction between fields and their metric: the fundamental difference between specifications concerning medium-independent fields and constitutive specifications concerning relations to the medium in which they exist. *Ann. Fondation Louis de Broglie*, 14, 37–75, 1989.

Barrett, T.W., Comments on 'Solutions of Maxwell's equations for general non-periodic waves in lossy media'. *IEEE Trans. Electromagn. Comput.*, 31, 197–199, 1989.

Barrett, T.W., Comments on 'Some comments on Harmuth and his critics'. *IEEE Trans. Electromagn. Comput.*, 31, 201–202, 1989.

Barrett, T.W., Maxwell's theory extended. Part 1. Empirical reasons for questioning the completeness of Maxwell's theory, effects demonstrating the physical significance of the A potentials. *Ann. Fondation Louis de Broglie*, 15, 143–183, 1990.

Barrett, T.W., Maxwell's theory extended. Part 2. Theoretical and pragmatic reasons for questioning the completeness of Maxwell's theory. *Ann. Fondation Louis de Broglie*, 15, 253–283, 1990.

Barrett, T.W., Tesla's non-linear oscillator-shuttle-circuit (OSC) theory. *Ann. Fondation Louis de Broglie*, 16, 23–41, 1991.

Barrett, T.W., Electromagnetic phenomena not explained by Maxwell's equations, pp. 8–86 in A. Lakhtakia (ed) *Essays on the Formal Aspects of Maxwell Theory*, World Scientific, 1993.

Barut, A.O., $E = \_h\omega$. *Phys. Lett. A*, 143, 349–352, 1991.

Beenakker, C.W.J., Theory of Coulomb-blockade oscillations in the conductance of a quantum dot. *Phys. Rev. B.*, 44, 1646–1656, 1991.

Beenakker, C.W.J., van Houton, H. and Staring, A.A.M., Influence of Coulomb repulsion on the Aharonov–Bohm effect in a quantum dot. *Phys. Rev. B*, 44, 1657–1667, 1991.

Belavin, A.A., Polyakov, A.M., Schwartz, A.S. and Tyupkin, Y.S., Pseudoparticle solutions of the Yang–Mills equations. *Phys. Lett. B*, 59, 85–87, 1975.

Bernido, C.S. and Inomata, A., Path integrals with a periodic constraint: the Aharonov-Bohm effect, *J. Math. Phys.*, 22(4), 715–718, 1981.

Benoit, A.D. Washburn, S., Umbach, C.P., Laibowitz, R.B. and Webb, R.A., Asymmetry in the magnetoconductance of metal wires and loops, *Phys. Rev. Lett.*, 57(14), 1765–1768, 1986.

Berry, M.V., Exact Aharonov–Bohm wavefunction obtained by applying Dirac's magnetic phase factor, *Eur. J. Phys.*, 1, 240–244, 1980.

Berry, M.V., Quantal phase factors accompanying adiabatic changes, *Proc. R. Soc. Lond. A*, 392, 45–57, 1984.

Berry, M.V., The adiabatic limit and the semiclassical limit, *J. Phys. A: Math. Gen.*, 17, 1225–1233, 1984.

Berry, M.V., Classical adiabatic angles and quantal adiabatic phase. *J. Phys. A: Math. Gen.*, 18, 15–27, 1985.

Berry, M.V., Interpreting the anholonomy of coiled light. *Nature*, 326, 277–278, 1987.

Berry, M.V., Quantum phase corrections from adiabatic iteration. *Proc. Roy. Soc. Lond. A*, A414, 31–46, 1987.

Bhandari, R. and Samuel, J., Observation of topological phase by use of a laser interferometer. *Phys. Rev. Lett.*, 60(13), 1211–1213, 1988.

Bialynicki-Birula, I. and Bialynicka-Birula, Z., Berry's phase in the relativistic theory of spinning particles. *Phys. Rev. D*, 35, 2383–2387, 1987.

Bialynicki-Birula, I. and Bialynicka-Birula, Z., *Quantum Electrodynamics*, Pergamon, New York, 1975.

Bishop, D.J., Licini, J.C. and Dolan, G.J., Lithium quenched-condensed microstructures and the Aharonov–Bohm effect. *Appl. Phys. Lett.*, 46(10), 1000–1002, 1985.

Bitter, T. and Dubbers, D., Manifestation of Berry's topological phase in neutron spin rotation. *Phys. Rev. Lett.*, 59, 251–254, 1987.

Bloch, F., Simple interpretation of the Josephson effect. *Phys. Rev.*, 21(17), 1241–1243, 1968.

Boersch,, H., Hamisch, H., Wohlleben, D. and Grohmann, K., Antiparallele Weissche Bereiche als Biprisma für Elektroneninterferenzen. *Zeit. Phys.*, 159, 397–404, 1960.

Bogomol'nyi, E.B., The stability of classical solutions. *Sov. J. Nucl. Phys.*, 24(4), 449–454, 1976.

Bohren, C.F., Scattering of electromagnetic waves by an optically active spherical shell. *J. Chem. Phys.*, 62, 1566–1570, 1975.

Bork, A.M., Maxwell and the vector potential, *Isis*, 58, 210–222, 1967.

Bork, A.M., Maxwell and the electromagnetic wave equation. *Am. J. Phys.*, 35, 844–849, 1967.

Born, M., Reciprocity theory of elementary particles. *Rev. Mod. Phys.*, 21(3), 463–473, 1949.

Botelho, L.C.L. and de Mello, J.C., A non-Abelian Aharonov–Bohm effect in the framework of pseudoclassical mechanics. *J. Phys. A: Math. Gen.*, 18, 2633–2634, 1985.

Boulware, D.G. and Deser, S., The Aharonov–Bohm effect and the mass of the photon, manuscript, 1989.

Brandt, R.A. and Primack, J.R., Dirac monopole theory with and without strings. *Phys. Rev.*, 15(4), 1175–1177, 1977.

Buchwald, J.Z., *From Maxwell to Microphysics*, University of Chicago Press, 1985.

Büttiker, M., Imry, Y. and Landauer, R., Josephson behavior in small normal one-dimensional rings. *Phys. Lett. A*, 96(7), 365–368, 1983.

Büttiker, M., Imry, Y., Landauer, R. and Pinhas, S., Generalized many-channel conductance formula with application to small rings. *Phys. Rev. B*, 31(10), 6207–6215, 1985.

Byers, N. and Yang, C.N., Theoretical considerations concerning quantized magnetic flux in superconducting cylinders. *Phys. Rev. Lett.*, 7(2), 46–49, 1961.

Cai, Y.Q., Papini, G. and Wood, W.R., On Berry's phase for photons and topology in Maxwell's theory. *J. Math. Phys.*, 31, 1942–1946, 1990.

Cavalloni, C. and Joss, W., Aharonov-Bohm effect and weak localization in cylindrical Ag films. *Phys. Rev. B*, 35(14), 7338–7349, 1987.

Chakravarty, S., Ablowitz, M.J. and Clarkson, P.A., Reductions of self-dual Yang–Mills fields and classical systems. *Phys. Rev. Lett.*, 65, 1085–1087, 1990.

Chambers, R.G., Shift of an electron interference pattern by enclosed magnetic flux. *Phys. Rev. Lett.*, 5, 3–5, 1960.

Chandrasekhar, V., Rooks, M.J., Wind, S. and Prober, D.E., Observation of Aharonov-Bohm electron interference effects with periods _h/e and _h/2e in individual micron-size, normal-metal rings. *Phys. Rev. Lett.*, 55(15), 1610–1613, 1985.

Chiao, R.Y. and Wu, Y-S. Manifestations of Berry's topological phase for the photon. *Phys. Rev. Lett.*, 57(8), 933–936, 1986.

Chiao, R.Y. and Tomita, A., Reply to Comments. *Phys. Rev. Lett.*, 59(15), 1789, 1987.

Chiao, R.Y., Antaramian, A., Ganga, K.M., Jiao, H., Wilkinson, S.R., and Nathel, Observation of a topological phase by means of a non-planar Mach–Zehnder interferometer. *Phys. Rev. Lett.*, 60(13), 1214–1217, 1988.

Chisholm, J.S.R. and Common, A.K., *Clifford Algebras and Their Applications in Mathematical Physics,* D. Reidel, Dordrecht, 1986.

Chow, W.W., Gea-Banacloche, J., Pedrotti, L.M., Sanders, V.E., Schleich, W. and Scully, M.O., The ring laser gyro. *Rev. Mod. Phys.*, 57(1), 61–104, 1985.

Clarke, J., Cleland, A.N., Devoret, M.H., Esteve, D. and Martinis, J.M., Quantum mechanics of a macroscopic variable: the phase difference of a Josephson junction. *Science*, 239, 992–997, 1988.

Craigie, N. (ed) *Theory and Detection of Magnetic Monopoles in gauge Theories (A Collection of Lecture Notes)*, World Scientific, Singapore, 1986.

Datta, S. and Bandyopadhyay, S., Aharonov-Bohm effect in semiconductor microstructures. *Phys. Rev. Lett.*, 5(7), 717–720, 1987.

Datta, S., Melloch, M.R., Bandyopadhyay, S., Noren, R., Vaziri, M., Miller, M. and Reifenberger, R., Novel interference effects between parallel quantum wells. *Phys. Rev. Lett.*, 55(21), 2344–2347, 1985.

Datta, S., Melloch, M.R., Bandyopadhyay, S. and Lundstrom, M.S., Proposed structure for large quantum interference effects. *Appl. Phys. Lett.*, 48(7), 487–489, 1986.

Delacrétaz, G., Grant, E.R., Whetten, R.L., W÷ste, L. and Zwanziger, J.W., Fractional quantization of molecular pseudorotation in Na_3. *Phys. Rev. Lett.*, 56(24), 2598–2601, 1986.

Dirac, P.A.M., Quantized singularities in the electromagnetic field. *Proc. Roy. Soc. Lond.*, A133, 60–72, 1931.

Dirac, P.A.M., Theory of magnetic monopoles. *Phys. Rev.*, 74, 817–830, 1948.

Djordjvic, A.R. and Sarkar, T., Comments on 'Correction of Maxwell's equations for signals I', 'Correction of Maxwell's equations for signals II' and 'Propagation velocity of electromagnetic signals'. *IEEE Trans. Electromagn. Compat.*, EMC-29(3), 255–256, 1987.

Dowker, J.S., Quantum mechanics and field theory on multiply connected and on homogeneous spaces. *J. Phys., A. Gen. Phys.*, 5, 936–943, 1972.

Dupuis, N. and Montambaux, G., Aharonov–Bohm flux and statistics of energy levels in metals. *Phys. Rev. B.*, 43, 14931–14937.

Dyson, F.J., Foreward to the 1st edition, Monastryrsky, M., *Riemann, Topology, and Physics*, Birkhäuser, 1st Edition, 1987, 2nd edition, 1999.

Edwards, J.T. and Thouless, D.J., Numerical studies of localization in disordered systems. *J. Phys. C*, 5, 807–820, 1972.

Ehrenberg, E. and Siday, R.E., The refractive index in electron optics and the principles of dynamics. *Proc. Phys. Soc., Lond. B*, 62(8), 8–21, 1949.

Eisenhart, L.P., *Continuous Groups of Transformations*, Princeton University Press, 1933.

Ellinas, D., Barnett, S.M. and Dupertius, M.A., Berry's phase in optical resonance. *Phys. Rev. A*, 39, 3228-3237, 1989.

Everitt, C.W.F., *James Clerk Maxwell*, Scribner's, New York, 1975.

Faraday, M., *Experimental Researches in Electricity*, 3 Volumes, 1839, 1844, 1855. Taylor and Francis, Vols. 1 and 3; Richard and John Edward Taylor, Vol. 2; Dover, New York, 1965.

Feynman, R.P., Leighton, R.B. and Sands, M., *The Feynman Lectures on Physics*, Vol. 1, Addison-Wesley, Reading, MA, 1964.

Forder, P.W., Ring gyroscopes: an application of adiabatic invariance. *J. Phys. A: Math. Gen.*, 17, 1343–1355, 1984.

Gates, S.J., Gauge invariance in nature: a simple view, pp. 31–80 in R.E. Mickens (ed) *Mathematical Analysis of Physical Systems*, Van Nostrand, NY, 1986.

Gerry, C.C., Berry's phase in the degenerate parametric amplifier. *Phys. Rev. A*, 39, 3204–3207, 1989.

Giavarini, G., Gozzi, E., Rohrlich, D. and Tacker, W.D., On the removability of Berry's phase. *Phys. Rev. D*, 39, 3007–3015, 1989.

Giavarini, G. and Onofri, E., Generalized coherent states and Berry's phase. *J. Math. Phys.*, 30, 659–663, 1989.

Gijs, M., Van Haesendonck, C. and Bruynseraede, Y., Resistance oscillations and electron localization in cylindrical Mg films. *Phys. Rev. Lett.*, 52(23), 2069–2072, 1984.

Goddard, P. and Olive, D.I., Magnetic monopoles in gauge field theories. *Rep. Progr. Phys.*, 41, 1357–1437, 1978.

Goldhaber, A.S. and Tower, W.P., Resource letter MM-1: Magnetic monopoles. *Am. J. Phys.*, 58, 429–439, 1990.

Graneau, P., *Ampére–Neumann Electrodynamics of Metals*, Hadronic Press, Nonantum, MA, 1985.

Gray, J.E. and Boules, R.N., Comment on a letter by E.F. Kuester. *IEEE Trans. Electromagn. Comput.*, EMC-29(4), 317, 1987.

Haldane, F.D.M., Comment on 'Observation of Berry's topological phase by use of an optical fiber'. *Phys. Rev. Lett.*, 59(15), 1788–1790, 1987.

Han, D. and Kim, Y.S., Special relativity and interferometers. *Phys. Rev. A*, 37, 4494–4496, 1988.

Hannay, J.H., Angle variable holonomy in adiabatic excursion of an integrable Hamiltonian. *J. Phys. A: Math. Gen.*, 18, 221–230, 1985.

Harmuth, H.F., Correction for Maxwell's equations for signals I. *IEEE Trans. Electromagn. Comput.*, EMC-28(4), 250–258, 1986.

Harmuth, H.F., Correction for Maxwell's equations for signals II. *IEEE Trans. Electromagn. Comput.*, EMC-28(4), 259–265, 1986.

Harmuth, H.F., Propagation velocity of Maxwell's equations. *IEEE Trans. Electromagn. Comput.*, EMC-28(4), 267–271, 1986.

Harmuth, H.F., Reply to Djordjvic and Sarkar. *IEEE Trans. Electromagn. Comput.*, EMC-29(3), 256, 1987.

Harmuth, H.F., Reply to wait. *IEEE Trans. Electromagn. Comput.*, EMC-29(3), 257, 1987.

Harmuth, H.F., Boles, R.N. and Hussain, M.G.M., Reply to Kuester's comments on the use of a magnetic conductivity. *IEEE Trans. Electromagn. Comput.* EMC-29(4), 318–320, 1987.

Hayden, H.C., On a recent misinterpretation of Sagnac experiment. *Galilean Electrodynamics*, 2, 57–58, 1991.

Hayden, H.C., Yes, moving clocks run slowly, but is time dilated? *Galilean Electrodynamics*, 2, 63–66, 1991.

Hendry, J., *James Clerk Maxwell and the Theory of the Electromagnetic Field*, Adam Hilger Ltd, Bristol, 1986.

Herzberg, G. and Longuet-Higgens, H.C., Intersection of potential energy surfaces in polyatomic molecules. *Disc. Faraday Soc.*, 35, 77–82, 1963.

Hestenes, D., *New Foundations for Classical Mechanics*, Reidel, Dordrecht, 1987.

Hestenes, D. and Sobczyk, G., *Clifford Algebra to Geometric Calculus*, Reidel, Dordrecht, 1984.

Higgs, P.W., Broken symmetries, massless particles and gauge fields. *Phys. Lett.*, 12(2), 132–133, 1964.

Higgs, P.W., Broken symmetries and the masses of gauge bosons. *Phys. Rev. Lett.*, 13(16), 508–509, 1964.

Higgs, P.W., Spontaneous symmetry breakdown without massless bosons. *Phys. Rev.*, 145(4), 1156–1163, 1966.

Hodge, W.V.D., *The Theory and Applications of Harmonic Integrals*, Cambridge University Press, 1959.

Horvathy, P.A., The Wu–Yang factor and the non-Abelian Aharonov–Bohm experiment. *Phys. Rev. D*, 33(2), 407–414, 1986.

Hunt, B.J., *The Maxwellians*, a dissertation submitted to the Johns Hopkins University with the requirements for the degree of Doctor of Philosophy, Baltimore, MD, 1984.

Hussain, M.G.M., Comments on the Comments of A.R. Djordjvic and T. Sarkar on 'Correction of Maxwell's equations for signals I and II' and 'Propagation velocity of electromagnetic signals,' *IEEE Trans. Electromagn. Comput.*, EMC-29(4), 317, 1987.

Imry, Y., Active transmission channels and universal conductance fluctuations. *Europhys. Lett.*, 1(5), 249–256, 1986.

Jackson, J.D., *Classical Electrodynamics*, 2nd edition, Wiley, New York, 1975.

Jaklevic, R.C., Lambe, J., Mercereau, J.E. and Silver, A.H., Macroscopic quantum interference in superconductors. *Phys. Rev.*, 140(5A), 1628–1637, 1965.

Jiao, H., Wilkinson, S.R., Chiao, R.Y. and Nathel, H., Two topological phases in optics by means of a non-planar Mach–Zehnder interferometer. *Phys. Rev. A*, 39, 3475–3486, 1989.

Josephson, B.D., Possible new effects in superconductive tunneling. *Phys. Lett.*, 1(7), 251–253, 1962.

Josephson, B.D., Coupled superconductors. *Rev. Mod. Phys.*, 36, 216–220, 1964.

Josephson, B.D., Supercurrents through barriers. *Adv. Phys.*, 14, 419–451, 1965.

Josephson, B.D., The discovery of tunneling supercurrents. *Science*, 184(4136), 527–530, 1974.

Kiritis, E., A topological investigation of the quantum adiabatic phase. *Commun. Math. Phys.*, 111, 417–437, 1987.

Kitano, M., Yabuzaki, T. and Ogawa, T., Comment on Observations of Berry's topological phase by use of an optical fiber. *Phys. Rev. Lett.*, 58(5), 523, 1987.

Konopinski, E.J., What the electromagnetic vector potential describes. *Am. J. Phys.*, 46, 499–502, 1978.

Korteweg, D.J. and de Vries, G., On the change of form of long waves advancing in a rectangular canal, and on a new type of long stationary wave. *Phil. Mag.*, 39, 422–443, 1895.

Kuester, E.F., Comments on 'Correction of Maxwell equations for signals I,' 'Correction of Maxwell equations for signals II,' and 'Propagation velocity of electromagnetic signals'. *IEEE Trans. Electromag. Comput.*, EMC-29(2), 187–190, 1987.

Lakhtakia, A., A case for magnetic sources. *Phys. Essays*, 4(1), 105–108, 1991.

Lakhtakia, A., Varadan, A. and Varadan, V.V., *Time Harmonic Electromagnetic Fields in Chiral Media*, Springer-Verlag, 1989.

Landauer, R., Electrical resistance of disordered one-dimensional lattices. *Phil. Mag.*, 21, 863–867, 1970.

Laughlin, R.B., Quantized Hall conductivity in two dimensions. *Phys. Rev. B*, B23(10), 5632–5633, 1981.

Layton, E., Huang, Y. and Chu, S-I., Cyclic quantum evolution and Aharonov–Anandan geometric phases in SU(2) spin-coherent states. *Phys. Rev.*, 41, 42–48, 1990.

Lebwohl, P. and Stephen, M.J., Properties of vortex lines in superconducting barriers. *Phys. Rev.*, 163(2), 376–379, 1967.

Lee, P.A., Stone, A.D. and Fukuyama, H., Universal conductance fluctuations in metals: effects of finite temperature interactions and magnetic fields. *Phys. Rev. B*, 35(3), 1039–1070, 1987.

Lenstra, D., Kamp, L.P.J. and Van Haeringen, W., Mode conversion, Bloch oscillations and Zener tunneling with light by means of the Sagnac effect. *Optics Com.*, 60, 339–344, 1986.

Lévay, P., Geometrical description of SU(2) Berry phases. *Phys. Rev. A*, 41, 2837–2840, 1990.

Longuet-Higgens, H.C., The intersection of potential energy surfaces in polyatomic molecules. *Proc. Roy. Soc. Lond. Ser. A*, 344, 147–156, 1975.

Martinis, J.M., Devoret, M.H. and Clarke, J., Experimental tests for the quantum behavior of a macroscopic degree of freedom: the phase difference across a Josephson junction. *Phys. Rev. B*, 35(10), 4682–4698, 1987.

Mason, L.J. and Sparling, G.A.J., Non-linear Schroedinger and Korteweg–de Vries are reductions of self-dual Yang-Mills. *Phys. Lett. A*, 137, 29–33, 1989.

Matteucci, G. and Pozzi, G., New diffraction experiment on the electrostatic Aharonov–Bohm effect. *Phys. Rev. Lett.*, 54(23), 2469–2472, 1985.

Maxwell, J.C., On Faraday's lines of force. *Trans. Camb. Phil. Soc.*, 10(Part I), 27–83, 1856.

Maxwell, J.C., A dynamical theory of the electromagnetic field. *Phil. Trans. Roy. Soc.*, 155, 459–512, 1865.

Maxwell, J.C., *A Treatise on Electricity and Magnetism*, 2 Vols., 1st edition 1873; 2nd edition 1881; 3rd edition, 1891, Oxford, Clarendon Press, republished in two volumes, Dover, New York, 1954.

Mead, C.A. and Truhlar, D.G., On the determination of Born–Oppenheimer nuclear motion wave functions including complications due to conical intersections and identical nuclei. *J. Chem. Phys.*, 70, 2284–2296, 1979.

Merzbacher, E., Single valuedness of wave functions. *Am. J. Phys.*, 30, 237–247, 1962.

Michelson, A.A., New measurement of the velocity of light. *Nature*, 114, 831, 1924.

Michelson, A.A., The effect of the earth's rotation on the velocity of light. Part I, *Astrophysical J.*, 61(3), 137–139, 1925.

Michelson, A.A., Gale, H.G. and Pearson, F., The effect of the earth's rotation on the velocity of light. Part II, *Astrophys., J.*, 61(3), 140–145, 1925.

Michelson, A.A. and Morley, E.W., Influence of motion of the medium on the velocity of light. *Am. J. Sci.,* (3), 31, 377–386, 1886.

Mikhailov, V.F., Observation of the magnetic charge effect in experiments with ferromagnetic aerosols. *Ann. Fondation Louis de Broglie*, 12, 491–523, 1987.

Mikhailov, V.F., On interpretation of the magnetic charge effect on ferromagnetic aerosols. *Academy of Sciences of Kazakh SSR*, Alma Ata, HEPI-88-05, 1988.

Mikhailov, V.F., Observation of the Dirac magnetic charges by virtue of diffusion chamber. *Academy of Sciences of Kazakh SSR*, Alma Ata, HEPI-88-17, 1988.

Mikhailov, V.F., Observation of apparent magnetic charges carried by ferromagnetic particles in water droplets. *J. Phys. A: Math. Gen.*, 24, 53–57, 1991.

Mikhailov, V.F. and Mikhailova, L.I., On some regularities of aerosol particle motion in electromagnetic fields. *J. Phys., A: Math. Gen.*, 23, 53–63, 1990.

Milliken, F.P., Washburn, S., Umbach, C.P., Laibowitz, R.B. and Webb, R.A., Effect of partial phase coherence on Aharonov–Bohm oscillations in metal loops. *Phys. Rev. B*, 36(8), 4465–4468, 1987.

Möllenstedt, G. and Bayh, W., Messung der kontinuierlichen Phasenschiebung von Elektronenwellen im kraftfeldfreien Raum durch das magnetische Vektorpotential einer Luftspule. *Die Naturwissenschaften*, 49(4), 81–82, 1962.

Montonen, C. and Olive, D., Magnetic monopoles as gauge particles. *Phys. Lett. B*, 72(1), 117–120, 1977.

Nahin, P.J., *Oliver Heaviside: Sage in Solitude*, IEEE Press, New York, 1988.

Neatrour, M.J., Comments on 'Correction of Maxwell's equations for Signals I', 'Correction for Maxwell's equations for signals II', and 'Propagation velocity of electromagnetic signals'. *IEEE Trans. Electromagn. Comput.*, EMC-29(3), 258–259, 1987.

Nikam, R.S. and Ring, P., Manifestation of Berry phase in diabolic pair transfer in rotating nuclei. *Phys. Rev. Lett.*, 58(10), 980–983, 1987.

Noether, E., Invariante variations probleme. *Nachr. Ges. Wiss. Goettingen, Math.-Phys.* Kl. 171, 235–257, 1918.

Oh, C.H., Soo, C.P. and Lai, C.H., The propagator in the generalized Aharonov–Bohm effect. *J. Math. Phys.*, 29, 1154–1157, 1988.

O'Hara, J.G. and Pricha, W., *Hertz and the Maxwellians: A Study and Documentation of the Discovery of Electromagnetic Wave Radiation, 1873–1894*, Peter Peregrinus Ltd, London, 1987.

Olariu, S. and Popescu, I.I., The quantum effects of electromagnetic fluxes. *Rev. Mod. Phys.*, 157(2), 349–436, 1985.

Palais, R.S., The symmetries of solitons. *Bull. Amer. Math. Soc.*, 34, 339–403, 1997.

Pancharatnam, S., Generalized theory of interference, and its applications. *Proc. Indian Acad. Sci. A*, 44, 247 (1956), reprinted in *Collected Works of S. Pancharatnam*, edited by G.W. Series, Oxford University Press, New York, 1975.

Pannetier, B., Chaussy, J., Rammal, R. and Gandit, P., Magnetic flux quantization in the weak-localization regime of a non-superconducting metal. *Phys. Rev. Lett.*, 53(7), 718–721, 1984.

Panofsky, W.K.H. and Phillips, M., *Classical Electricity and Magnetism*, 2nd edition, Addison-Wesley, Reading, MA, 1962.

Paranjape, M.B., Induced angular momentum and gauge invariance. *Phys. Rev. D*, 36(12), 3766–3771, 1987.

Penrose, R. and Rindler, W., *Spinors and Spacetime*, 2 vols., Cambridge University Press, 1984 & 1986.

Peshkin, M., The Aharonov–Bohm effect: why it cannot be eliminated from quantum mechanics. *Phys. Rep.*, 80, 375–386, 1981.

Peshkin, M. and Tonomura, A., *The Aharonov–Bohm Effect*, Springer-Verlag, New York, 1989.

Polyakov, A.M., Particle spectrum in quantum field theory. *JETP Lett.*, 20(6), 194–195, 1974.

Polyarkov, Y.B., Kontarev, V.Y., Krylov, I.P. and Sharvin, Y.V., Observation of quantum oscillations of the magnetoresistance of multiply connected objects with a hopping conductivity. *JETP Lett.*, 44, 373–376, 1986.

Post, E.J., *Formal Structure of Electromagnetics*, North-Holland, Amsterdam, 1962.

Post, E.J., Sagnac effect. *Rev. Mod. Phys.*, 39(2), 475–493, 1967.

Post, E.J., The constitutive map and some of its ramifications. *Ann. Phys.*, 71, 497–518, 1972.

Post, E.J., Interferometric path-length changes due to motion. *J. Opt. Soc. Am.*, 62, 234–239, 1972.

Post, E.J., Ramifications of flux quantization. *Phys. Rev. D*, 9(9), 3379–3386, 1974.

Post, E.J., The Gaussian version of Ampere's law. *J. Math. Phys.*, 19(1), 347, 1978.

Post, E.J., Can microphysical structure be probed by period integrals? *Phys. Rev. D*, 25(12), 3223–3229, 1982.

Post, E.J., Electroflux quantization. *Phys. Lett.*, 92(5), 224–226, 1982.

Post, E.J., On the quantization of the Hall impedance. *Phys. Lett. A*, 94(8), 343–345, 1983.

Prasad, M.K. and Sommerfeld, C.M., Exact solution for the t'Hooft monopole and the Julia-Zee dyon. *Phys. Rev. Lett.*, 35(12), 760–762, 1975.

Richardson, D.J., Kilvington, A.I., Green, K. and Lamoreaux, S.K., Demonstration of Berry's phase using stored ultracold neutrons. *Phys. Rev. Lett.*, 61, 2030–2033, 1988.

Sagnac, G., L'éther lumineux demonstré par l'effet du vent relatif d'éther dans un interféromêtre en rotation uniforme. *Comptes Rendus*, 157, 708–710, 1913.

Sagnac, G., Sur la preuve de la réalité de l'Éther lumineaux par l'expérience de l'interférographe tournant. *Comptes. Rendus*, 157, 1410–1413, 1913.

Sagnac, G., Circulation of the ether in rotating interferometer. *Le J. Phys. Radium*, 5(4), 177–195, 1914.

Sandesara, N.B. and Stark, R.W., Macroscopic quantum coherence and localization for normal-state electrons in Mg. *Phys. Rev. Lett.*, 53(17), 1681–1683, 1984.

Schulman, L.S., Approximate topologies. *J. Math. Phys.*, 12(2), 304–308, 1971.

Schulman, L.S., *Techniques and Applications of Path Integration*, Wiley, New York, 1981.

Schwinger, J., Magnetic charge and quantum field theory. *Phys. Rev.*, 144(4), 1087–1093, 1966.

Schwinger, J., A magnetic model of matter. *Science*, 165(3895), 757–761, 1969.

Segert, J., Photon Berry's phase as a classical topological effect. *Phys. Rev.*, 36(1), 10–15, 1987.

Segert, J., Non-Abelian Berry's phase, accidental degeneracy, and angular momentum. *J. Math. Phys.*, 28(9), 2102–2114, 1987.

Shaarawi, A.M., Besieris, I.M. and Ziolkowski, R., A novel approach to the synthesis of non-dispersive wave packet solutions to the Klein–Gordon and Dirac equations. *J. Math. Phys.*, 31, 2511–2519, 1990.

Sharvin, D.Y. and Sharvin, Y.V., Magnetic flux quantization in a cylindrical film of a normal metal. *JETP Lett.*, 34 272–275, 1981.

Silvertooth, E.W., Experimental detection of the ether. *Speculations Sci. Technol.*, 10(1), 3–7, 1986.

Simon, B., Holonomy, the quantum adiabatic theorem, and Berry's phase. *Phys. Rev. Lett.*, 51(24), 2167–2170, 1983.

Stone, A.D., Magnetoresistance fluctuations in mesoscopic wires and rings. *Phys. Rev. Lett.*, 54(25), 2692–2695, 1985.

Stone, A.D. and Imry, Y., Periodicity of the Aharonov–Bohm effect in normal-metal rings. *Phys. Rev. Lett.*, 56(2), 189–192, 1986.

Stormer, H.L. and Tsui, D.C., The quantized Hall effect. *Science*, 220, 1241–1246, 1983.

Sundrum, R. and Tassie, L.J., Non-Abelian Aharonov–Bohm effects, Feynman paths, and topology. *J. Math. Phys.*, 27(6), 1566–1570, 1986.

Suter, D., Chingas, G.C., Harris, R.A. and Pines, A., Berry's phase in magnetic resonance. *Mol. Phys.*, 61(6), 1327–1340, 1987.

Suter, D., Mueller, K.T. and Pines, A., Study of the Aharonov–Bohm quantum phase by NMR interferometry. *Phys. Rev. Lett.*, 60(13), 1218–1220, 1988.

Tellegen, B.D.H., The gyrator: a new electric network element. *Phillips Res. Rep.*, 3, 81, 1948.

Thirring, W., A soluble relativistic field theory. *Ann. Phys.*, (NY), 3, 91–112, 1958.

Thomas, T.F., Direct calculation of the Berry phase for spins and helicities, manuscript, 1988.

't Hooft, G., Renormalizable Lagrangians for massive Yang–Mills fields. *Nucl. Phys. B*, 35, 167–188, 1971.

't Hooft, G., Magnetic monopoles in unified gauge theories. *Nucl. Phys. B*, 79, 276–284, 1974.

Timp, G., Chang, A.M., Cunningham, J.E., Chang, T.Y., Mankiewich, P., Behringer, R. and Howard, R.E., Observation of the Aharonov–Bohm effect for $\omega ct > 1$. *Phys. Rev. Lett.*, 58(26), 2814–2817, 1987.

Timp, G., Mankiewich, P.M., deVegvar, Behringer, R., Cunningham, J.E., Howard, R.E., Baranger, H.U. and Jain, J.K., Suppression of the Aharonov–Bohm effect in the quantized Hall regime. *Phys. Rev. B*, 39, 6227–6230, 1989.

Tomita, A. and Chiao, R.Y., Observation of Berry's topological phase by use of an optical fiber. *Phys. Rev. Lett.*, 57(8), 937–940, 1986.

Tonomura, A., Matsuda, T., Suzuki, R., Fukuhara, A., Osakabe, N., Umezaki, H., Endo, J., Shinagawa, K., Sugita, Y. and Fujiwara, H., Observation of Aharonov–Bohm effect by electron microscopy. *Phys. Rev. Lett.*, 48, 1443–1446, 1982.

Tonomura, A., Umezaki, H., Matsuda, T., Osakabe, N., Endo, J. and Sugita,Y., Is magnetic flux quantized in a toroidal ferromagnet? *Phys. Rev. Lett.*, 51(5), 331–334, 1983.

Tonomura, A., Osakabe, N., Matsuda, T., Kawasaki, T., Endo, J., Yano, S. and Yamada, H., Evidence for Aharonov–Bohm effect with magnetic field completely shielded from electron wave. *Phys. Rev. Lett.*, 56(8), 792–795, 1986.

Tonomura, A. and Callen, E., Phase, electron holography, and a conclusive demonstration of the Aharonov–Bohm effect. *ONRFE Sci. Bul.*, 12(3), 93–104, 1987.

Tsui, D.C., Stormer, H.L. and Gossard, A.C., Two-dimensional magneto-transport in extreme quantum limit. *Phys. Rev. Lett.*, 48(22), 1559–1562, 1982.

Umbach, C.P., Van Haesendonck, C., Laibowitz, R.B., Washburn, S. and Webb, R.A., Direct observations of ensemble averaging of the Aharonov–Bohm effect in normal metal loops. *Phys. Rev. Lett.*, 56(4), 386–389, 1986.

van Dantzig, D., The fundamental equations of electromagnetism, independent of metrical geometry. *Proc. Cambridge Philos. Soc.*, 30, 421–427, 1934.

von Klitzing, K., Dorda, G. and Pepper, M., New method for high-accuracy determination of the fine-structure constant based on quantized Hall resistance. *Phys. Rev. Lett.*, 45(6), 494–497, 1980.

Wait, J.R., Comments on Correction of Maxwell's equations for signals I and II. *IEEE Trans. Electromagn. Comput.*, EMC-29(3), 256–257, 1987.

Wang, S-J., Nonadiabatic Berry's phase for a spin particle in a rotating magnetic field. *Phys. Rev. A*, 42, 5107–5110, 1990.

Ward, R.S. and Wells, R.O., *Twistor Geometry & Field Theory*, Cambridge University Press, 1990.

Washburn, S., Umbach, C.P., Laibowitz, R.B. and Webb, R.A., Temperature dependence of the normal-metal Aharonov-Bohm effect. *Phys. Rev. B*, 32(7), 4789–4792, 1985.

Washburn, S. and Webb, R.A., Aharonov-Bohm effect in normal metal: quantum coherence and transport. *Adv. Phys.*, 35(4), 375–422, 1986.

Washburn, S., Schmid, H., Kern, D. and Webb, R.A., Normal-metal Aharonov–Bohm effect in the presence of a transverse electric field. *Phys. Rev. Lett.*, 59(16), 1791–1794, 1987.

Webb, R.A., Washburn, S., Umbach, C.P. and Laibowitz, R.B., Observation of _h/e Aharonov-Bohm oscillations in normal-metal rings. *Phys. Rev. Lett.*, 54(25), 2696–2699, 1985.

Webb, R.A., Washburn, S., Benoit, A.D., Umbach, C.P. and Laibowitz, R.B., The Aharonov–Bohm effect and long-range phase coherence in normal metal rings. *Proc. 2nd Int. Symp. Foundations of Quantum Mechanics 1986*, M. Namiki *et al.* (eds), The Physical Society of Japan, Tokyo, pp. 193–206, 1987.

Weinberg, E.J., Fundamental monopoles and multimonopole solutions for arbitrary simple gauge fields. *Nucl. Phys. B*, 167, 500–524, 1980.

Weyl, H., Gravitation und Elektrizität. *Sitzungsberichte der Koeniglich Preussischen Akad. Wissenschaften zu Berlin*, 465–480, 1918.

Weyl, H., Eine neue Erweiterung der RelativitΣts theorie. *Ann. Phys.*, 59, 101–133, 1919.

Weyl, H., *Gruppen Theorie und Quantenmechanik, S. Hirzel, Leipzig, 1928* (*The Theory of Groups and Quantum Mechanics*), Dover, 1949.

Weyl, H., *The Classical Groups, Their Invariants and Representations*, Princeton University Press, 1939.

Whittaker, E.T., On the partial differential equations of mathematical physics. *Math. Ann.*, 57, 333–355, 1903.

Whittaker, E.T., On an expression of the electromagnetic field due to electrons by means of two scalar potential functions. *Proc. Lond. Math. Soc.*, 1, 367–372, 1904.

Wignall, J.W.G., Frame dependence of the phase of de Broglie waves. *Am. J. Phys.*, 57, 415–416, 1989.

Wilczck, F. and Zee, A., Appearance of gauge structure in simple dynamical systems. *Phys. Rev. Lett.*, 52(24), 2111–2114, 1984.

Wilkinson, M., An example of phase holonomy in WKB theory. *J. Phys. A: Math. Gen.*, 17, 3459–3476, 1984.

Wilkinson, M., Critical properties of electron eigenstates in incommensurate systems. *Proc. R. Soc. Lond. A*, 391, 305–350, 1984.

Woodruff, A.E., The contributions of Hermann von Helmholtz to electrodynamics. *Isis*, 59, 300–311, 1968.

Wu, T.T. and Yang, C.N., Concept of non-integrable phase factors and global formulation of gauge fields. *Phys. Rev. D*, 12(12), 3843–3857, 1975.

Wu, T.T. and Yang, C.N., Dirac monopole without strings: monopole harmonics. *Nuclear Phys. B*, 107, 365–380, 1976.

Wu, T.T. and Yang, C.N., Dirac's monopole without strings: classical Lagrangian theory. *Phys. Rev. D*, 14(2), 437–445, 1976.

Xie, X.C. and DasSarma, S., Aharonov–Bohm effect in the hopping conductivity of a small ring. *Phys. Rev. B*, 36, 9326–9328, 1987.

Yang, C.N., Charge quantization, compactness of the gauge group, and flux quantization. *Phys. Rev. D*, 1(8), 2360, 1970.

Yang, C.N., Integral formalism for gauge fields. *Phys. Rev. Lett.*, 33(7), 445–447, 1974.

Yang, C.N., Gauge fields, electromagnetism and the Bohm–Aharonov effect. *Proc. Int. Symp. Foundations of Quantum Mechanics*, Tokyo, pp. 5–9, 1983.

Yang, C.N., Hermann Weyl's contribution to physics, in K. Chandrasekharan (ed), *Hermann Weyl, 1885–1985*, Springer, New York, pp. 7–21, 1986.

Yang, C.N. and Mills, R.L., Conservation of isotopic spin and isotopic gauge invariance. *Phys. Rev.*, 96, 191–195, 1954.

Yurke, B., McCall, S. L. and Klauder, J.R., SU(2) and SU(1,1) interferometers. *Phys. Rev. A*, 33(6), 4043–4054, 1986.

Zeleny, W.B., Symmetry in electrodynamics: a classical approach to magnetic monopoles. *Am. J. Phys.*, 59, 412–415, 1991.

Chapter 2

Sagnac Effect: A Consequence of Conservation of Action Due to Gauge Field Global Conformal Invariance in a Multiply-Joined Topology of Coherent Fields[1]

2.1. Overview

The Sagnac effect underlying the ring laser gyro is a coherent field effect and described here as a global, not a local, effect in a multiply-joined, not a simply-joined, topology of those fields. Given a Yang–Mills or gauge field formulation of the electromagnetic field (Barrett, 1993), the measured quantity in the Sagnac effect is the phase factor. Gauge field formulation of electromagnetism requires in many cases uncoupling the electromagnetic field from the Lorentz group algebra. As conventionally interpreted, the Lorentz group is the defining algebraic topology for the concepts of inertia and acceleration from an inertialess state. However, those concepts find new definitions here in gauge theory and new group theory descriptions. The explanation for the origins of the Sagnac effect offered here lies in the generation of a *constraint* or *obstruction* in the interferometer's field topology under conditions of conserved action, that constraint or obstruction being generated *only* when the platform of the interferometer is rotated. Whereas previous explanations of the Sagnac effect have

[1]Based on Barrett (1995).

left the conventional Maxwell equations inviolate but seen a need to change the constitutive relations, here we see a need to do the opposite. Just as the Lorentz group description appears only as a limiting (*zero rotation or stationary*) case in this new explanation, so Minkowski spacetime is also viewed as a limiting case appropriate for the Sagnac interferometer in the *stationary* platform situation, with Cartan–Weyl spacetime appropriate for *rotated* platform situations. We attribute the existence of a measurable phase factor in the Sagnac interferometer with rotated platform as due to the conformal invariance of the action in the presence of the creation of a topological obstruction by the rotation.

2.2. Sagnac Effect Phenomenology

The Lorentz group algebra is the defining field algebra for the set of all *inertial* frames and the spacetime symmetry. Any frame of reference that is not an inertial frame is an "accelerated" frame and experienced as a force field. However, we shall attempt to show that for *noninertial* frames the Lorentz group is *not* the defining algebra. Such situations are measured by rotation sensors. Of these, Sagnac (1913a,b, 1914) first demonstrated a ring interferometer which indicates the state of rotation of a frame of reference, i.e., a ring interferometer as a rotation rate sensor. The ring interferometer performs the same function as a mechanical gyroscope. When a laser is used as the source of radiation in the interferometer, it is called a ring laser gyro.

Figure 2.1 shows the basic Sagnac interferometer. One light beam circulates a loop in a clockwise direction, and another beam circulates a loop in a counterclockwise direction. When the interferometer is set in motion, interference fringes (phase difference) are observed at the overlap area H, i.e., in the heterodyned counterpropagating beams. Details of the interferometer can be found in Post (1967) and Chow *et al.* (1985) and reviews of the Sagnac effect have been given by von Laue (1920), Zernicke (1947) and Metz (1952). There are two basic kinds of ring interferometers sensing rotation: the passive ring resonator and the active ring laser gyro. The general theory of the

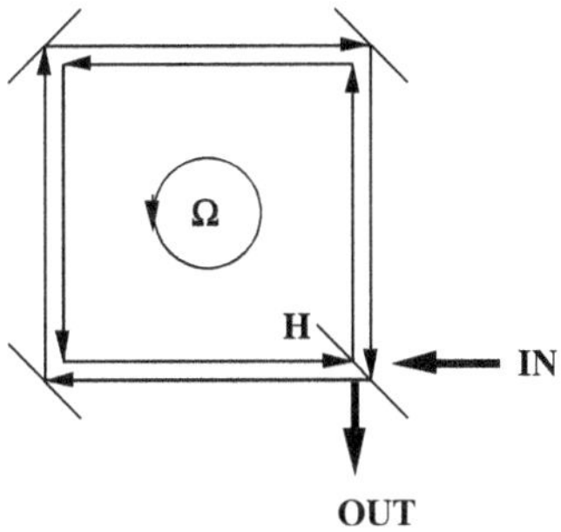

Fig. 2.1. The basic Sagnac interferometer. The clockwise and counterclockwise beams are not separated in reality and are only shown so here for purposes of exposition.

ring laser gyro is addressed in Gyorffi and Lamb (1965), Aronowitz (1971) and Menegozzi and Lamb (1973).

Here, we shall quickly cover the main descriptive features and move on to address the central issue: explanations of the effect. In the case of the passive ring resonator, the interference fringes are described by:

$$\Delta\phi = \frac{4\Omega \cdot \mathbf{A}}{\lambda c}, \tag{2.1}$$

where $\Delta\phi$ is the phase difference between clockwise and counter-clockwise propagating beams, Ω is angular rate in rad/sec., λ is the vacuum wavelength, $\mathbf{A}$ is area enclosed by the light path, and a velocity field $\mathbf{v}$ defines the angular velocity:

$$\nabla \times \mathbf{v}\Omega = 2\Omega. \tag{2.2}$$

Rosenthal (1962) suggested a self-oscillating version of the Sagnac ring interferometer which was demonstrated by Macek and Davis (1963). In this version, the clockwise and counterclockwise modes occur in the same optical cavity. In the case of the laser version of the self-oscillating version of the Sagnac interferometer, i.e., the ring laser gyroscope, and in contrast to the Sagnac ring interferometer, a comoving optical medium in the laser beam affects the beat frequency, which, rather than the phase difference, is the measured variable. In this version, the frequency difference, $\Delta\omega$, of the clockwise and counterclockwise propagating beams with respect to the resonant frequency, ω, is described by:

$$\left|\frac{\Delta\omega}{\omega}\right| = \frac{2\oint n^2(1-\alpha)\mathbf{v}\cdot dr}{c\oint nds}, \tag{2.3}$$

where $\mathbf{n}$ is the index of refraction of the stationary medium; and α is a coefficient of drag. The same frequency difference with respect to the angular velocity is given by

$$\Delta\omega = \sqrt{\frac{\mathbf{A}}{\pi}\frac{2\Omega}{\lambda}} = \frac{4\mathbf{A}\Omega}{\mathbf{P}\lambda}, \tag{2.4}$$

where $\mathbf{P}$ is the perimeter of the light path. The Sagnac interferometer path length change in terms of phase reversals is independent of the waveguide mode and completely independent of the optical properties of the path (Post, 1972a).

There are details of ring laser gyroscope operation, such as lock-in and scale factor variation, which necessitate the amendment of the above descriptions, but these operational details will not be addressed here, and we move directly to consider explanations of the effect.

There are three current explanations/descriptions of the Sagnac effect:

(a) the kinematic description;
(b) the physical-optical description;
(c) the dielectric metaphorical description.

It is the intention of the present essay to introduce a fourth:

(d) the gauge field explanation.

We shall examine each of these approaches in turn.

2.2.1. *Kinematic description*

The force field exerted on the fields of the Sagnac interferometer can be either due to gravitational, linear acceleration or rotational velocity (kinematic acceleration) field effects. The kinematic acceleration field is due to Coriolis force contributions, but the gravitational field and linear acceleration field do not have such contributions.

There is also a distinction between the different fields with respect to the convergence/divergence of those fields. For example, the lines of gravitational force converge to a nonlocal point, e.g., the center of the earth, in the case of a platform at rest on the earth, but in the case of a linearly accelerated platform the lines of force converge to a nonlocal point at infinity. No Coriolis force is present in either of these cases. In contrast to these, with a platform with rotational velocity, the lines of force diverge from the local axis of rotation and a Coriolis force is present. If the platform with rotational velocity is also located on or near the earth, it will also experience the gravitational force of the earth, besides the Coriolis force. However, only the platform undergoing kinematic acceleration (rotational velocity) is in a state of motion with respect to *all* inertial frames.

The kinematic description thus primarily implicates the Coriolis (acceleration) force and a state of kinematic acceleration is associated with a state of absolute motion with respect to all inertial frames. For example, Konopinski (1978) defined the electromagnetic vector potential as field momentum exchanged with the kinetic momenta of charged particles. According to this author most defining relations between potentials and fields, such as, e.g., in equations of motion, are defined in a *static condition*. In these static or *local conditions*, $q\phi$ can be defined as a "store" of field energy, and $\frac{q\mathbf{A}}{c}$ a "store" of momentum energy. That is the conventional interpretation of the four vector potential.

However, the subject of Konopinski's paper is a second condition which is *global* in character. The model he considers is not a *point* but a *volume* $\oint dV(r)$ of an electromagnetic field with an energy and vector momentum defined as in the Coulomb gauge:

$$w(r,t) = \frac{(\mathbf{E}^2 + \mathbf{B}^2)}{8\pi}, \tag{2.5}$$

$$\mathbf{g}(r,t) = \frac{(\mathbf{E} \times \mathbf{B})}{4\pi c}, \tag{2.6}$$

and with a total mass

$$W = \oint \frac{dVw}{c^2}, \tag{2.7}$$

which is constant in the *absence* of fluxes through the surface enclosing the volume. The test condition for this situation consists of a point charge q at a fixed position $\mathbf{r}_q$ in a static external field $\mathbf{E}_0 = -\nabla\phi$, $\mathbf{B}_0 = -\nabla \times \mathbf{A}(\mathbf{r})$. The fact that the model considered is a volume introduces *global* (across the volume) and *local* (within the volume, e.g., pointlike) conditions.

The equation of motion for a point charge is given as follows:

$$\frac{d(Mv + \frac{q}{c}A)}{dt} = -\nabla q\left(\varphi - \left(\frac{v}{c}\right) \cdot A\right), \tag{2.8}$$

which describes changes in conjugate momentum (left side) + interaction energy (right side). With q constant (the left hand side), any variation of $\mathbf{v}$ causes $\mathbf{A}$ to vary and vice versa. The total field momentum that changes when the position $\mathbf{r}_q$ is changed is derived as

$$\mathbf{P}(\mathbf{r}_q) \equiv \oint dV(\mathbf{r})\frac{\mathbf{E}_q(\mathbf{r}-\mathbf{r}_q) \times \mathbf{B}_0(\mathbf{r})}{4\pi c}. \tag{2.9}$$

Introducing source terms gives the field momentum $\mathbf{P}(\mathbf{r}_q)$ as

$$\mathbf{P}(\mathbf{r}_q) = \frac{q\mathbf{A}(\mathbf{r}_q)}{c}. \tag{2.10}$$

The result is that under the total momentum conditions expressed by Eq. (2.8), changes in the total field momentum when the position of the particle at $\mathbf{r}_q$ is changed, must result in changes in the kinetic momentum of the particle $\mathbf{Mv}$.

Considered from a different perspective, one may state the equilibrium condition in the reverse causal condition to that considered by Konopinski, namely: changes in the velocity of the system defining the kinetic energy of a "particle", $\mathbf{v}$, result in changes in the vector potential in the total field momentum, $\mathbf{A}$. Moreover, in the Sagnac effect there are *two* vector potential components with respect to clockwise and counterclockwise beams. The measured quantity, as will be explained more fully below, is then *the phase factor* or the integral of the *potential difference* between those beams and related to the *angular velocity difference* between the two beams. Therefore, as the vector potential measures the momentum gain and the scalar potential measures the kinetic energy gain, the photon will acquire "mass".[2]

Konopinski using variational principles formulated a Lagrangian for this global field situation with A_ν as "generalized coordinates" and $\partial_\mu A_\nu$ as "generalized velocities".

$$L = -\frac{(\partial_\mu A_\nu)^2}{8\pi} + \frac{j_\nu A_\nu}{c}. \tag{2.11}$$

With gauge invariance (the Lorenz gauge) there are no source terms, $j_\nu \equiv 0$, so:

$$L = -\frac{(\partial_\mu A_\nu)^2}{8\pi}. \tag{2.12}$$

Thus, in our adaptation of this argument, Eq. (2.12) describes the field conditions of the Sagnac interferometer *when its platform is stationary*, but conveys no more information than the field tensor $F_{\mu\nu} = \partial_\mu A_\nu - \partial_\nu A_\mu$. On the other hand, (i) the conservation condition expressed by Eqs. (2.8) and (2.11) describe the Sagnac interferometer platform *in rotation and kinetically*; and (ii) it is relevant that Eq. (2.12), but *not* Eq. (2.11), is determined by the Lorenz gauge.[3] Therefore, the field algebraic logic underlying the Sagnac effect, i.e., the Sagnac interferometer platform *in rotation*, is *not* that of the Lorenz gauge.

2.2.2. *Physical-optical description*

Post's physical optical theory of the Sagnac effect (Post, 1967, 1972b) demands the loosening of the ties of the theory of electromagnetism to a Lorentz[4] invariant structure. Post correctly observed that it is

---

[2]Moyer (1987), addressing the combining of electromagnetism and general relativity, identified charge with a Lagrange multiplier and the Hamiltonian with the self-energy mc^2, using an optimal control argument rather than the calculus of variations. The Lagrangian identified is a function of the electromagnetic scalar potential and the vector potential, i.e., the four potential.

[3]Evidently, this gauge is due to L. Lorenz (1829–1891) of Copenhagen, not H.A. Lorentz (1853–1928) of Leiden (cf. Whittaker, 1951, Vol. 1, pp. 267–268, and Penrose and Rindler, 1984, Vol. 1, p. 321 (footnote)).

[4]H.A. Lorentz (1853–1928).

customary to assume no distinction in free space between dielectric displacement, **D**, and the electric field, **E**, nor between magnetic induction, **B**, and the magnetic field, **H**. This identification is called the *Gaussian identification* and is justified by the supposed absence of polarization mechanisms in free space. The Gaussian identification, together with the Maxwell equations leads to the d'Alembertian equation, which is a Lorentz invariant structure. As the d'Alembertian does not permit mixed spacetime derivatives, it cannot account for the nonreciprocal asymmetry between clockwise and counterclockwise beam rotation in the Sagnac interferometer. Therefore, Post suggested that in order to account for this asymmetry, either the Gaussian field identification is incorrect in a rotating frame, or the Maxwell equations are affected by the rotation. However, in offering this choice Post tacitly assumed no linkage between the Gaussian field identification and the Maxwell equations (i.e., their exclusivity was assumed). He also assumed that the solution to the asymmetry effect must be a *local* effect, because he was convinced that both the Gaussian field identification and the conventional Maxwell equations describe *local* effects. Below, on the other hand, we argue that the field arrangements in the interferometer should be described as a *global* situation and, as a result, the occurrence of an asymmetry does not warrant a change in the *local* Abelian Maxwell equations (also rejected by Post), but to the required use of *nonlocal*, non-Abelian Maxwell equations in a *multiply-connected interferometric* situation (neglected by Post); and also does not warrant a change in the *local* Gaussian field identification (adopted by Post), but warrants the use of *nonlocal* non-Abelian field-metric interactions (neglected by Post). Both *nonlocal* non-Abelian equations and *nonlocal* interactions are required because the asymmetry under discussion arises in the Sagnac interferometer, that interferometer being a *global*, i.e., nonlocal, situation; and the amendment suggested by Post, whether of the local equations or the local identification, is inappropriately *ad hoc* in the presence of that *global* situation. Rather, the field topology of the Sagnac interferometer requires drastic redefinition of those equations and the full interaction logic, rather than a topologically inappropriate amendment of their local form regardless of field topology.

Nonetheless, Post understood that there was, and is, a problem in defining field-metric relations in empty space and referred to, among others, the Pegram (1917) experiment as illustrative. Quickly stated, in this experiment, simultaneously and around the same axis, a coaxial cylindrical condenser and solenoid are rotated. The rotation produces a magnetic field in the solenoid in the axial direction and between the plates of the condensor. The condensor can then be charged by shortening the plates of the condenser. Post observed that the experiment indicates a cross relation between electric and magnetic fields in a vacuum, a relation which is denied by a Lorentz transformation.

Post (1967) also noticed correctly that Weyl and Cartan were aware of the metric independence of Maxwell's equations. But Post (1974) presented the view that the asymmetry of the conventional Maxwell equations (absence of magnetic monopoles) is compatible with a certain topological symmetry, which he then used to suggest that the law of flux quantization, $\oint \mathbf{F} = 0$, is a fundamental law. This suggestion, however, is contradicted by the law of flux quantization being *global*, rather than *local,* in nature, and being based on a *multiply-connected* symmetry. Post's suggestion was motivated evidently by the assumption incorrect from our perspective — that the spacetime situation of the Sagnac interferometer is *simply connected.* He also distinguished two points of view:

(1) Topology enters physics through the families of integration manifolds that are generated by physical fields and spacetime is the arena in which these integration manifolds are embedded.
(2) Spacetime is endowed with a topological structure relating to its physics.

The first point of view brings *the field* to center stage; the second, *the metric*. However, from a gauge field perspective these two points of view are neither unconnected nor exclusive and thus do not constitute an exclusive choice. Under a gauge field formulation, there *can be* interaction between field and metric. The exclusive choice offered by Post is understandable in that Post did not distinguish between force-fields and related gauge-fields. From a gauge theory point of view, however, one is not forced to choose exclusively between these two alternatives.

Post also proposed that the global condition for the **A** potential to exist is that all cyclic integrals of **F** vanish. However, this statement is again based on consideration of *simply-connected* domains, for which the local Maxwell theory is, indeed, described by $\oint \mathbf{F} = 0$. He stated that flux quantization is formally incompatible with the magnetic-monopole hypothesis, because of his belief that a global **A** exists only if $\oint \mathbf{F} = 0$. But as we shall show below, in a *multiply-connected* domain in the presence of a topological obstruction, a *global phase factor* defined over local **A** fields exists. Therefore in this global, multiply-connected situation, $\oint \mathbf{F} \neq 0$.

Nonetheless, Post's physical optical theory was a major advance in understanding and is based on the following valid observations:

(1) The Maxwell equations have no specific constitutive relations to free space. The traditional equalities in free space, $\mathbf{E} = \mathbf{D}$ and $\mathbf{B} = \mathbf{H}$, assume the Gaussian approximation (absence of detected polarization mechanisms in free space) discussed above, and define the properties of free space *only as seen from inertial frames*. This is because those relations and the Maxwell equations lead to the standard free-space d'Alembertian wave equation, and the d'Alembertian, as Post pointed out, is a Lorentz invariant structure. (The Gaussian approximation corresponds to the Minkowskian metric $(c^2, -1, -1, -1)$, or $(1, -1, -1, -1)$ with $(c^2 = 1)$, defining the Lorentz group as a symmetry property of the spacetime continuum.)

(2) Only in the case of *uniformly translating* systems does the mutual motion of observer and platform *completely* define the physical situation.

Post's solution to the problems raised by these observations is, as we have seen, to modify the constitutive relations, but not to modify the field or Maxwell equations. This solution, we have suggested, inappropriately assumes that the Sagnac interferometer is a simply-connected geometry. The Pegram experiment, described above, indicates the presence of cross coupling between electric and magnetic fields. According to Post this cross coupling is responsible for the Sagnac effect and is due to the constitutive relations on a rotating frame. However, it is ironic that it is not possible to assume the physicality of such cross terms without modification of the Maxwell equations which Post left untouched. Therefore, while we may agree that the cross coupling is related to the Sagnac effect, below we ascribe its existence to the presence of non-Abelian Maxwell relations (Barrett, 1993), rather than amended constitutive relations.

We can also agree that "A nonuniform motion produces a real and intrinsic physical change in the object in motion; the motion of the frame of reference by contrast produces solely a difference in the observational viewpoint" (Post, 1967, p. 488). However, we would reword the distinction as follows: whereas the *linear motion* of the frame of reference of an interferometer incorporating area **A** produces solely a *local* difference in the observational viewpoint describable by the Lorentz gauge, *nonuniform motion* of an interferometer incorporating area **A**, as on a rotating platform, produces a *global* difference describable by the Ampère gauge.

Despite these, considered here incorrect, theoretical positions, Post (1967) greatly improved and advanced the understanding of the Sagnac effect by squarely addressing the physical issues. We shall return to these physical issues later. He was also a leader in realizing the necessity of distinguishing local versus global approaches to physics (Post, 1962, 1971, 1974, 1982).

2.2.3. *Dielectric metaphor description*

Chow *et al.* (1985) commenced their dielectric metaphor for the gravitational field with the Plebanski (1960) observation that it is possible to write Maxwell's equations in an arbitrary gravitational field in a form in which they resemble electrodynamic equations in a dielectric medium. Thus, the gravitational field is in some sense equivalent to a dielectric medium and represented by the metric $g_{\mu\nu} = g_{\nu\mu}$ of the form:

$$g_{\mu\nu} = \eta_{\mu\nu} + h_{\mu\nu} = \begin{bmatrix} 1 & 0 & 0 & 0 \\ 0 & -1 & 0 & 0 \\ 0 & 0 & -1 & 0 \\ 0 & 0 & 0 & -1 \end{bmatrix} + \begin{bmatrix} 0 & h_{01} & h_{02} & h_{03} \\ h_{10} & 0 & 0 & 0 \\ h_{20} & 0 & 0 & 0 \\ h_{30} & 0 & 0 & 0 \end{bmatrix}, \tag{2.13}$$

where $\eta_{\mu\nu}$ is the metric of special relativity and $h_{\mu\nu}$ is the effect of the gravitational field. Using the definition

$$\mathbf{h} \equiv (h_{01}, h_{02}, h_{03}), \tag{2.14}$$

Chow *et al.* (1985) defined amended constitutive relations. That is, just like Post (1967) examined above, these authors chose to introduce any metric influences on the fields into the constitutive relations rather than into amended Maxwell equations. The amended constitutive relations offered by these authors are as follows:

$$\mathbf{D} = \mathbf{E} - c(\mathbf{B} \times \mathbf{h}), \tag{2.15}$$

$$\mathbf{B} = \mathbf{H} + \frac{1}{c}(\mathbf{E} \times \mathbf{h}). \tag{2.16}$$

These authors then proceeded to derive an equation of motion for the electric field. However, as we have seen above, because an equation of motion (d'Alembertian) *assumes* a Lorenz gauge[5] (i.e., the special theory of relativity), and as $h_{\mu\nu}$ is introduced *as a correction to the special theory components*, $\eta_{\mu\nu}$, such a derivation must be at the price of *group algebraic inconsistency.* Furthermore, if, as is claimed by many (commencing with Heaviside, 1893) there is a formal analogy between the gravitational potential, **h**, and the vector potential, **A**, and the field $\nabla \times \mathbf{h}$ and the magnetic field **B**, that analogy can nonetheless be introduced into the electromagnetic field in ways *other than* by the amended constitutive relations, Eqs. (2.15) and (2.16). The next section addresses this other way.

[5]L. Lorenz (1829–1891) of Copenhagen, not H.A. Lorentz (1853–1928) of Leiden.

2.2.4. *The gauge field explanation*

Having found the physical optical and the dielectric metaphorical explanations of the Sagnac effect wanting, we now introduce a gauge theory[6] explanation, but before doing so we examine Forder's (1984) analysis of ring gyroscopes. A presupposition of gauge theory is constant action and Forder (1984) showed that the adiabatic invariance of the action implies the invariance of the flux enclosed by the contour, which is Lenz's law. There is a common framework for treatment of (1) the ring laser gyro, (2) conductors, and (3) superconductors. The treatment of (1) is according to the adiabatic invariance of the quantum phase:

$$\Delta\phi = 0 = L\Delta k + \omega\Delta T, \tag{2.16}$$

where L is the length of the contour; and the treatments of (2) and (3) are according to the adiabatic invariance of the magnetic flux:

$$\Delta\Phi = 0 = L_0\Delta i + \frac{E}{q}\Delta T, \tag{2.17}$$

where L_0 is the inductance of the contour.

Forder's thesis implies:

(1) treatment of the action as an adiabatic invariant when the gyro is subject to a slow angular acceleration;
(2) generalization of the action integral to a noninertial frame of reference which requires the general theory of relativity;
(3) defining particles on the contour (of a platform), which provide angular momentum and the action of their motion in the rotating frame, that angular momentum and action involving not only (A) the particles' (linear) momentum, but also (B) a term proportional to the particles' energy.

[6]The concept of gauge was originally introduced by Weyl (1918a, 1929) to describe a local scale or metric invariance for a global theory (the general theory of relativity). This use was discontinued. Later, it was applied to describe a local phase invariance in quantum theory.

Forder's claim is that (B) distinguishes rotation sensitive from rotation-insensitive devices. We quickly outline the main points of this claim.

With the intrinsic angular momentum defined:

$$I = \oint pdq, \tag{2.18}$$

(and the action = integral over one complete cycle), Forder's model is one of particles at the periphery of the frame at a distance l. If Hamilton's principal function is $S(t, l)$, the energy is $\mathbf{E}$ and momentum is p for each particle, then

$$I = \oint \frac{\partial S}{\partial l} dl = \oint pdl, \tag{2.19}$$

and the transit time around the contour is

$$T = \frac{dI}{dE}. \tag{2.20}$$

Free propagation is assumed for both particles and light propagating between perfectly reflecting surfaces arranged around the contour. Then

$$S(t, l) = pl - Et, \tag{2.21}$$

and

$$I = pL, \tag{2.22}$$

where $L = \oint dl$ is the length of the contour. The period of motion is then

$$T = \frac{dI}{dE} = L\frac{dp}{dE} = \frac{L}{v}, \tag{2.23}$$

where $v = \frac{dE}{dp}$ is the velocity of the particle.

If the contour and observer are accelerated to an angular velocity Ω, the action changes to

$$I^A = \oint \frac{\partial S^A}{\partial x^i} dx^i = -\oint p_i dx^i, \tag{2.24}$$

$(i = 1, 2, 3)$ and where the p_i components are the spatial parts of the four-momentum

$$p_\mu = -\frac{\partial S^A}{\partial x^\mu}. \tag{2.25}$$

Forder's argument is that:

$$I^{\mathbf{A}} = I, \tag{2.26}$$

i.e., the action is an adiabatic invariant under the acceleration, and that:

$$I^{\mathbf{A}} = p^{\mathbf{A}} L + E^{\mathbf{A}} \Delta T, \tag{2.27}$$

where ΔT is a synchronization discrepancy between clocks at different points in a rotating frame. The action for a ring laser gyro is then simply:

$$I^{\mathbf{A}} = E^{\mathbf{A}} T^{\mathbf{A}}, \tag{2.28}$$

where

$$T^A = \oint \frac{dx^0}{c} = \frac{dI}{dE^A} = \frac{L dp^A}{dE^A} + \Delta T = \frac{L}{v^A} + \Delta T \tag{2.29}$$

is the transit time around the contour, and

$$v^A = \frac{dE^A}{dp^A} \tag{2.30}$$

is the proper velocity of the particle. Forder then claimed that although the proper velocity is c for all observers, in a rotating frame a photon takes a different length of time to traverse the contour. However, it is difficult to understand how this can be claimed. If the clockwise and counterclockwise beams in the Sagnac interferometer traverse *different* paths, then one might state that, under platform rotation, the physical distance changed to compensate for changes in ΔT, the time taken to traverse the length of the rotating interferometer back to point H (see Fig. 2.1). But this is not the case. It is the *same* interferometric path for both beams. So how it can shorten for one beam and lengthen for another is mysterious.

Leaving that aside: depending on the direction of propagation, the period is increased or decreased by ΔT defined:

$$\Delta T = \frac{1}{c^2} \oint (\Omega \wedge \mathbf{r}) \cdot dl = \frac{2}{c^2} \Omega \cdot S, \tag{2.31}$$

where S is the contour area, and a link was established by Forder between the Sagnac effect and action constancy — a constancy which underlies gauge theory to which we now turn.

In the case of conventional electromagnetism, phase is arbitrary (there is gauge invariance) and fields (of force) are described completely by the electromagnetic field tensor, f_{uv}. This is the case when the theoretical model addresses only *local* effects. In the case of *parallel transport*, however, the phase of a wave function ψ representing a particle of charge e at point x is parallel to the phase at another point $x + dx^\mu$ if the local values differ by

$$ea_\mu(x)dx^\mu, \tag{2.32}$$

where $\mu = 0, 1, 2, 3$ and $a_\mu(x)$ represents a set of functions. A *gauge transformation* at x with a phase change $e\alpha(x)$ is written as

At x,

$$\psi(x) \rightarrow \psi'(x) = \exp[ie\alpha(x)]\psi(x), \tag{2.33a}$$

but at $x + dx^\mu$ it is written as

$$\psi(x + dx^\mu) \rightarrow \psi'(x + dx^\mu) = \exp[ie\{\alpha(x) + \partial_\mu\alpha(x)dx^\mu\}]\psi(x). \tag{2.33b}$$

So in the case of a phase change *with phase parallelism* (Hong-Mo and Tsun, 1993):

$$a'_\mu(x) = a_\mu(x) + \partial_\mu(x)\alpha(x). \tag{2.34}$$

where ∂_μ signifies $\partial/\partial x^\mu$.

Figures 2.2–2.4 are examples of representations defining topologically parallel transport for two counterpropagating beams. Referring

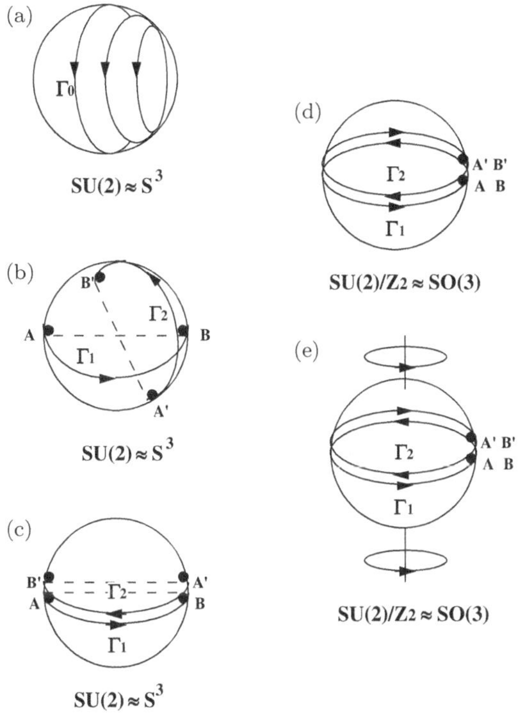

Fig. 2.2. Generation of a Topological Obstruction (source of magnetic flux) in the Sagnac interferometer. (a) parallel transport along two or more different paths; (b) parallel transport along two different paths with a path reversal; (c) parallel transport along the same path; (d) parallel transport along the same path with a path reversal around an obstruction; (e) parallel transport along the same path with a path reversal around an obstruction and with twist. Paths taken by two counterpropagating beams are separated in (c), (d) and (e) for purposes of exposition. The counterpropagating beams are superposed in reality with $A = B'$ and $B = A'$ (c) and $A = B = A' = B'$ in (d) and in (e). SU(2) is a 3-sphere S^3 in 4-space; Z_2 are the integers modulo 2; $SU(2)/Z_2 = SO(3)$ is a 3-sphere in 4-space with identity of pairs of opposite signs, e.g., $|\pm a| = \xi$. The twist in E corresponds to a "patch" condition and in the Sagnac interferometer is caused by the presence of angular velocity (+ linear acceleration) as shown in (e).

to Fig. 2.2, along either of the paths, a change in phase is given by

$$e \int_{\Gamma}^{Q} a_{\mu}(x) dx^{\mu}. \tag{2.35}$$

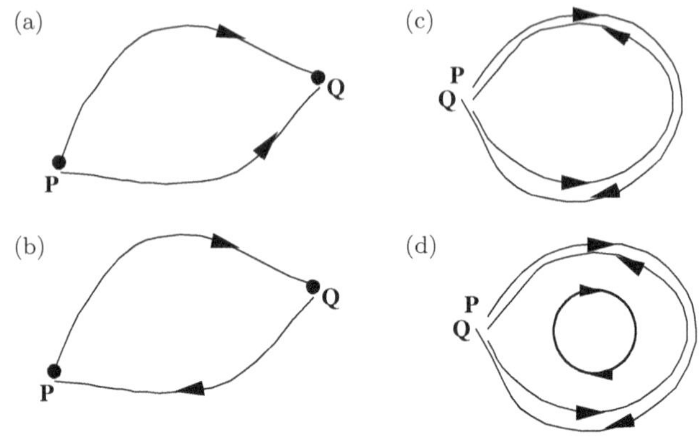

Fig. 2.3. Generation of a Topological Obstruction (source of magnetic flux) in the Sagnac interferometer. (a) two beam parallel transport along two different paths from P to Q around an obstruction with progress in the same direction along the paths; (b) two beam parallel transport along two different paths around an obstruction with a path reversal; (c) two beam parallel transport along one path around an obstruction with $P = -Q$ and $Q' = -P$; (d) two beam parallel transport along one path around an obstruction with $P = -Q$ and $Q' = -P$ and with a twist. Paths taken by two counterpropagating beams are separated in (c) and (d) for purposes of exposition. The twist in (d) corresponds to a "patch" condition and in the Sagnac interferometer is caused by angular velocity (+ linear acceleration). The counterpropagating beams are superposed in reality. Figure 2.3(c) corresponds to Fig. 2.2(d); and Fig. 2.3(d) to Fig. 2.2(e).

The difference between the phases at Q along two distinct paths is

$$e\int_{\Gamma_2}^{Q} a_\mu(x)dx^\mu - e\int_{\Gamma_1}^{Q} a_\mu(x)dx^\mu = e\oint_{\Gamma_2-\Gamma_1} a_\mu(x)dx^\mu \quad (2.36)$$

Use of Stokes' theorem obtains the surface integral:

$$-e\iint f_{\mu\nu}(x)d\sigma^{\mu\nu} \quad (2.37)$$

over any surface Σ bounded by the closed curve $\Gamma_2 - \Gamma_1$ with

$$f_{\mu\nu} = \partial_\nu a_\mu(x) - \partial_\mu a_\nu(x). \quad (2.38)$$

If $f_{\mu\nu}$ is nonzero, then parallel-transport of phases is path-dependent. But $f_{\mu\nu}$ is gauge invariant (because as a difference, it is independent of any phase rotations at that point). Therefore $f_{\mu\nu}$ has the defining characteristics of the electromagnetic field tensor and the $a_i(x)$ of

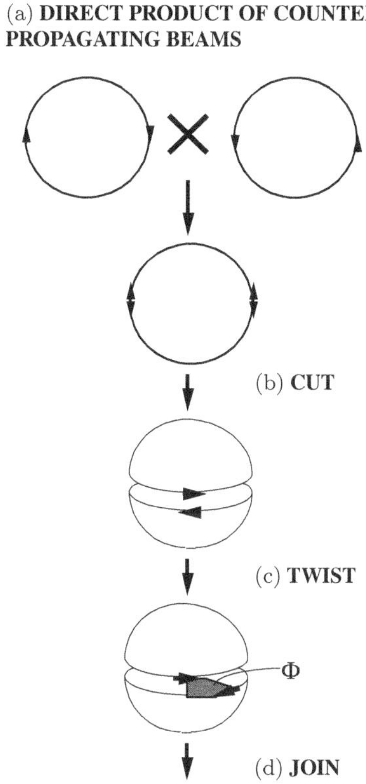

Fig. 2.4. Generation of a Topological Obstruction (source of magnetic flux) in the Sagnac interferometer. Commencing with a direct product of two counter-propagating beams, on S^3, (a) the sphere S^3 is cut in (b) given a twist in (c) and joined in (d). The twist and join in (d) corresponds to a "patch" condition and in the Sagnac interferometer is caused by angular velocity (+ linear acceleration). Figure 2.4(c) corresponds to Figs. 2.2(d) and 2.3(c); and Fig. 2.4(d) to Figs. 2.2(e) and 2.3(d). Adapted from Hong-Mo and Tsun (1993).

gauge potentials. This is as far as conventional Maxwell theory takes us, leaving the situation of a rotated platform shown in Figs. 2.2–2.4 underdescribed. To progress further, requires Yang–Mills theory (Yang and Mills, 1954a,b) to which we now turn.

Yang–Mills theory is a generalization of electromagnetism in which ψ, a wave function with two components, e.g., $\psi = \psi'(x)$, $i = 1, 2$, is the focus of interest, instead of a complex wave function of a charged particle as in conventional Maxwell theory. In the case of the Sagnac effect, we assign $i = (1)$ to clockwise, and $i = (2)$ to *counterclockwise* propagating beams, but the clockwise and counterclockwise propagating beams in the interferometric situation we are considering can only be distinguished after parallel transport following a *patch condition*, which is a condition initiated when the Sagnac interferometer platform undergoes an angular rotation. A change in phase then means a change in the orientation in *"internal space"* under the transformation: $\psi \to S\psi$, where S is a unitary matrix with a unitary determinant.

With the gauge potential now defined as a matrix, $A_\mu(x)$, with g a generalized charge, and with parallel transport along separate arms of the interferometer, we have

- beam traveling *clockwise* and platform *rotated* (at x):

$$\psi(x) \to \psi'(x) = S(x)\psi(x), \tag{2.39a}$$

- beam traveling *counterclockwise* and platform *rotated* (at $x+dx^\mu$):

$$\hat{\psi}(x) \to \hat{\psi}'(x) = S(x + dx^\mu)\hat{\psi}(x), \tag{2.39b}$$

or more explicitly:

- beam traveling *clockwise* and platform *rotated* (at x):

$$\begin{aligned} \psi_P(x) \to_{\Gamma_{\text{clockwise}}} \psi_Q(x) &= \exp\left[igA_\nu(x + dx^\mu)dx^\nu\right] \\ &\times \exp\left[igA_\mu(x)dx^\mu\right]\psi_P(x), \end{aligned} \tag{2.40a}$$

- beam traveling *counterclockwise* and platform *rotated* (at $x+dx^\mu$):

$$\begin{aligned} \psi_P(x) \to_{\Gamma_{\text{counterclockwise}}} \psi_Q(x) &= \exp\left[igA_\mu(x + dx^\nu)dx^\mu\right] \\ &\times \exp\left[igA_\nu(x)dx^\nu\right]\psi_P(x). \end{aligned} \tag{2.40b}$$

Taking the difference and using Stokes' theorem gives:

$$F_{\mu\nu}(x) = \partial_\nu A_\mu(x) - \partial_\mu A_\nu(x) + ig\left[A_\mu(x), A_\nu(x)\right]. \tag{2.41}$$

This is the description we seek. $F_{\mu\nu}(x)$ (Eq. 2.41) is *gauge covariant* due to the phase-direction in internal symmetry space, which is not the case with $f_{\mu\nu}$ (Eq. (2.38)). The field at H (cf Fig. 2.1) in the Sagnac interferometer with the platform *under rotation* is then described by

$$F'_{\mu\nu} = S(x)F_{\mu\nu}(x)S'(x) \tag{2.42}$$

where $S(x)$, and its converse, $S'(x)$, is a gauge group to be defined.

It should be emphasized that this description applies only to the platform *under rotation*. On cessation of rotation, the Sagnac interferometer platform no longer exhibits a patch condition, whereupon:

$$\begin{aligned}
&A_\mu \to a_\mu,\\
&A_\nu \to a_\nu,\\
&g \to 0,\\
&(\partial_\nu A_\mu(x) - \partial_\mu A_\nu(x) + ig\,[A_\mu(x), A_\nu(x)]) \to (\partial_\nu a_\mu(x) - \partial_\mu a_\nu(x)),\\
&F_{\mu\nu} \to f_{\mu\nu},
\end{aligned} \tag{2.43}$$

and we recapture conventional Maxwell theory. Upon rerotation of the frame of reference of the platform of the Sagnac interferometer, we again have

$$\begin{aligned}
&a_\mu \to A_\mu,\\
&a_\nu \to A_\nu,\\
&g > 0,\\
&(\partial_\nu a_\mu(x) - \partial_\mu a_\nu(x)) \to (\partial_\nu A_\mu(x) - \partial_\mu A_\nu(x) + ig\,[A_\mu(x), A_\nu(x)]),\\
&f_{\mu\nu} \to F_{\mu\nu},
\end{aligned} \tag{2.44}$$

and we recapture Yang–Mills theory once again. Stated differently, with rotation of the platform the gauge symmetry is $SU(2)/Z_2 \cong SO(3)$ and on stabilization of the platform the gauge symmetry is U(1).[7]

[7]Gauge transformations of U(1) are obtained by means of multiplication by complex scalar fields of unit modulus — or by fields of elements of the Lie group $U(1)$ (cf. Penrose and Rindler, 1984, p. 342).

2.3. The Lorentz Group and The Lorenz Gauge Condition

The Lorentz group expresses a spacetime symmetry in which free space exhibits the same physical properties in all inertial frames. Accelerated frames are deviations from inertial frames and the acceleration can be of gravitational or kinematic in origin. Gyroscopes measure deviation of frames from an inertial frame. Therefore rotation of the Sagnac interferometer platform causes a deviation from the Lorentz group description of spacetime.

If **W** is a vector defined over the vector space of real numbers and in the Minkowskian metric, then a *Lorentz norm*[8] is defined as

$$\|W\| = W \cdot W = W^i W^j \eta_{ij} = \left(W^0\right)^2 + \left(W^1\right)^2 + \left(W^2\right)^2 + \left(W^3\right)^2 \tag{2.45}$$

If **V** is a vector similarly defined, then (Penrose and Rindler, 1984):

$$\mathbf{W} \cdot \mathbf{V} = \frac{1}{2}\left\{\|\mathbf{W} + \mathbf{V}\| - \|\mathbf{W}\| - \|\mathbf{V}\|\right\} \tag{2.46}$$

is the *inner product* defined in terms of the Lorentz norm. Therefore, a *Lorentz transformation* on **W** in vector space can be defined as preserving the Lorentz norm, the inner product, the spatial orientation and time orientation of **W**.

If S and T are points in the Minkowski metric and ST is the position vector of T relative to S then:

$$\chi(S, T) = \|ST\| \tag{2.47}$$

is the *squared interval* for any pair of points S and T in Minkowski space. A transformation of Minkowski space which preserves the squared interval is a *Poincaré transformation.*

The Lorenz[9] gauge condition is

$$\nabla^a a_a = 0, \tag{2.48}$$

---

[8] H.A. Lorentz (1853–1928) of Leiden.

[9] L. Lorenz (1829–1891) of Copenhagen.

which guarantees invariance of the Maxwell equations under transformations. The Lorenz gauge is conformally invariant. However, the Maxwell equations with the Lorenz gauge *are not conformally invariant due to a conformal invariance mismatch* (cf. Penrose and Rindler, 1984, p. 373). Yet this is a minor point. The major one is that the Lorenz gauge and the Lorentz group are not necessary choices *for all platform conditions.* According to the present analysis, they are also *incorrect choices* when the Sagnac interferometer platform is *in rotation.*

2.4. The Phase Factor Concept

In conventional electromagnetism, on the one hand, $f_{\mu\nu}$ *underdescribes* the system because phase is undetermined; on the other hand, $a_\mu(x)$ *overdescribes* the system because different values of $a_\mu(x)$ correspond to the same physical condition (Wu and Yang, 1975). But the Dirac phase factor defined:

$$\Phi(C) = \exp\left[ie \oint_C a_\mu(x)dx^\mu\right], \tag{2.49}$$

where e is electric charge and a_μ is the electromagnetic potential, *completely* describes the system in the case of the Sagnac interferometer fields when the interferometer platform is *at rest.*

For the case of the Sagnac interferometer in rotation and multiply connected in a generated A_μ matrix field of specific symmetry, Φ may be said to be in an excited state, Φ^*. Accordingly, in Yang–Mills theory:

$$\Phi^*(C) = \mathbf{P}\exp\left[ig \oint_C A_\mu(x)dx^\mu\right], \tag{2.50}$$

where $\Phi^*(C)$ specifies parallel phase transport over any loop C in rotation, g is generalized charge, P specifies path dependence of the integral and A_μ is a matrix variable. As $F_{\mu\nu}$ specifies only infinitesimal loops (i.e., local conditions), $\Phi^*(C)$ is the global version of $F_{\mu\nu}$ and differs from $F_{\mu\nu}$ *if the region is multiply connected.* The phase factor, $\Phi^*(C)$, describes the Sagnac interferometer fields when the interferometer platform *is rotated.*

One may then ask: if the surface of the Sagnac interferometer, Σ, is continuously deformed in space, do the paths Γ_1 and Γ_2 deform to a point in a gauge group? Our answer is: *yes*, if the platform of the interferometer *is not* rotated, but *no*, if the platform of the interferometer *is* rotated. The counterpropagation of the two beams around an obstruction permits deformation to a point only in the non-rotated condition. When rotated, a "patch" condition (cf. Figs. 2.2–2.4) exists in the multiply-connected topology. The measured phase difference of the rotated Sagnac interferometer is then:

$$\oint \left(A_\mu^{\text{clockwise}} - A_\mu^{\text{anticlockwise}}\right) dx^\mu = \oint (\Delta\varphi)\, dx^\mu = 4\pi g_m, \tag{2.51}$$

where g_m is "magnetic charge". The counterpropagating beams on a rotating platform are an instance of a patching condition which precludes shrinkage to a single point. Therefore, Eq. (2.51) describes the dynamic explaining the Sagnac effect. On the other hand, the *unrotated* Sagnac interferometer has no measured phase difference and for that stabilized state:

$$\oint \left(a_\mu^{\text{clockwise}} - a_\mu^{\text{anticlockwise}}\right) dx^\mu = 0. \tag{2.52}$$

Whereas in conventional electromagnetics a_μ and $f_{\mu\nu}$ are labeled only at *points* in spacetime, in gauge theory $\Phi(C)$ and $\Phi^*(C)$ are labeled by C, a *closed curve* in spacetime. $\Phi^*(C)$ is a functional of the function:

$$A = \{A^\mu(s); s = 0 \to 2\pi\}, \tag{2.53}$$

and the relation to the field tensor, F_μ, is

$$F_\mu(A\,[s]) = \Phi *_A^{-1} (s,0) F_{\mu\nu}(A(s)) \Phi *_A (s,0) \frac{dA^\nu(s)}{ds}. \tag{2.54}$$

2.4.1. SU(2) *Group Algebra*

The explanation we have sought for the Sagnac effect pivots on an understanding of the SU(2) continuous group. The following is an account of how that group relates to the U(1) group of rotations.

A general rotation about some arbitrary axis is designated $R(\alpha, \beta, \gamma)$ in group O(3) for the Euler angles α, β and γ. The SU(2) group has the following group elements which refer to a complex two-dimensional vector $(\mathbf{u}, \mathbf{v})$:

$$\begin{matrix}\mathbf{u}' \\ \mathbf{v}'\end{matrix} = \begin{bmatrix} a & b \\ c & d \end{bmatrix} \begin{matrix}\mathbf{u} \\ \mathbf{v}\end{matrix}, \tag{2.55}$$

where a, b, c and d are complex numbers. With the additional requirement that the determinant is ± 1, making the group unitary, and $ad - bc = +1$, making the group special (S) (so special unitary is a subclass of unitary group), the transformation rules simplify to the matrix:

$$\begin{bmatrix} a & b \\ -b^* & a^* \end{bmatrix}, \tag{2.56}$$

which is the defining matrix for the SU(2) group of continuous transformations. In terms of the $R(\alpha, \beta, \gamma)$ rotation, and as an example, we could choose $a = \exp[\frac{-i\alpha}{2}]$ and $b = 0$, which gives a rotation $R(\alpha, 0, 0)$ about the z-axis, and $a = \cos\left[\frac{\beta}{2}\right]$ and $b = \sin\left[\frac{\beta}{2}\right]$ which gives the rotation $R(0, \beta, 0)$ about the y-axis. Then rotations $R(\alpha, \beta, \gamma)$ would then be associated with the SU(2) matrix:

$$\begin{bmatrix} \cos\left[\dfrac{\beta}{2}\right] \exp\left[\dfrac{i(\alpha+\gamma)}{2}\right] & \sin\left[\dfrac{\beta}{2}\right] \exp\left[\dfrac{-i(\alpha-\gamma)}{2}\right] \\ -\sin\left[\dfrac{\beta}{2}\right] \exp\left[\dfrac{i(\alpha-\gamma)}{2}\right] & \cos\left[\dfrac{\beta}{2}\right] \exp\left[\dfrac{-i(\alpha+\gamma)}{2}\right] \end{bmatrix}. \tag{2.57}$$

These SU(2) transformations define the relations between the Euler angles of group O(3) with the parameters of SU(2). Each of two fields, $(\mathbf{u}, \mathbf{v}) \rightarrow (\mathbf{u}', \mathbf{v}')$, can be associated with the rotation matrix in O(3) for $R(\alpha, \beta, \gamma)$:

$$\begin{bmatrix} \cos(\alpha)\cos(\beta)\cos(\gamma) - \sin(\alpha)\sin(\beta) & \sin(\alpha)\cos(\beta)\cos(\gamma) + \cos(\alpha)\sin(\gamma) & -\sin(\beta)\cos(\gamma) \\ -\cos(\alpha)\cos(\beta)\sin(\gamma) - \sin(\alpha)\cos(\gamma) & -\sin(\alpha)\cos(\beta)\sin(\gamma) + \cos(\alpha)\cos(\gamma) & \sin(\beta)\sin(\gamma) \\ \cos(\alpha)\sin(\beta) & \sin(\alpha)\sin(\beta) & \cos(\beta) \end{bmatrix} \tag{2.58}$$

Then matrix (2.56) defining SU(2) algebra can be related to O(3) with algebra defined by matrix (2.58). This relationship is called a *homomorphism.*

The SU(2) group is a Lie algebra such that for the angular momentum generators, J_i, the commutation relations are $[J_i, J_j] = i\varepsilon_{ijk}J_k$, $i, j, k = 1, 2, 3$, where the ε_{ijk} are "structure constants". The four dimensions *can be the three Euclidean spatial dimensions and time* (*or Minkowski spacetime*), *but need not be. Here they are related to a complex spacetime with a holomorphic metric* (*see below*).

As an example of an SU(2) transformation, we show the following. An isotropic parameter, w, can be defined:

$$w = \frac{x - iy}{z}, \tag{2.59}$$

where x, y, z are the spatial coordinates. If w is written as the quotient of μ_1 and μ_2, or the homogeneous coordinates of the bilinear transformation, then, corresponding to (2.51) and using the matrix (2.57) we have the following example of a vector undergoing an SU(2) transformation:

$$\left|\mu_1^{'}\mu_2^{'}\right\rangle = \begin{bmatrix} \cos\left[\frac{\beta}{2}\right]\exp\left[\frac{i(\alpha+\gamma)}{2}\right] & \sin\left[\frac{\beta}{2}\right]\exp\left[\frac{-i(\alpha-\gamma)}{2}\right] \\ -\sin\left[\frac{\beta}{2}\right]\exp\left[\frac{i(\alpha-\gamma)}{2}\right] & \cos\left[\frac{\beta}{2}\right]\exp\left[\frac{-i(\alpha+\gamma)}{2}\right] \end{bmatrix} |\mu_1\mu_2\rangle \tag{2.60}$$

The *sourceless* condition for such transformations is that the action:

$$L = \oint\!\!\!\oint\!\!\!\oint\!\!\!\oint f^i_{\mu\nu} f^i_{\nu\mu} dS = \text{an absolute minimum}, \tag{2.61}$$

where the integral is taken over a closed four-dimensional surface. The solutions to this equation are gauge fields (nonintegrable phase factors) implying conformal invariance.

Let us define further restrictions or boundary conditions. By defining the coordinates:

$$
\begin{aligned}
x &= \frac{1}{2}(u^2 - v^2), \\
y &= \frac{1}{2i}(u^2 + v^2), \\
z &= uv,
\end{aligned}
\tag{2.62}
$$

then:

$$
x^2 + y^2 + z^2 \text{ is invariant.} \tag{2.63}
$$

Suppose we let $\alpha = \gamma = 0$ and choose $\beta = 0$ or $\beta = 2\pi$. For $\beta = 0$ the SU(2) matrix is

$$
\begin{bmatrix} 1 & 0 \\ 0 & 1 \end{bmatrix}. \tag{2.64}
$$

However, for $\beta = 2\pi$ the SU(2) matrix is

$$
\begin{bmatrix} -1 & 0 \\ 0 & -1 \end{bmatrix}. \tag{2.65}
$$

Therefore, for zero rotation in three-dimensional space there corresponds *two distinct* SU(2) elements depending on the value of β. The two-to-one relationship of a U(1) group to a SU(2) form is an indication of compactification of degrees of freedom.

Previously, the conventional U(1) symmetry Maxwell theory was placed in a Yang–Mills context and generalized to a SU(2) symmetry form (Barrett, 1988, 1993). Table 2.1 compares the two formulations.

In Table 2.1, the Noether currents are as follow:

$$
\begin{aligned}
\mathbf{J}_x &= (\mathbf{A} \times \mathbf{B} - \mathbf{B} \times \mathbf{A}), \\
\mathbf{J}_y &= (\mathbf{A} \cdot \mathbf{B} - \mathbf{B} \cdot \mathbf{A}), \\
\mathbf{J}_z &= (\mathbf{A} \times \mathbf{E} - \mathbf{E} \times \mathbf{A}), \\
\mathbf{iJ}_z &= (\mathbf{A} \cdot \mathbf{E} - \mathbf{E} \cdot \mathbf{A}).
\end{aligned}
\tag{2.66}
$$

Table 2.1. Variables in U(1) and SU(2) symmetries.

U(1)	SU(2) Symmetry
$\rho_e = J_0$	$\rho_e = J_0 - iq(A \cdot E - E \cdot A) = J_0 + qJ_z$
$\rho_m = 0$	$\rho_m = -iq(\mathbf{A} \cdot \mathbf{B} - \mathbf{B} \cdot \mathbf{A}) = -iq\mathbf{J}_y$
$g_e = \mathbf{J}$	$g_e = iq[A_0, \mathbf{E}] - iq(\mathbf{A} \times \mathbf{B} - \mathbf{B} \times \mathbf{A}) + \mathbf{J} = iq[A_0, \mathbf{E}] - iq\mathbf{J}_x + \mathbf{J}$
$g_m = 0$	$g_m = iq[A_0, \mathbf{B}] - iq(\mathbf{A} \times \mathbf{E} - \mathbf{E} \times \mathbf{A}) = iq[A_0, \mathbf{B}] - iq\mathbf{J}_z$
$\sigma = \mathbf{J}/\mathbf{E}$	$\sigma = \frac{\{iq[A_0,\mathbf{E}] - iq(\mathbf{A}\times\mathbf{B} - \mathbf{B}\times\mathbf{A}) + \mathbf{J}\}}{\mathbf{E}} = \frac{\{iq[A_0,\mathbf{E}] - iq\mathbf{J}_x + \mathbf{J}\}}{\mathbf{E}}$
$s = 0$	$s = \frac{\{iq[A_0,\mathbf{B}] - iq(\mathbf{A}\times\mathbf{E} - \mathbf{E}\times\mathbf{A})\}}{\mathbf{H}} = \frac{\{iq[A_0,\mathbf{B}] - iq\mathbf{J}_z\}}{\mathbf{H}}$

Table 2.2. Maxwell equations in U(1) and SU(2) symmetries.

	U(1)	SU(2)
Gauss's law	$\nabla \cdot \mathbf{E} = J_0$	$\nabla \cdot \mathbf{E} = J_0 - iq(\mathbf{A} \cdot \mathbf{E} - \mathbf{E} \cdot \mathbf{A})$
Ampère's law	$\frac{\partial \mathbf{E}}{\partial t} - \nabla \times \mathbf{B} + \mathbf{J} = 0$	$\frac{\partial \mathbf{E}}{\partial t} - \nabla \times \mathbf{B} + \mathbf{J} + iq[A_0, \mathbf{E}] - iq(\mathbf{A} \times \mathbf{B} - \mathbf{B} \times \mathbf{A}) = 0$
Coulomb's law	$\nabla \cdot \mathbf{B} = 0$	$\nabla \cdot \mathbf{B} + iq(\mathbf{A} \cdot \mathbf{B} - \mathbf{B} \cdot \mathbf{A}) = 0$
Faraday's law	$\nabla \times \mathbf{E} + \frac{\partial \mathbf{B}}{\partial t} = 0$	$\nabla \times \mathbf{E} + \frac{\partial \mathbf{B}}{\partial t} + iq[A_0, \mathbf{B}] + iq(\mathbf{A} \times \mathbf{E} - \mathbf{E} \times \mathbf{A}) = 0$

Table 2.2 compares the Maxwell equations formulated for U(1) group symmetries, i.e., the conventional form, and the Maxwell equations formulated for SU(2) group symmetries.

2.4.2. *Short primer of topological concepts*

We have attempted to explain a physical effect, the Sagnac effect, using topological concepts *as necessary descriptive forms*, besides using well-known mathematical analysis techniques. Such topological concepts are not commonly in the armamentarium, i.e., the toolkit, of most physicists and engineers. It is a fact of human experience that without the correct mathematical "filter", a physical problem cannot be correctly defined. A major reason for the use of topology is that it is a study of continuity and continuous deformation, but, above all, it provides *justification for the existence or nonexistence of a qualitative, rather than a quantitative, object.* The conventional mathematical toolkit is generally a toolkit for quantitative prediction and *such methods already assume the existence of the qualitative objects they address.* The present approach proposes that the correct mathematical filter or descriptive tool is a topological filter or a topological toolkit (see the following chapter). That is, the physical problem offered by the Sagnac effect cannot be defined without topological concepts. Therefore, the following section is a very short introduction of definitions of some useful topological concepts.

Topological invariants: In a topological space, X, the number of connected parts and the number of holes are topological invariants if the number does not change under homeomorphism. Topological invariants uniquely specify equivalence classes.

Homeomorphism: If $\phi_x : X \rightarrow X'$ is continuous and if ϕ_x^{-1} exists and is also continuous, so that X and X' have the same number of components and holes, then ϕ_x is a homeomorphism and a *coordinate mapping* and X is a *coordinate chart.*

Manifold or differentiable manifold: A manifold of real dimension is a topological space which is locally homeomorphic to an open set of the real numbers. A differential manifold is a manifold with the property that if X and Y are coordinate charts which have a nonempty intersection, then the mapping $\phi_X \circ \phi_Y$ is a differential mapping. In other words, a differentiable manifold is the primitive topological space for the study of differentiability.

Homotopy classes: If Γ defines closed curves on a gauge group G, if Γ cannot be shrunk to a point and if members of a class of Γ cannot be continuously deformed into each other, then that class is the homotopy class of Γ. Stated differently, two coterminous paths in parameter space are *homotopic* if they can be continuously deformed into each other. Thus, whereas homeomorphisms define equivalence classes of topological *spaces*, homotopy defines equivalence classes of continuous *maps*.

Connectedness:

(a) If
  (1) the zeroth homotopy set of X, or $\Pi_0(X)$, is the set of path connected components of X;
  (2) $\Pi_1(X)$ is the set of loops which cannot be continuously deformed into each other, and
  (3) $\Pi_1(X) = 0$, then the space X is *simply connected.*

(b) If $\Pi_1(X) \neq 0$, then the space X is said to be *multiply connected.*

(c) If $\Pi_n(X) = \Pi_{n-1}(\Omega X), n \geq 1$, and for $n \geq 1$, $\Pi_n(X)$ is a *group*, then $\Pi_0(X)$ is not a group.

If X is a Lie group, then $\Pi_0(X)$ inherits a group structure from X, because it can be identified with the quotient group of X by its identity-connected component. A *connection* can always be defined independently of the choice of metric on the space.

Topological obstruction: If X and X$'$ are two equivalence classes, and X cannot be deformed continuously into X', then there exists a topological obstruction preventing a mapping, which is a topological invariant of X.

Fiber bundles: A fiber bundle is a twisted product of two spaces, X and Y, where Y is acted on by a group G, and the twist in the product has been effected by the group action. X is the base space and Y is the *fiber* on which the group structure acts. A rotating Sagnac interferometer is thus a nontrivial fiber bundle over X with *total space* or *topological space* E such that $\Pi : E \to X$. In that interferometric situation, X, the base space is identified with the Sagnac interferometer *with platform at rest* and E is identified with the Sagnac interfometer *with platform in motion.* The clockwise and counterclockwise propagating beams play the role of two disjoint arcs on E. With the Sagnac interferometer at rest, coordinates in A are equivalent to coordinates in B. However, for the Sagnac interferometer platform in motion changing coordinates in A to coordinates in B(or vice versa) requires a transfer function or group element G. For the platform in motion, but not at rest, the fiber Y is acted on by the group element G. The fiber Y in the Sagnac interferometer is the local propagating beam segment for both clockwise and counterclockwise beams and common to both the Sagnac interferometer at rest, as well as in motion.

Holomorphic: A differential function defined on an open set of complex numbers is said to be holomorphic if it satisfies the Cauchy–Rieman equations. A holomorphic transformation is a complex analytic transformation for which the real and imaginary parts are Taylor expandable. It is also conformal and orientation preserving.

Holonomy: The holonomy of a given closed curve is the parallel transport considered as an element of the structure group under an embedding.

Homeomorphic: If two figures can be transformed into each other by continuous deformations without cutting and pasting, the figures are said to be homeomorphic.

Homology and Homotopy: Homology theory is the study of physical bodies through their internal structure; homotopy is the study of physical bodies through their interaction with known objects.

Conformal: A conformal structure on a manifold is the prescription of a null cone defined by a quadratic function in the tangent space at each point of the manifold. A conformal mapping is an angle-preserving mapping.

Conformal structure: The concept of *conformal structure* is related to that of *conformal rescaling*. In the case of conventional electromagnetism, significance is only given to an equivalence class of fields which can be obtained from a given metric, g_{ab}, by a conformal rescaling:

$$g_{ab} \mapsto \hat{g}_{ab} = \Omega^2 g_{ab}, \tag{2.67}$$

where Ω is any scalar field. No transformation of *points* is involved, *but information is lost when a field undergoes conformal rescaling in conventional electromagnetism.* However, if a spinor description is used, we have

$$\varepsilon_{AB} \mapsto \hat{\varepsilon}_{AB} = \Omega\varepsilon_{AB}. \tag{2.68}$$

Therefore, whereas the argument of the inner product between two spin-vectors is conformally invariant, *the modulus of the inner product is altered under conformal rescaling.* Maxwell's equations of conventional electromagnetism are conformally invariant and the Lorenz gauge is also conformally invariant. However, Maxwell's equations with the Lorenz gauge is not conformally invariant (cf. Penrose and Rindler, 1984, p. 373). The Sagnac interferometer exhibits conformal invariance under rotation.

Conformal invariance: A system of fields and field equations is conformally invariant if it is possible to attach conformal weights to all field quantities in such a way that the field equations remain true after conformal rescaling. The Euler–Lagrange equation

$$D{*}F = 0, \tag{2.69}$$

(where D is the covariant exterior derivative), the self-dual and anti-self-dual equaions

$$F = \mp{}^{*}F \tag{2.70}$$

for the 2-form F, and the action:

$$S = \frac{-1}{16}\int d^4x f_{\mu\nu}(x) f^{\mu\nu}(x), \tag{2.71}$$

are all conformally invariant.

Cohomology: Cohomology groups are dual to homology groups. A homology group can be formed from a class C of p-chains, which is a formal finite linear combination. If C_pC_p is the set of all $C^{\infty}p$, then ω is a map from $\mathbf{C_p}$ to $\mathbf{R}$:

$$\omega : \mathbf{C_p} \to \mathbf{R} \tag{2.72}$$

or to an element of the dual of C_p. Therefore, if $H_p(M;\mathbf{Z})$ is a homology group, $H^p(M;\mathbf{R})$ is the corresponding cohomology group, where the de Rham cohomology group (de Rham, 1960) is

$$H^p = \{\text{closed}\, p - \text{form}\}/\{\text{exact}\, p\text{ - form}\}.$$

In the case of Stokes' theorem:

$$\int_M D\omega = \int_{\partial M} \omega, \tag{2.73}$$

where ∂ is the boundary operator and D is the exterior derivative. If

$$\langle \omega, C\rangle = \int_C \omega, C \in C_p, \tag{2.74}$$

then Stokes' theorem is

$$\langle D\omega, C\rangle = \langle \omega, \partial C\rangle. \tag{2.75}$$

de Rham's theorem is (Ward and Wells, 1990, p. 161):

$$H^q(M, \mathbf{C}) \cong H^q_{DR}(M), \tag{2.76}$$

or, in words, the complex cohomology group, $H^q(M, \mathbf{C})$, is equivalent to the qth de Rham cohomology group of thc manifold, M, $H^q_{DR}(M)$.

Differential forms: In Euclidean space $\mathbf{R}^n$ the coordinates $x^1, \ldots, x^n$ are *O-forms* and differentiation is the gradient. In $\mathbf{R}^n$ the differentiation on covariant vector fields, e.g., $A_\mu dx^\mu$ on $\mathbf{R}^n$, is the curl and those fields are 1-forms. So the differential symbol can represent a linear map from O-forms to 1-forms and correspond to line integrals.

In $\mathbf{R}^3$ the area $A_{ij}dx^i dx^j$ for a two-dimensional surface is a 2-form and differentiation then is an exterior differentiation or divergence.

For every form ξ in a Riemannian manifold, there is a dual form $^*\xi$:

$$^*\xi \to \xi^{n-p}, \tag{2.77}$$

e.g.,

$$^*f_{\mu\nu} \to -\frac{1}{2}\varepsilon_{\mu\nu\rho\sigma} f^{\rho\sigma}, \tag{2.78}$$

where f is the electromagnetic tensor. If $d\xi = 0$, then ξ is said to be *closed*. If $\xi = df$, where f is any function, then ξ is said to be *exact*.

Self-duality and anti-self-duality: If we commence with the spinor form of the Yang–Mills field tensor, we have[10]:

$$F_{ab} = \varphi_{AB}\varepsilon_{A'B'} + \varepsilon_{AB}\bar{\varphi}_{A'B'} \tag{2.79}$$

The *dual* of F_{ab}, $^*F_{ab}$, is

$$^*F_{ab} = -i\varphi_{AB}\varepsilon_{A'B'} + i\varepsilon_{AB}\psi_{A'B'}, \tag{2.80}$$

where[11]

$$\varphi_{AB} = \varphi_{(AB)} = \frac{1}{2}F_{ABC'}^{C'} \tag{2.81}$$

and

$$\psi_{A'B'} = \psi_{(A'B')} = \frac{1}{2}F_{C}^{C}{}_{A'B'}. \tag{2.82}$$

---

[10]In this equation: (1) a prime on a label indicates complex conjugation; (2) the ε-spinor is used, which is *antisymmetrical*: $\varepsilon_{AB} = -\varepsilon_{BA}$ or $\varepsilon_A^B = -\varepsilon_A^B$ and $x_B = x^A\varepsilon_{AB}$; $x^A = \varepsilon^{AB}x_B$; and (3) a bar on a spinor indicates complex conjugation; (4) φ_{AB} is a *symmetric* spinor; (5) the indices, AB, are abstract indices without any reference to any basis or coordinate system; and (6) the clumped pair of indices are defined $a = AA'; b = BB', c = CC', \ldots, z = ZZ'$ (Penrose and Rindler, 1984).

[11]Round brackets indicate symmetrization and square brackets indicate antisymmetrization, e.g.:

$$\begin{aligned}
U_{A(BC)D} &= \tfrac{1}{2!}(U_{ABCD} + U_{ACBD});\\
U_{A(BCD)E} &= \tfrac{1}{3!}(U_{ABCDE} + U_{ACBDE} + U_{ACDBE}\\
&\quad + U_{ABDCE} + U_{ADBCE} + U_{ADCB});\\
U_{A[BC]D} &= \tfrac{1}{2!}(U_{ABCD} - U_{ACBD});\\
U_{A[BCD]E} &= \tfrac{1}{3!}(U_{ABCDE} - U_{ACBDE} + U_{ACDBE} - U_{ABDCE}\\
&\quad + U_{ADBCE} - U_{ADCB}).
\end{aligned}$$

If[12]

$$^{*}F_{ab} = iF_{ab}, \text{ i.e., } \varphi_{AB} = 0, \tag{2.83}$$

the bivector field F_{ab} is *self-dual.* If

$$^{*}F_{ab} = -iF_{ab}, \text{ i.e., } \psi_{A'B'} = 0, \tag{2.84}$$

the bivector field F_{ab} is *anti-self-dual.*

In other words, every complex bivector F_{ab} is a sum of self-dual, $^{+}F_{ab}$, and anti-self-dual, $^{+}F_{ab}$, components:

$$F_{ab} = {}^{+}F_{ab} + {}^{-}F_{ab}, \tag{2.85}$$

where

$$^{+}F_{ab} = \frac{1}{2}(F_{ab} - i * F_{ab}) = \varepsilon_{AB}\psi_{A'B'}, \tag{2.86}$$

and

$$^{-}F_{ab} = \frac{1}{2}(F_{ab} + i * F_{ab}) = \varphi_{AB}\varepsilon_{A'B'}. \tag{2.87}$$

In the case of Yang–Mills fields we have[13]:

$$^{+}F^{\Psi}_{ab} = \varepsilon_{AB}\chi^{\Psi}_{A'B'\Theta}, \tag{2.88}$$

and

$$F^{\Psi}_{ab} = \varphi_{AB\Theta}\varepsilon_{A'B'}. \tag{2.89}$$

---

[12]The clumped pairs of spinor indices are defined:
$a = AA'; b = BB', c = CC', \ldots, z = ZZ'$. Therefore:

$$\psi^{AA'}{}_{BB'}{}^{Q} = \psi^{a}{}_{BB'}{}^{Q} = \psi^{a}{}_{b}{}^{Q} = \psi^{A'A}{}_{b}{}^{Q} = \psi^{a}{}_{B}{}^{Q}{}_{B'} \text{ and}$$

$$\psi^{a}{}_{AA'}{}^{Q} = \psi^{AA'}{}_{a}{}^{Q} = \psi^{a}{}_{a}{}^{Q} = \psi^{AA'}{}_{AA'}{}^{Q} \text{and}$$

$$\psi^{AA'}{}_{BB'}{}^{B} = \psi^{a}{}_{b}{}^{B} = \psi^{AA'}{}_{b}{}^{B} = -\psi^{aB}{}_{b} = -\psi^{aB}{}_{BB'}$$

(Penrose and Rindler, 1984, Vol. 1, p. 116).

[13]Capital Greek labels are used to indicate elements of vector spaces. That is, they are bundle indices. This relabeling indicates that a component λ^{Φ} *locally* assigned to a Yang–Mills field has a basis which corresponds to the basis of the Yang–Mills field in the *global* manifold. The set of fields α^{Ψ}_{Ψ} is a gauge for $\mathcal{I}^{\Psi}$, the tensor describing a ring of scalar fields. (Penrose and Rindler, 1984, Vol. 1, p. 345).

Affine transformation. If $\mathbf{A}$ is a connection form, and if the local coordinates are changed:

$$(x, g) \to (x', g'); \quad g' = hg, \tag{2.90}$$

then

$$\mathbf{A}' = hdh^{-1} + h\mathbf{A}h^{-1} \tag{2.91}$$

is an affine transformation. It is affine because it *both translates* (by the amount hdh^{-1}) *and rotates* $\mathbf{A}$ (according to $h\mathbf{A}h^{-1}$). Due to the presence of the translational quantity, $\mathbf{A}$ cannot be represented in tensorial form (Nash and Shen, 1983, p. 178).

Action invariance and gauge transformations: Because of the assumption of local gauge invariance, the action is invariant under local gauge transformations. The Euler–Lagrange equations and the action are conformally invariant. The dual operator $*$ is also conformally invariant.

2.5. Minkowski Spacetime Vs. Cartan–Weyl Form

Minkowski space, or vector space, is the spacetime of *special relativity.* In the curved spacetime of *general relativity*, Minkowski vector spaces occur as the *tangent spaces* of spacetime events. Minkowski space is a four-dimensional vector space over the field of real numbers possessing (a) an orientation, (b) a bilinear inner product of signature (+−) and (c) a time orientation. The signature refers to a tetrad — or four linearly independent vectors — t, x, y, z, such that:

$$\begin{aligned} &t \cdot t = 1, \\ &x \cdot x = y \cdot y = z \cdot z = -1, \\ &t \cdot x = t \cdot y = t \cdot z = x \cdot y = x \cdot z = y \cdot z = 0. \end{aligned} \tag{2.92}$$

If

$$t = g_0, x = g_1, y = g_2, z = g_3, \tag{2.93}$$

then

$$g_i \cdot g_j = \eta_{ij}, \tag{2.94}$$

where

$$(\eta_{ij}) = \left(\eta^{ij}\right) = \begin{pmatrix} 1 & 0 & 0 & 0 \\ 0 & -1 & 0 & 0 \\ 0 & 0 & -1 & 0 \\ 0 & 0 & 0 & -1 \end{pmatrix}. \tag{2.95}$$

The Minkowski spacetime or group refers to a well-defined metric and is the underlying algebraic logic for the special theory of relativity. However, as is well known, the special theory of relativity is only valid locally in a frame in free fall, which disconnects it from any gravitational effects (Kenyon, 1990, p. 46) — hence the reason for the appellative "special".

Compare now Minkowski space to the Weyl group. The Weyl group has a well-defined conformal structure but no preferred metric. The Riemann tensor of of general relativity is composed of two components: the Ricci tensor and the Weyl tensor (Weyl, 1918b). The Ricci tensor reflects the distribution of matter fields and the Weyl tensor describes space curvature which is not *locally* determined by the matter density. In terms of the *Weyl conformal tensor*, $C^{\cdot d}_{abc}$, the *Weyl tensor in spinor form*, Ψ_{ABCD}, is[14]:

$$C_{AA'BB'CC'DD'} = \Psi_{ABCD}\varepsilon_{A'B'}\varepsilon_{C'D'} + \varepsilon_{AB}\varepsilon_{CD}\bar{\Psi}_{A'B'C'D'}. \tag{2.96}$$

where $\bar{\Psi}_{A'B'C'D'}$ is the complex conjugate of the four-valent spinor Ψ_{ABCD}.

[14]As before, in the following equation: (1) a prime on a label indicates complex conjugation, so that spinors come in pairs; (2) the ε-spinor is used, which is *antisymmetrical*: $\varepsilon_{AB} = -\varepsilon_{BA}$ or $\varepsilon^{B}_{A} = -\varepsilon^{B}_{A}$ and $x_B = x^A\varepsilon_{AB}$; $x^A = \varepsilon^{AB}x_B$, i.e., the ε-spinors are used to raise and lower indices on other spinors; (3) a bar on a spinor indicates complex conjugation; and (4) the indices, **AB** are abstract indices without any reference to any basis or coordinate system (Penrose and Rindler, 1984, Vol. 1).

In a spacetime with Lorentzian signature, the self-dual and the anti-self-dual parts of the Weyl tensor (namely C^{+}_{abcd} and C^{-}_{abcd}), or of the Weyl spinor ($\bar{\Psi}_{A'B'C'D'}$, and Ψ_{ABCD}), are complex conjugates of each other.[15] Therefore an anti-self-dual spacetime (one with $C^{+}_{abcd} = 0$) is necessarily conformally flat ($C_{abcd} = 0$). However, this restriction does not apply to positive-definite four-spaces, or to complex spacetimes.[16] By a "complex spacetime" is meant a four-dimensional complex manifold $\mathbf{M}$, equipped with a *holomorphic* metric g_{ab}. In other words: with respect to a holomorphic coordinate basis $x^a = (x^0, x^1, x^2, x^3)$ the metric is a 4×4 matrix of holomorphic functions of x^a, and its determinant is nowhere vanishing. The Ricci tensor (which before was real) becomes complex-valued and the self-dual and anti-self-dual parts of the Weyl tensor (which before were complex-valued but conjugate to each other) become *independent* holomorphic tensors. The vanishing of the Weyl tensor implies that the spacetime is locally Minkowskian.

The origins of the Weyl group extend back to the discovery of spinors by Cartan in 1913 (Cartan, 1913, 1981). Although the development of spinors has proceeded formally (Brauer and Weyl, 1935; Weyl, 1939, 1968), the original Cartan approach provides geometrical definitions of the mathematical entities.

Whatever the global conditions of this complex spacetime, the following Riemann manifolds are possible (Ward and Wells, 1990): (1) $\mathbf{M}$ is compact without boundary; (2) $(\mathbf{M}, g_{ab})$ is asymptotically locally Euclidean (ALE); and (3) $(\mathbf{M}, g_{ab})$ is asymptotically locally flat (ALF). In empty space the Ricci tensor vanishes, and the Bianchi

[15] As before, the clumped pairs of spinor indices are defined: $a = AA^{'} b = BB^{'}; c = CC^{'}, \ldots, z = ZZ^{'}$. Therefore:

$$\psi^{AA'}{}_{BB'}{}^{Q} = \psi^{a}{}_{BB'}{}^{Q} = \psi^{a}{}_{b}{}^{Q} = \psi^{A'A}{}_{b}{}^{Q} = \psi^{a}{}_{B}{}^{Q}{}_{B'} \text{ and}$$

$$\psi^{a}{}_{AA'}{}^{Q} = \psi^{AA'}{}_{a}{}^{Q} = \psi^{a}{}_{a}{}^{Q} = \psi^{AA'}{}_{AA'}{}^{Q} \text{ and}$$

$$\psi^{AA'}{}_{BB'}{}^{B} = \psi^{a}{}_{b}{}^{B} = \psi^{AA'}{}_{b}{}^{B} = -\psi^{aB}{}_{b} = -\psi^{aB}{}_{BB'}$$

(Penrose and Rindler, 1984, Vol. 1, p. 116).

[16] Ward and Wells (1990, p. 293).

identity equation is:

$$\nabla^{AA'}\Psi_{ABCD} = 0, \tag{2.97}$$

which is of the same form as the Maxwell equations' spinor description:

$$\nabla^{AA'}\varphi_{AB} = 0. \tag{2.98}$$

This symmetric spinor object defines the Maxwell field tensor[17]:

$$f_{AA'BB'} = \varphi_{AB}\varepsilon_{A'B'} + \varepsilon_{AB}\bar{\varphi}_{A'B'} \tag{2.99}$$

which, according to our account above, describes the Sagnac interferometer *when the platform is at rest.*

The electromagnetic potential, $\mathbf{a}$, can be defined in terms of the covariant derivative as:

$$\nabla_{\mathbf{a}} \equiv \partial_a - ie\mathbf{a}_a, \tag{2.100}$$

where $\nabla_{\mathbf{a}}$ is a covariant derivative, ∂_a is the flat-space covariant derivative, and e is the charge of the field; or as:

$$\mathbf{a}_a \equiv \frac{i}{\varepsilon\alpha}\nabla_\alpha\alpha, \tag{2.101}$$

where α is a gauge, and the charge $e = n\varepsilon$, or as[18]:

$$\varphi_{AB} = \nabla_{A'(A}\mathbf{a}_{B)}^{A'} = \frac{1}{2}\left(\nabla_{A'A}\mathbf{a}_B^{A'} + \nabla_{A'B}\mathbf{a}_A^{A'}\right). \tag{2.102}$$

---

[17] φ_{AB} is a *symmetric* spinor.

[18] As before, round brackets indicate symmetrization and square brackets indicate antisymmetrization, e.g.:

$$\begin{aligned}
U_{A(BC)D} &= \tfrac{1}{2!}(U_{ABCD} + U_{ACBD});\\
U_{A(BCD)E} &= \tfrac{1}{3!}(U_{ABCDE} + U_{ACBDE} + U_{ACDBE} + U_{ABDCE}\\
&\quad + U_{ADBCE} + U_{ADCB});\\
U_{A[BC]D} &= \tfrac{1}{2!}(U_{ABCD} - U_{ACBD});\\
U_{A[BCD]E} &= \tfrac{1}{3!}(U_{ABCDE} - U_{ACBDE} + U_{ACDBE} - U_{ABDCE}\\
&\quad + U_{ADBCE} - U_{ADCB}).
\end{aligned}$$

Using:

$$f_{AB} \equiv \frac{i}{\varepsilon\alpha}\Delta_{AB}\alpha, \tag{2.103}$$

where the commutator is defined:

$$\Delta_{AB} = \nabla_A\nabla_B - \nabla_B\nabla_A = 2\nabla_{[A}\nabla_{B]}, \tag{2.104}$$

we can then obtain:

$$f_{AB} = \nabla_A \mathbf{a}_B - \nabla_B \mathbf{a}_A, \tag{2.105}$$

which, in terms of the flat-space covariant derivative, recaptures Eq. (2.38)

$$f_{AB} = \partial_B a_A(x) - \partial_A a_B(x), \tag{2.106}$$

again describing the Sagnac interferometer with *platform at rest.*

In the case of the Sagnac interferometer with *platform in rotation,* and using again Yang–Mills fields, the matrix potential, **A**, is defined as a matrix of convectors:

$$\mathbf{A}_{a\Psi}{}^{\Theta} = i\alpha_{\Psi}{}^{\Theta}\nabla_a\alpha_{\Psi}{}^{\Psi}, \tag{2.107}$$

and the Yang–Mills field, with obstruction, is defined in spinor form[19] as (Penrose and Rindler, 1986, Vol. 2, p. 34):

$$F_{ab\Theta}{}^{\Psi} = \varphi_{AB\Theta}{}^{\Psi}\varepsilon_{A'B'} + \varepsilon_{AB}\chi_{A'B'\Theta}{}^{\Psi}, \tag{2.108}$$

where

$$\varphi_{AB\Theta}{}^{\Psi} = \varphi_{(AB)\Theta}{}^{\Psi} = \frac{1}{2}F_{ABC'}{}^{C'}{}_{\Theta}{}^{\Psi}, \tag{2.109}$$

is the Yang–Mills potential and

$$\chi_{A'B'\Theta}{}^{\Psi} = \chi_{(A'B')\Theta}{}^{\Psi} = \frac{1}{2}F_{C}{}^{C}{}_{A'B'\Theta}{}^{\Psi}; \tag{2.110}$$

[19] $\varphi_{AB} = \varphi_{(AB)}$ (Penrose and Rindler, 1986, Vol. 2, p. 32).

and in vector potential form as (Penrose and Rindler, 1984, Vol. 1, p. 349):

$$\frac{1}{2}F_{ab\Theta}{}^{\Psi} = \nabla_{[a}\mathbf{A}_{b]\Theta}{}^{\Psi} - i\mathbf{A}_{\Lambda}{}^{\Psi}{}_{[a}\mathbf{A}_{b]\Theta}{}^{\Lambda} = \frac{1}{2!}(\nabla_a A_{b\Theta}{}^{\Psi} - \nabla_b A_{a\Theta}{}^{\Psi} - i\left[A_{a\Lambda}{}^{\Psi}, A_{b\Theta}{}^{\Lambda}\right]), \tag{2.111}$$

which, in terms of the flat-space covariant derivative and but for a generalized charge $-g$, recaptures Eq. (2.41):

$$F_{\mu\nu}(x) = \partial_\nu A_\mu(x) - \partial_\mu A_\nu(x) + ig\left[A_\mu(x), A_\nu(x)\right], \tag{2.112}$$

describing the Sagnac interferometer with *platform in rotation.* Unlike the case of the unrotated platform, the rotated platform is Yang–Mills charged.

In summary, a comparison can be made (Table 2.3) between descriptions of the Sagnac interferometer with platform at rest and with platform in motion.

Turning now to other algebraic approaches to field description: the mathematical algebra offered by *twistors* sheds light on the

Table 2.3. Sagnac interferometer at rest and in motion.

Platform at rest	$\leftrightarrow$	Platform in motion
Minkowski spacetime locally and globally	$\leftrightarrow$	Minkowski spacetime only locally; Minkowski vector spaces as the tangent spaces of spacetime events
Self-dual and anti-self-dual Weyl spacetimes are complex conjugates	$\leftrightarrow$	Weyl anti-self-dual spacetime independent of self-dual spacetime
Conformally flat spacetime	$\leftrightarrow$	Conformally curved spacetime
Abelian Maxwell equations apply locally and globally	$\leftrightarrow$	Abelian Maxwell equations apply locally and Non-Abelian Maxwell equations apply globally
Absence of Weyl tensor	$\leftrightarrow$	Presence of Weyl tensor
Real spacetime	$\leftrightarrow$	Complex spacetime
Twistor space algebra applies	$\leftrightarrow$	Curved twistor space algebra applies
Fields of SO(3) gauge	$\leftrightarrow$	Fields of $SU(2)/Z_2$ gauge

structure of energy–momentum/angular–momentum of systems.[20] A twistor is a conformally invariant structure and in Minkowski spacetime is a pair:

$$Z = (\omega^A, \pi_{A'}) \tag{2.113}$$

consisting of a spinor field, ω^A, and a complex conjugate spinor field, $\pi_{A'}$, satisfying the twistor equation:

$$\partial_{AA'}\omega^B = -i\varepsilon_A^B \pi_{A'}. \tag{2.114}$$

However, the extrapolation of twistor algebra to non-Minkowski spacetime (as required for a description of the Sagnac interferometer platform in motion) is beyond the scope of the present book. Such a program is addressed by the *ambitwistor* program[21] and covers the topics of curved anti-self-dual spacetimes, curved twistor spaces and complex spacetimes.

2.6. Discussion

We have argued that Coriolis acceleration results in an unmeasurable change in the gauge potentials and a measurable change in the phase factor. Since the phase factor is defined by the gauge potentials, this implies a new definition of the gauge concept. Originally introduced by Weyl as referencing a change in length, the concept was later adapted to quantum mechanical requirements and henceforth referenced a change in phase. In a further extension, the present usage implicates a causal relationship between Coriolis acceleration and the gauge potentials. That being so, there seems to be no reason to discriminate between the *effects* of the electromagnetic **A** potential field on a test particle (e.g., as in the Aharonov–Bohm effect) and the *effects* of acceleration (kinematic, linear or gravitational) on test fields in an interferometer. In all cases, a measurable change in a phase factor results. Because (1) the only measurable indication of

---

[20] Penrose and Rindler (1984, 1986), Huggett and Tod (1985) and Ward and Wells (1990).

[21] cf. the Ward construction in Penrose and Rindler (1986, Vol. 2, pp. 164–168) and Ward and Well (1990, Chapter 9).

the presence of the forces in these cases, is indicated by the change the phase factor of test particles/waves in a global interferometric situation, and (2) we have attributed this phase factor change to a conservation of action law and topological group constraints, there seems reason enough to conclude that the electromagnetic test particle/wave does not, and cannot, discriminate between the different causal (force) origins of a change in phase factor in (1) and (2). That is, and considering diverse causal origins of the qualitatively identical result, if either (a) some change in the spacetime metric, or (b) the nearby presence of a mass (gravitational attraction), or (c) linear acceleration, or (d) kinematic acceleration, *all* result in the *same* qualitative effect, namely, electromagnetic phase factor changes, then there is no need to search for the *unification* of, e.g., gravitational and electromagnetic *forces*. Rather, what should be appreciated is that if electromagnetic force fields and the spacetime metric gauge fields are bound by conservation of action and topological group constraints, the electromagnetic group, of whatever symmetry, can be perturbed in a number of ways of which (a), (b), (c) and (d) are examples. Furthermore, the electromagnetic gauge group does not discriminate between (a), (b), (c) and (d), for all effect the qualitatively same change — a change in the phase factor. It is inappropriate, then, to require unification of electromagnetic force with the "force" of gravity in that gravity is one manifestation of the electromagnetic gauge field registering a perturbation of the force field under condition (b), both force and gauge fields being confined by conservation of action bounded by topological degrees of freedom. Bound by global conformal invariance or conservation of action, *any change* in the conformal structure of *e.m. fields* results in a spacetime metric change; but, *pari passu*, *any change* in *the spacetime metric* results in a change in the conformal structure of e.m. fields. Therefore, rather than seeking to unite the *forces* of gravity and electromagnetism, the electromagnetic *force* fields and *gauge* fields (gravity) are already united due to preservation of action and topological degrees of freedom.

The viewpoint presented here is in conformity with the general theory of relativity. It is also in conformity with the special theory of relativity, *when the platform is at rest, but not when the platform is accelerated or subjected to a Coriolis force, as it should be if the epithet "special" is justified.* The viewpoint presented also implies that the source of the metric fields (of spacetime) are the electromagnetic fields and the source of the electromagnetic fields are the metric fields, both being subject to conditions of global conformal invariance.

The present approach to the Sagnac effect, a coherent field effect, is in the spirit of Rainich–Wheeler–Misner's "already unified field theory" considering that RWM adopts the stance that spacetime is not just an arena for the electromagnetic fields to play out their dynamic interactions, but both spacetime and electromagnetism, together, form a dynamic interactive entity.[22] However, the present view has taken advantage of the major conceptual advances made in both physics and mathematics since RWM was first proposed. For example, advances have occurred in application of gauge theory and specifically, Yang–Mills theory, with the discovery of the instanton concept, as well as with the recognition that the underlying algebraic logic of fields, as defined by group theory, together with the contributions of Cartan, Weyl and Lie, prescribes those fields' dynamic behavior. We are now able, therefore, to distinguish between gauge potentials and phase factors and even between the gauge potentials a_μ and $\mathbf{A}_\mu$ and the phase factors Φ and Φ^*, and, moreover, to ascribe physical meaning to these distinctions. Drawing on group theory concepts, the present approach does not seek gravitational unity with the *conventional* U(1) formulation of Maxwell's theory, but with the *"compactified"* SU(2) and higher-order versions. And again: whereas for RWM the concept of *force* remains central, here *the gauge field* is at least as important. But the major difference is that whereas RWM do not use the electromagnetic potential, the use of this potential is pivotal to the present approach. Nonetheless, despite these fundamental differences and other incompatibilities, the

[22]Rainich (1924, 1925), Wheeler (1955, 1962), Misner and Wheeler (1957), Power and Wheeler (1957).

strategical theoretical orientation towards fields, their metric and the dynamic interaction of the two, is in a similar spirit.

Taken to its logical conclusion, the approach adopted here requires that under the special topological conditions described above, and only under those conditions, the photon associated with the Φ^* field will acquire mass and propagate as a disturbance of the gravitational metric. Using field conversion, a Φ^* field-based mechanism would efficiently propagate energy as well as communications, and penetrate media normally impenetrable to force field photons.

Bibliography

Aronowitz, F., The laser gyro. In M. Ross (ed.) *Laser Applications, Volume 1*, Academic, New York, 1971.

Barrett, T.W., Comments on the Harmuth Ansatz: use of a magnetic current density in the calculation of the propagation velocity of signals by amended Maxwell theory. *IEEE Trans. Electromagn. Compat.*, 30, 419–420, 1988.

Barrett, T.W., Electromagnetic phenomena not explained by Maxwell's equations. pp. 6–86 in A. Lakhtakia (ed.) *Essays on the Formal Aspects of Electromagnetic Theory*, World Scientific, Singapore, 1993.

Barrett, T.W., Sagnac effect: A consequence of conservation of action due to gauge field global conformal invariance in a multiply-joined topology of coherent fields. pp. 278–279, in Barrett, T.W. and Grimes, D.M. (eds.) *Advanced Electromagnetism: Foundations, and Applications*, World Scientific, Singapore, 1995.

Brauer, R. and Weyl, H., Spinors in n dimensions. *Am. J. Math.*, 57, 425–449, 1935.

Cartan, E., Les groupes projectifs qui ne laissent invariante aucune multiplicité plane. *Bull. Soc. Math. France*, 41, 53–96, 1913.

Cartan, E., *The Theory of Spinors*, Dover, New York, 1981 (*Leçons sur la Théorie des spineurs* (2 volumes), Hermann, Paris, 1966).

Chow, W.W., Gea-Banacloche, J., Pedrotti, L.M., Sanders, V.E., Schleich, W. and Scully, M.O., The ring laser gyro. *Rev. Mod. Phys.*, 57, 61–104, 1985.

de Rham, G., *Variétés Differentiable*, Hermann, Paris, 1960.

Forder, P.W., Ring gyroscopes: an application of adiabatic invariance. *J. Phys. A: Math. Gen.*, 17, 1343–1355, 1984.

Gyorffi, G.L. and Lamb, W.E., *Theory of a Ring Laser*, Michigan Microfilm, Yale University, 1965.

Huggett, S.A. and Tod, K.P., *An Introduction to Twistor Theory*, Cambridge University Press, 1985.

Kenyon, I.R., *General Relativity*, Oxford University Press, 1990.

Heaviside, O., *Electromagnetic Theory,* 1893, Chelsea, New York, 1971.

Hong-Mo, C. and Tsun, T.S., *Some Elementary Gauge Theory Concepts*, World Scientific, Singapore, 1993.

Konopinski, E.J., What the electromagnetic vector potential describes. *Am. J. Phys.*, 46, 499–502, 1978.

Macek, W.M. and Davis, D.T.M., Rotation rate sensing with traveling wave ring lasers. *Appl. Phys. Lett.*, 2, 67–68, 1963.

Menegozzi, L.M. and Lamb, W.E., Theory of a ring laser. *Phys. Rev.*, A8, 2103–2125, 1973.

Metz, A., The problems relating to rotation in the theory of relativity. *J. Phys. Radium,* 13, 224–238, 1952.

Metz, A., Théorie relativiste de l'expérience de Sagnac avec interposition de tubes réfringents immobiles. *Compt Rend.*, 234, 597–599, 1952.

Metz, A., Théorie relativiste d'une expérience de Dufour et Prunier. *Compt Rend.*, 234, 705–707, 1952.

Misner, C.W. and Wheeler, J.A., Classical physics as geometry: Gravitation, electromagnetism, unquantized charge, and mass as properties of curved empty space. *Ann. Phys.*, 2, 525–603, 1957.

Moyer, H.G., An action principle combining electromagnetism and general relativity. *J. Math. Phys.*, 28, 705–710, 1987.

Nash, C. and Shen, S., *Topology and Geometry for Physicists*, Academic Press, New York, 1983.

Pegram, G.B., Unipolar induction and electron theory. *Phys. Rev.*, 10, 591–600, 1917.

Penrose, R. and Rindler, W., *Spinors and Space–Time*, Volume 1, Cambridge University Press, 1984.

Penrose, R. and Rindler, W., *Spinors and Space–Time*, Volume 2, Cambridge University Press, 1986.

Plebanski, J., Electromagnetic waves in gravitational fields. *Phys. Rev.*, 118, 1396–1408, 1960.

Post, E.J., *Formal Structure of Electromagnetics: General Covariance and Electromagnetics,* North-Holland Publishing Co., Amsterdam, John Wiley & Sons, New York, 1962.

Post, E.J., Sagnac effect. *Rev. Mod. Phys.*, 39, 475–493, 1967.

Post, E.J., Geometry and physics: A global approach. pp. 57–78 in M. Bunge (ed.) *Problems in the Foundations of Physics,* Springer-Verlag, New York, 1971.

Post, E.J., Interferometric path-length changes due to motion. *J. Opt. Soc. Am.*, 62, 234–239, 1972a.

Post, E.J., The constitutive map and some of its ramifications. *Ann. Phys.* 71, 497–518, 1972b.

Post, E.J., Ramifications of flux quantization. *Phys. Rev.*, D9, 3379–3385, 1974.

Post, E.J., Can microphysical structure be probed by period integrals? *Phys. Rev.*, D25, 3223–3229, 1982.

Power, E.A. and Wheeler, J.A., Thermal geons. *Rev. Mod. Phys.*, 29, 480–495, 1957.

Rainich, G.Y., Electrodynamics in the general relativity theory. *Proc. Natl. Acad. Sci.*, 10, 124–127, 1924.

Rainich, G.Y., Electrodynamics in the general relativity theory. *Trans. Am. Math. Soc.*, 27, 106–136, 1925.

Rosenthal, A.H., Regenerative circulatory multiple-beam interferometry for the study of light-propagation effects. *J. Opt. Soc. Am.*, 52, 1143, 1962.

Sagnac, G., L'éther lumineux demonstré par l'effet du vent relatif d'éther dans un interférometre en rotation uniforme. *Comptes Rendus Acad. Sci.*, 157, Séance du 27 Octobre, 708–710, 1913a.

Sagnac, G., Sur la preuve de la réalité de l'éther lumineaux par l'expérience de l'interférographe tournant. *Comptes Rendus Acad. Sci.*, Séance du 22 Décembre, 157, 1410–1413, 1913b.

Sagnac, G., Circulation of the ether in rotating interferometer. *J. Phys. Radium.*, (5), 4, 177–195, 1914.

von Laue, M., Velocity of light in moving bodies. *Ann. Phys.*, 62, 448–463, 1920.

Wald, R.M., *General Relativity*, University of Chicago Press, 1984.

Ward, R.S. and Wells, R.O., *Twistor Geometry and Field Theory*, Cambridge University Press, 1990.

Weyl, H., Gravitation und electrizität. *Sitz. Ber. Preuss. Ak. Wiss.*, 465–480, 1918a.

Weyl, H., Reine infinitesimalgeometrie. *Math. Zeit.*, 2, 384–411, 1918b.

Weyl, H., Gravitation and the electron. *Proc. Natl. Acad. Sci. USA*, 15, 323–334, 1929.

Weyl, H., *The Classical Groups, Their Invariants and Representations*, Princeton University Press, 1939 and 1946.

Weyl, H., *Gesammelte Abhandlungen*, in 4 Volumes, K. Chandrasekharan (ed.), Springer-Verlag, Heidelberg, 1968.

Wheeler, J.A., Geons. *Phys. Rev.*, 97, 511–536, 1955.

Wheeler, J.A., *Geometrodynamics*, Academic Press, 1962.

Whittaker, E.T., *A History of the Theories of Aether and Electricity*, 2 Volumes, 1951 and 1953, Dover, 1989.

Wu, T.T. and Yang, C.N., Concept of nonintegrable phase factors and global formulation of gauge fields. *Phys. Rev.*, D12, 3845, 1975.

Yang, C.N. and Mills, R., Isotopic spin conservation and generalized gauge invariance. *Phys. Rev.*, 95, M7, 631, 1954a.

Yang, C.N. and Mills, R.N., Conservation of isotopic spin and isotopic gauge invariance. *Phys. Rev.*, 96, 191–195, 1954b.

Zernike, F., The convection of light under various circumstances with special reference to Zieman's experiments. *Physica*, 13, 279–288, 1947.

Chapter 3

Topological Approaches to Electromagnetism[1]

3.1. Overview

Topology addresses those properties, often associated with *invariant qualities*, which are not altered by continuous deformations. Objects are topologically equivalent, or *homeomorphic*, if one object can be changed into another by bending, stretching, twisting, or any other continuous deformation or mapping. Continuous deformations are allowed, but prohibited are foldings which bring formerly distant points into direct contact or overlap, and cutting — unless followed by a regluing, reestablishing the preexisting relationships of continuity.

The continuous deformations of topology are commonly described in differential equation form and the quantities conserved under the transformations commonly described by differential equations exemplifying an algebra of operations which preserve that algebra. Évariste Galois[2] first gave the criteria that an algebraic equation must satisfy in order to be solvable by radicals. This branch of mathematics came to be known as Galois or *group theory*.

[1]Based on: Barrett, T.W. Topological approaches to electromagnetism. in *Modern Nonlinear Optics Part 3*, 2nd edition, Wiley, pp. 669–734, 2001.
[2]Évariste Galois (1811–1832).

Beginning with Leibniz[3] in the 17th, Euler[4] in the 18th, Reimann,[5] Listing[6] and Möbius[7] in the 19th and Poincaré[8] in the 20th centuries, "*analysis situs*" (Riemann) or "*topology*" (Listing) has been used to provide answers to questions concerning what is most fundamental in physical explanation. That question itself implies a further question concerning what mathematical structures one uses with confidence to adequately "paint" or describe physical models built from empirical facts. For example, differential equations of motion cannot be fundamental, because they are dependent on boundary conditions which must be justified — usually by group theoretical considerations. Perhaps, then, group theory[9] is fundamental.

Group theory certainly offers an austere shorthand for fundamental transformation rules. But it appears to the present writer that the final judge of whether a mathematical group structure can, or cannot, be applied to a physical situation is the topology of that physical situation. Topology dictates and justifies the group transformations.

So for the present writer, the answer to the question of what is the most fundamental physical description is that it is a description of the topology of the situation. With the topology known, the group theory description is justified and equations of motion can then also be justified and defined in specific differential equation form. If there is a requirement for an understanding more basic than the topology of the situation, then all that is left is verbal description of visual images. So we commence an examination of electromagnetism

[3]Gottfried Wilhelm Leibnitz (1646–1716).

[4]Leonhard Euler (1707–1783).

[5]Bernhard Reimann (1826–1866).

[6]Johann Benedict Listing (1808–1882). See his *Vorstudien zur Topologie*, 1847, where, for the first time, the title "topology" (in German) appeared in print.

[7]August Ferdinand Möbius (1790–1868).

[8]Henri Poincaré (1854–1912).

[9]Here we address the kind of groups addressed in Yang–Mills theory, which are *continuous* groups (as opposed to *discrete* groups). Unlike discrete groups, continuous groups contain an infinite number of elements and can be differentiable or analytic. Cf. Yang, C.N. and Mills, R.L., Conservation of isotopic spin and isotopic gauge invariance. *Phys. Rev.*, 96, 191–195, 1954.

under the assumption that topology defines group transformations and the group transformation rules justify the algebra underlying the differential equations of motion.

Differential equations or a set of differential equations describe a *system* and its evolution. Group symmetry principles summarize both invariances and the laws of nature independent of a system's specific dynamics. It is necessary that the symmetry transformations be continuous or specified by a set of parameters which can be varied continuously. The symmetry of continuous transformations leads to conservation laws.

There are a variety of special methods used to solve ordinary differential equations. It was Sophus Lie[10] in the 19th century who showed that all the methods are special cases of integration procedures which are based on the invariance of a differential equation under a continuous group of symmetries. These groups became known as Lie groups.[11] A symmetry group of a system of differential equations is a group which transforms solutions of the system to other solutions.[12] In other words, there is an invariance of a differential equation under a transformation of independent and dependent variables. This invariance results in a diffeomorphism on the space of independent and dependent variables, permitting the mapping of solutions to solutions.[13]

---

[10]Sophus Lie (1842–1899).

[11]***Lie Group Algebras*:** If a topological group is a group and also a topological space in which group operations are continuous, then Lie groups are topological groups which are also analytic manifolds on which the group operations are analytic.

In the case of Lie algebras, the parameters of a product are analytic functions of the parameters of each factor in the product. For example, $L(\gamma) = L(\alpha)L(\beta)$ where $\gamma = f(\alpha, \beta)$. This guarantees that the group is differentiable. The Lie groups used in Yang–Mills theory are *compact* groups, i.e., the parameters range over a closed interval.

[12]Cf. Olver, P.J., *Applications of Lie Groups to Differential Equations*, Springer-Verlag, 1986.

[13]Baumann, G., *Symmetry Analyis of Differential Equations with Mathematica*, Springer-Verlag, 1998.

The relationship was made more explicit by Noether (1882–1935)[14] in theorems now known as *Noether's theorems*,[15] which related symmetry groups of a variational integral to properties of its associated Euler–Lagrange equations. The most important consequences of this relationship are that (i) conservation of energy arises from invariance under a group of time translations; (ii) conservation of linear momentum arises from invariance under (spatial) translational groups; and (iii) conservation of angular momentum arises from invariance under (spatial) rotational groups; and (iv) conservation of charge arises from invariance under change of phase of the wave function of charged particles. Conservation and group symmetry laws have been vastly extended to other systems of equations, e.g., the standard model of modern high-energy physics, and also, of importance to the present interest: soliton equations. For example, the *Korteweg–de Vries* "soliton" equation[16] yields a symmetry algebra spanned by the four vector fields of (i) space translation; (ii) time translation; (iii) Galilean translation; and (iv) scaling.

An aim of the present book is to show that the spacetime topology defines electromagnetic field equations[17] — whether the fields be of force or of phase. That is to say, the premise of this enterprise is that a set of field equations are only valid with respect

[14]Emmy (Amalie) Noether (1882–1935).

[15]Noether, E., Invariante variations probleme. *Nachr. Ges. Wiss. Goettingen, Math.-Phys.* Kl. 171, 235–257, 1918.

[16]Korteweg, D.J. and de Vries, G. On the change of form of long waves advancing in a rectangular canal, and on a new type of long stationary wave, *Philos. Mag.*, 39, 422–443, 1895.

[17]Barrett, T.W. Maxwell's theory extended. Part I. Empirical reasons for questioning the completeness of Maxwell's theory — effects demonstrating the physical significance of the A potentials, *Ann. Fondation Louis de Broglie*, 15, 143–183, 1990.

———, Maxwell's theory extended. Part II. Theoretical and pragmatic reasons for questioning the completeness of Maxwell's theory, *Ann. Fondation Louis de Broglie*, 12, 253–283, 1990.

———, The Ehrenhaft–Mikhailov effect described as the behavior of a low energy density magnetic monopole instanton, *Ann. Fondation Louis de Broglie*, 19, 291–301, 1994.

to a set defined topological description of the physical situation. In particular, the writer has addressed demonstrating that the A_μ potentials, $\mu = 0, 1, 2, 3$, are not just a mathematical convenience, but — *in certain well-defined situations* — are measurable, i.e., physical. Those situations in which the A_μ potentials are measurable possess a topology, the transformation rules of which are describable by the SU(2) group[18] or lower symmetry groups; and those situations

———, Electromagnetic phenomena not explained by Maxwell's equations, in Lakhtakia, A. (Ed.) *Essays on the Formal Aspects of Maxwell's Theory*, World Scientific, Singapore, pp. 6–86, 1993.

———, Sagnac effect. pp. 278–313 in Barrett, T.W. and Grimes, D.M., (Eds) *Advanced Electromagnetism: Foundations, Theory, Applications*, World Scientific, Singapore, 1995.

———, The toroid antenna as a conditioner of electromagnetic fields into (low energy) gauge fields, *Speculations Sci. Technol.*, 21(4), 291–320, 1998.

[18]**SU(n) Group Algebra:** Unitary transformations, $U(n)$, leave the modulus squared of a complex wavefunction invariant. The elements of a $U(n)$ group are represented by $n \times n$ unitary matrices with a determinant equal to ± 1. Special unitary matrices are elements of unitary matrices which leave the determinant equal to $+1$. There are $n^2 - 1$ independent parameters. *SU(n)* is a subgroup of $U(n)$ for which the determinant equals $+1$.

SL(2,*C*) Group Algebra: The special linear group of 2×2 matrices of determinant 1 with complex entries is SL(2,C).

SU(2) Group Algebra: SU(2) is a subgroup of SL(2,C). The are $2^2 - 1 = 3$ independent parameters for the special unitary group SU(2) of 2×2 matrices. SU(2) is a Lie algebra such that for the angular momentum generators, J_i, the commutation relations are $[J_i, J_j] = i\varepsilon_{ijk} J_k; i, j, k = 1, 2, 3$. The SU(2) group describes rotation in three-dimensional space with two parameters (see below). There is a well-known SU(2) matrix relating the Euler angles of $O(3)$ and the complex parameters of SU(2) is:

$$\begin{matrix} \cos\left[\frac{\beta}{2}\right]\exp\left[\frac{i(\alpha+\gamma)}{2}\right], & \sin\left[\frac{\beta}{2}\right]\exp\left[\frac{-(\alpha-\gamma)}{2}\right], \\ -\sin\left[\frac{\beta}{2}\right]\exp\left[\frac{i(\alpha-\gamma)}{2}\right], & \cos\left[\frac{\beta}{2}\right]\exp\left[\frac{-i(\alpha+\gamma)}{2}\right], \end{matrix}$$

where α, β, γ are the Euler angles. It is also well known that a homomorphism exists between $O(3)$ and SU(2), and the elements of SU(2) can be associated with rotations in $O(3)$; and SU(2) is the *covering group* of $O(3)$. Therefore, it is easy to show that SU(2) can be obtained from $O(3)$. These SU(2) transformations define the relations between the Euler angles of group $O(3)$ with the parameters of SU(2). For comparison with the above, if the rotation matrix $R(\alpha, \beta, \gamma)$ in $O(3)$

---

is represented as

$$\begin{pmatrix} \cos[\alpha]\cos[\beta]\cos[\gamma]-\sin[\alpha]\sin[\gamma] & \sin[\alpha]\cos[\beta]\cos[\gamma]+\cos[\alpha]\sin[\gamma] & -\sin[\beta]\cos[\gamma] \\ -\cos[\alpha]\cos[\beta]\sin[\gamma]-\sin[\alpha]\cos[\gamma] & -\sin[\alpha]\cos[\beta]\sin[\gamma]+\cos[\alpha]\cos[\gamma] & \sin[\beta]\sin[\gamma] \\ \cos[\alpha]\sin[\beta] & \sin[\alpha]\sin[\beta] & \cos[\beta] \end{pmatrix}$$

then the orthogonal rotations about the coordinate axes are:

$$R_1(\alpha)=\begin{pmatrix} \cos[\alpha] & \sin[\alpha] & 0 \\ -\sin[\alpha] & \cos[\alpha] & 0 \\ 0 & 0 & 1 \end{pmatrix}, \quad R_2(\beta)=\begin{pmatrix} \cos[\beta] & 0 & -\sin[\beta] \\ 0 & 1 & 0 \\ \sin[\beta] & 0 & \cos[\beta] \end{pmatrix},$$

$$R_3(\gamma)=\begin{pmatrix} \cos[\gamma] & \sin[\gamma] & 0 \\ -\sin[\gamma] & \cos[\gamma] & 0 \\ 0 & 0 & 1 \end{pmatrix}.$$

An isotropic parameter, ϖ, can be defined:

$$\varpi=\frac{x-iy}{z},$$

where x, y, z are the spatial coordinates. If ϖ is written as the quotient of μ_1 and μ_2, or the homogeneous coordinates of the bilinear transformation, then:

$$\left|\mu_1'\mu_2'\right\rangle=\begin{bmatrix} \cos\left[\frac{\beta}{2}\right]\exp\left[\frac{i(\alpha+\gamma)}{2}\right] & \sin\left[\frac{\beta}{2}\right]\exp\left[\frac{-(\alpha-\gamma)}{2}\right] \\ -\sin\left[\frac{\beta}{2}\right]\exp\left[\frac{i(\alpha-\gamma)}{2}\right] & \cos\left[\frac{\beta}{2}\right]\exp\left[\frac{-i(\alpha+\gamma)}{2}\right] \end{bmatrix}|\mu_1\mu_2\rangle,$$

which is the relation between the Euler angles of $O(3)$ and the complex parameters of SU(2). However, there is not a unique one-to-one relation, for 2 rotations in $O(3)$ correspond to 1 direction in SU(2). There is thus a many-to-one or homomorphism between $O(3)$ and SU(2).

In the case of a complex two-dimensional vector (u, v):

$$\begin{pmatrix} u' \\ v' \end{pmatrix}=\begin{pmatrix} \cos\left[\frac{\beta}{2}\right]\exp\left[\frac{i(\alpha+\gamma)}{2}\right] & \sin\left[\frac{\beta}{2}\right]\exp\left[\frac{-(\alpha-\gamma)}{2}\right] \\ -\sin\left[\frac{\beta}{2}\right]\exp\left[\frac{i(\alpha-\gamma)}{2}\right] & \cos\left[\frac{\beta}{2}\right]\exp\left[\frac{-i(\alpha+\gamma)}{2}\right] \end{pmatrix}\begin{pmatrix} u \\ v \end{pmatrix}.$$

If we define:

$$a=\cos\left[\frac{\beta}{2}\right]\exp\left[\frac{i(\alpha+\gamma)}{2}\right],$$

$$b=\sin\left[\frac{\beta}{2}\right]\exp\left[\frac{-(\alpha-\gamma)}{2}\right],$$

then

$$\left|\mu_1'\mu_2'\right\rangle = \begin{bmatrix} a & b \\ -b^* & a^* \end{bmatrix} |\mu_1\mu_2\rangle,$$

where

$$\begin{bmatrix} a & b \\ -b^* & a^* \end{bmatrix}$$

are the well-known SU(2) transformation rules. Defining: $c = -b^*$ and $d = a^*$, we have the determinant:

$$ad - bc = 1 \text{ or } aa^* - b(-b^*) = 1.$$

Defining the (x, y, z) coordinates with respect to a complex two-dimensional vector (u, v) as:

$$x = \frac{1}{2}\left(u^2 - v^2\right), \quad y = \frac{1}{2i}\left(u^2 + v^2\right), \quad z = uv$$

then SU(2) transformations leave the squared distance $x^2 + y^2 + z^2$ invariant.

Every element of SU(2) can be written as:

$$\begin{bmatrix} a & b \\ -b^* & a^* \end{bmatrix}, \quad |a|^2 + |b|^2 = 1.$$

Defining

$$a = y_1 - iy_2, \quad b = y_3 - iy_4,$$

the parameters y_1, y_2, y_3, y_4 indicate positions in SU(2) with the constraint:

$$y_1^2 + y_2^2 + y_3^2 + y_4^2 = 1,$$

which indicates that the group SU(2) is a three-dimensional unit sphere in the four-dimensional y-space. This means that any closed curve on that sphere can be shrunk to a point. In other words, SU(2) is *simply-connected.*

It is important to note that SU(2) is the quantum mechanical "rotation group".

***Homomorphism of O*(3) *and SU*(2)**: There is an important relationship between $O(3)$ and SU(2). The elements of SU(2) are associated with rotations in three-dimensional space. To make this relationship explicit, new coordinates are defined:

$$x = \frac{1}{2}\left(u^2 - v^2\right); \quad y = \frac{1}{2i}\left(u^2 + v^2\right); \quad z = uv$$

Explicitly, the SU(2) transformations leave the squared three-dimensional distance $x^2 + y^2 + z^2$ invariant, and invariance which relates three-dimensional in which the $\mathbf{A}_\mu$ potentials are not measurable possess a topology, the transformation rules of which are describable by the $U(1)$ group.[19]

---

rotations to elements of SU(2). If a, b of the elements of SU(2) are defined:

$$a = \cos\frac{\beta}{2}\exp\frac{i(\alpha+\gamma)}{2}, b = \sin\frac{\beta}{2}\exp\frac{-i(\alpha-\gamma)}{2},$$

then the general rotation matrix $R(\alpha,\beta,\gamma)$ can be associated with the SU(2) matrix:

$$\begin{pmatrix} \cos\frac{\beta}{2}\exp\frac{i(\alpha+\gamma)}{2} & \sin\frac{\beta}{2}\exp\frac{-i(\alpha-\gamma)}{2} \\ -\sin\frac{\beta}{2}\exp\frac{i(\alpha-\gamma)}{2} & \cos\frac{\beta}{2}\exp\frac{-i(\alpha+\gamma)}{2} \end{pmatrix}$$

by means of the Euler angles.

It is important to indicate that this matrix does not give a unique one-to-one relationship between the general rotation matrix $R(\alpha,\beta,\gamma)$ and the SU(2) group. This can be seen if (i) we let $\alpha = 0, \beta = 0, \gamma = 0$, which gives the matrix:

$$\begin{pmatrix} 1 & 0 \\ 0 & 1 \end{pmatrix},$$

and (ii) $\alpha = 0, \beta = 2\pi, \gamma = 0$, which gives the matrix:

$$\begin{pmatrix} -1 & 0 \\ 0 & -1 \end{pmatrix}.$$

Both matrices define zero rotation in three-dimensional space, so we see that this zero rotation in three-dimensional space corresponds to two different SU(2) elements depending on the value of β. There is thus a *homomorphism*, or many-to-one mapping relationship between $O(3)$ and SU(2) — where "many" is 2 in this case — but not a one-to-one mapping.

SO(2) Group Algebra: The collection of matrices in Euclidean two-dimensional space (the plane) which are orthogonal and moreover for which the determinant is +1 is a subgroup of $O(2)$. SO(2) is the special orthogonal group in two variables.

The rotations in the plane is represented by the SO(2) group:

$$R(\theta) = \begin{pmatrix} \cos[\theta] & -\sin[\theta] \\ \sin[\theta] & \cos[\theta] \end{pmatrix} \theta \in \mathscr{R}$$

where $R(\theta)R(\gamma) = R(\theta+\gamma)$. S^1, or the unit circle in the complex plane with multiplication as the group operation is an SO(2) group.

[19] ***U(n)* Group Algebra:** Unitary matrices, U, have a determinant equal to ± 1. The elements of $U(n)$ are represented by $n \times n$ unitary matrices.

***U*(1) Group Algebra:** The one-dimensional unitary group, or $U(1)$, is characterized by one continuous parameter. $U(1)$ is also differentiable and the derivative is also an element of $U(1)$. A well-known example of a $U(1)$ group is that of all

Historically, electromagnetic theory was developed for situations described by the $U(1)$ group. The dynamic equations describing the transformations and interrelationships of the force field are the well known Maxwell equations, and the group algebra underlying these equations is $U(1)$. There was a need to extend these equations to describe SU(2) situations and to derive equations whose underlying algebra is SU(2). These two formulations were provided in previous chapters and are shown again here in Table 3.1. Table 3.2 shows the electric charge density, ρ_e, the magnetic charge density, ρ_m, the electric current density, g_e, the magnetic current density, g_m, the electric conductivity, σ, and the magnetic conductivity, s.

In the following sections, four topics are addressed: The mathematical entities, or waves, called *solitons*; the mathematical entities called *instantons*; a beam — an electromagnetic wave — which is *polarization modulated over a set sampling interval*; and the *Aharonov–Bohm effect*. Our intention is to show that these entities, waves or effects, can only be adequately characterized and differentiated, and thus understood, by using topological characterizations. Once characterized, the way becomes open for engineering control of these entities, waves and effects for useful activities.

Table 3.1. Maxwell equations in U(1) and SU(2) symmetry forms.

	U(1) symmetry form (traditional Maxwell equations)	**SU(2) symmetry form**
Gauss's law	$\nabla \cdot \mathbf{E} = J_0$	$\nabla \cdot \mathbf{E} = J_0 - iq(\mathbf{A} \cdot \mathbf{E} - \mathbf{E} \cdot \mathbf{A})$
Ampère's law	$\frac{\partial \mathbf{E}}{\partial t} - \nabla \times \mathbf{B} + \mathbf{J} = 0$	$\frac{\partial \mathbf{E}}{\partial t} - \nabla \times \mathbf{B} + \mathbf{J} + iq\,[A_0, \mathbf{E}] - iq(\mathbf{A} \times \mathbf{B} - \mathbf{B} \times \mathbf{A}) = 0$
Coulomb's law	$\nabla \cdot \mathbf{B} = 0$	$\nabla \cdot \mathbf{B} + iq(\mathbf{A} \cdot \mathbf{B} - \mathbf{B} \cdot \mathbf{A}) = 0$
Faraday's law	$\nabla \times \mathbf{E} + \frac{\partial \mathbf{B}}{\partial t} = 0$	$\nabla \times \mathbf{E} + \frac{\partial \mathbf{B}}{\partial t} + iq\,[\mathbf{A}_0, \mathbf{B}] + iq(\mathbf{A} \times \mathbf{E} - \mathbf{E} \times \mathbf{A}) = 0$

the possible phases of a wave function, which are angular coordinates in a two-dimensional space. When interpreted in this way — as the internal phase of the $U(1)$ group of electromagnetism — the $U(1)$ group is merely a circle $(0 - 2\pi)$.

Table 3.2. The U(1) and SU(2) symmetry forms of the major variables.

U(1) symmetry form (traditional maxwell theory)	SU(2) Symmetry Form
$\rho_e = \mathbf{J}_0$	$\rho_e = \mathbf{J}_0 - iq(\mathbf{A}\cdot\mathbf{E} - \mathbf{E}\cdot\mathbf{A}) = \mathbf{J}_0 + q\mathbf{J}_z$
$\rho_m = 0$	$\rho_m = -iq(\mathbf{A}\cdot\mathbf{B} - \mathbf{B}\cdot\mathbf{A}) = -iq\mathbf{J}_y$
$g_e = \mathbf{J}$	$g_e = iq\,[A_0, \mathbf{E}] - iq(\mathbf{A}\times\mathbf{B} - \mathbf{B}\times\mathbf{A}) + \mathbf{J}$ $= iq\,[A_0, \mathbf{E}] - iq\mathbf{J}_x + \mathbf{J}$
$g_m = 0$	$g_m = iq\,[A_0, \mathbf{B}] - iq(\mathbf{A}\times\mathbf{E} - \mathbf{E}\times\mathbf{A}) = iq\,[A_0, \mathbf{B}] - iq\mathbf{J}_z$
$\sigma = \mathbf{J}/\mathbf{E}$	$\sigma = \frac{\{iq[A_0,\mathbf{E}] - iq(\mathbf{A}\times\mathbf{B}-\mathbf{B}\times\mathbf{A})+\mathbf{J}\}}{\mathbf{E}} = \frac{\{iq[A_0,\mathbf{E}]-iq\mathbf{J}_x+\mathbf{J}\}}{\mathbf{E}}$
$s = 0$	$s = \frac{\{iq[A_0,\mathbf{B}] - iq(\mathbf{A}\times\mathbf{E}-\mathbf{E}\times\mathbf{A})\}}{\mathbf{H}} = \frac{\{iq[A_0,\mathbf{B}]-iq\mathbf{J}_z\}}{\mathbf{H}}$

3.2. Solitons[20]

Soliton solutions to differential equations require complete integrability and integrable systems conserve geometric features related to symmetry. Unlike the equations of motion for conventional Maxwell theory, which are solutions of $U(1)$ symmetry systems, solitons are solutions of SU(2) symmetry systems. These notions of group symmetry are more fundamental than differential equation descriptions. Therefore, although a complete exposition is beyond the scope of the present review, we develop some basic concepts in order to place differential equation descriptions within the context of group theory.

Within this context, *ordinary differential equations are viewed as vector fields on manifolds or configuration spaces.*[21] For example, Newton's equations are second-order differential equations describing

[20] A soliton is a solitary wave which preserves its shape and speed in a collision with another solitary wave. Cf. Barrett, T.W., 404–413 in Taylor, J.D. (ed.) *Introduction to Ultra-Wideband Radar Systems*, CRC Press, Boca Raton, 1995; Infeld, E. and Rowlands, G., *Nonlinear Waves, Solitons and Chaos*, 2nd edition, Cambridge University Press, 2000.

[21] Cf. Olver, P.J., *Applications of Lie Groups to Differential Equations*, Springer-Verlag, 1986.

smooth curves on Riemannian manifolds. Noether's theorem[22] states that a diffeomorphism,[23] ϕ, of a Riemannian manifold, C, indices a diffeomorphism, $D\phi$, of its tangent[24] bundle,[25] TC. If ϕ is a symmetry of Newton's equations, then $D\phi$ preserves the Lagrangian, i.e.,

$$L \circ D\varphi = L. \tag{3.1}$$

As opposed to equations of motion in conventional Maxwell theory, *soliton flows are Hamiltonian flows*. Such Hamiltonian functions define *symplectic structures*[26] for which there is an absence of *local* invariants but an infinite dimensional group of diffeomorphisms which preserve *global* properties. In the case of solitons, the global properties are those permitting the matching of the nonlinear and dispersive characteristics of the medium through which the wave moves.

In order to achieve this match, two linear operators, L and A, are postulated associated with a partial differential equation (PDE). The two linear operators are known as the *Lax pair*. The operator L

---

[22] Noether, E., Invariante variations probleme. *Nachr. Ges. Wiss. Goettingen, Math.-Phys. Kl.* 171, 235–257, 1918.

[23] A ***diffeomorphism*** is an elementary concept of topology and important to the understanding of differential equations. It can be defined in the following way: If the sets U and V are open sets both defined over the space R^m, i.e., $U \subset R^m$ is open and $U \subset R^m$ is open, where open means nonoverlapping, then the mapping $\psi : U \to V$ is an infinitely differentiable map with an infinitely differential inverse, and objects defined in U will have equivalent counterparts in V. The mapping ψ is a diffeomorphism. It is a smooth and infinitely differentiable function. The important point is: conservation rules apply to diffeomorphisms, because of their infinite differentiability. Therefore, diffeomorphisms constitute fundamental characterizations of differential equations.

[24] A vector field on a manifold M gives a ***tangent vector*** at each point of M.

[25] A ***bundle*** is a structure consisting of a manifold E, and manifold M, and an onto map: $\pi : E \to M$.

[26] ***Symplectic topology*** is the study of the global phenomena of symplectic symmetry. Symplectic symmetry structures have no local invariants. This is a subfield of topology: for example: McDuff, D. and Salamon, D., *Introduction to Symplectic Topology*, Oxford: Clarendon Press, 1995.

is defined by

$$L = \frac{\partial^2}{\partial x^2} + u(x,t), \tag{3.2}$$

with a related eigenproblem:

$$L\psi + \lambda\psi = 0. \tag{3.3}$$

The temporal evolution of ψ is defined as

$$\psi_t = -A\psi, \tag{3.4}$$

with the operator of the form:

$$A = a_0 \frac{\partial^n}{\partial x^n} + a_1 \frac{\partial^{n-1}}{\partial x^{n-1}} + \cdots + a_n,$$

where a_0 is a constant and the n coefficients a_i are functions of x and t. Differentiating Eq. (3.3) gives

$$L_t\psi + L\psi_t = -\lambda_t\psi - \lambda\psi_t.$$

Inserting (3.4):

$$L\psi_t = -LA\psi,$$

or

$$\lambda\psi_t = AL\psi.$$

Using (3.3) again

$$[L, A] = LA - AL = L_t + \lambda_t, \tag{3.5}$$

and for a time-independent λ:

$$[L, A] = L_t.$$

This equation provides a method for finding A.

Translating the above into a group theory formulation: in order to relate the three major soliton equations to group theory it is necessary to examine the *Lax equation*[27] (3.5) above as a the *zero-curvature condition* (*ZCC*). The *ZCC* expresses the flatness of a connection by the commutation relations of the covariant derivative operators and in terms of the Lax equation is

$$L_t - A_x - [L, A] = 0,$$

or[6]:

$$\left[\frac{\partial}{\partial x} - L\frac{\partial}{\partial t} - A\right] = 0,$$

or:

$$\left(\frac{\partial}{\partial x} - L\right)_t = \left[A\frac{\partial}{\partial x} - L\right].$$

Following this development, Palais[28] showed that the generic cases of soliton — the *Korteweg–de Vries Equation* (*KDV*), the *Nonlinear Schrödinger Equation* (*NLS*), the *Sine-Gordon Equation* (*SGE*) — can be given as an SU(2) formulation. In each of the three cases considered below, V is a one-dimensional space that is embedded in the space of off-diagonal complex matrices, $\left(\begin{smallmatrix} 0 & b \\ c & 0 \end{smallmatrix}\right)$ and in each case $L(u) - a\lambda + u$, where u is a potential, λ is a complex parameter, and **a** is the constant, diagonal, trace zero matrix

$$a = \begin{pmatrix} -i & 0 \\ 0 & i \end{pmatrix}.$$

The matrix definition of **a** links these equations to an SU(2) formulation. (Other matrix definitions of **a** could, of course, link **a** to lower group symmetries.)

[27] Lax, P.D., Integrals of nonlinear equations of evolution and solitary waves. *Comm. Pure Appl. Math.*, 21, 467–490, 1968;
Lax, P.D., Periodic solutions of the KdV equations, in *Nonlinear Wave Motion, Lectures in Applied Math.*, 15, American Mathematical Society, pp. 85–96, 1974.

[28] Palais, R.S., The symmetries of solitons. *Bull. Amer. Math. Soc.* 34, 339–403, 1997.

To carry out this objective, an inverse scattering theory function is defined:

$$B(\xi) = \sum_{n=1}^{N} c_n^2 \exp[-\kappa_n \xi] + \frac{1}{2\pi} \int_{-\infty}^{+\infty} b(k) \exp[ik\xi] dk,$$

where $-\kappa_1^2, \ldots, -\kappa_N^2$ are discrete eigenvalues of $u, c_1, \ldots, c_N$ are normalizing constants, and $b(k)$ are reflection coefficients.

Therefore, in a ***first case*** (the *KdV*), if $u(x) = \begin{pmatrix} 0 & q(x) \\ -1 & 0 \end{pmatrix}$ and

$$B(u) = a\lambda^3 + u\lambda^2 + \begin{pmatrix} \frac{i}{2}q & \frac{i}{2}q_x \\ 0 & -\frac{i}{2}q \end{pmatrix} \lambda + \begin{pmatrix} \frac{q_x}{4} & \frac{-q^2}{2} \\ \frac{q}{2} & \frac{-q_x}{4} \end{pmatrix},$$

and the *ZCC* (*Lax equation*) is satisfied if and only if. q satisfies the *KdV* in the form $q_t = -\frac{1}{4}(6qq_x + q_{xxx})$.

In a ***second case*** (the *NLS*), if

$$u(x) = \begin{pmatrix} 0 & q(x) \\ -\bar{q}(x) & 0 \end{pmatrix}$$

and

$$B(u) = a\lambda^3 + u\lambda^2 + \begin{pmatrix} \frac{i}{2}|q|^2 & \frac{i}{2}q_x \\ -\frac{i}{2}\bar{q}_x & -\frac{i}{2}|q|^2 \end{pmatrix},$$

then *ZCC* (*Lax equation*) is satisfied if and only if $q(x,t)$ satisfies the *NLS* in the form $q_t = \frac{i}{2}\left(q_{xx} + 2|q|^2 q\right)$

In a ***third case*** (the *SGE*), if

$$u(x) = \begin{pmatrix} 0 & -\frac{q_x(x)}{2} \\ \frac{q_x(x)}{2} & 0 \end{pmatrix}$$

and

$$B(u) = \frac{i}{4\lambda} \begin{pmatrix} \cos[q(x)] & \sin[q(x)] \\ \sin[q(x)] & -\cos[q(x)] \end{pmatrix},$$

then *ZCC* (*Lax equation*) is satisfied if and only if q satisfies the *SGE* in the form $q_t = \sin[q]$.

With the connection of PDEs, and especially soliton forms, to group symmetries established, then one can conclude that *if* the Maxwell equation of motion which includes electric *and* magnetic conductivity is in soliton (*SGE*) form, the group symmetry of the Maxwell field is SU(2). Furthermore, because solitons define Hamiltonian flows, their energy conservation is due to their *symplectic structure.*

In order to clarify the difference between conventional Maxwell theory which is of $U(1)$ symmetry, and Maxwell theory extended to SU(2) symmetry, we can describe both in terms of mappings of a field $\psi(x)$. In the case of $U(1)$ Maxwell theory, a mapping $\psi \to \psi'$ is:

$$\psi(x) \to \psi'(x) = \exp\left[ia(x)\right]\psi(x),$$

where $a(x)$ is the conventional vector potential. However, in the case of SU(2) extended Maxwell theory, a mapping $\psi \to \psi'$ is

$$\psi(x) \to \psi'(x) = \exp\left[iS(x)\right]\psi(x),$$

where $S(x)$ is the action and an element of an SU(2) field defined:

$$S(x) = \int \mathbf{A}dx,$$

and $\mathbf{A}$ is *the matrix form of the vector potential.* Therefore, we see the necessity *to adopt a matrix formulation of the vector potential when addressing* SU(2) *forms of Maxwell theory.*

3.3. Instantons

Instantons[29] correspond to the minima of the Euclidean action and are pseudo-particle solutions[30] of SU(2) Yang–Mills equations in

[29] Cf. Jackiw, R, Nohl, C. and Rebbi, C. *Classical and Semi-classical Solutions to Yang–Mills Theory.* Proceedings 1977 Banff School, Plenum Press.

[30] Belavin, A., Polyakov, A., Schwartz, A. and Tyupkin, Y., Pseudoparticle solutions of the Yang–Mills equations. *Phys. Lett.*, 59B, 85–87, 1975.

Euclidean 4 space.[31] A complete construction for any Yang-Mills group is also available.[32] In other words:

"It is reasonable... to ask for the determination of the classical field configurations in Euclidean space which minimize the action, subject to appropriate asymptotic conditions in 4-space. These classical solutions are the instantons of the Yang–Mills theory."[33]

In the light of the intention of this book to further the use of topology in electromagnetic theory, we quote further:

"If one were to search *ab initio* for a nonlinear generalization of Maxwell's equation to explain elementary particles, there are various symmetry group properties one would require. These are

(i) *external symmetries* under the Lorentz and Poincare groups and under the conformal group if one is taking the rest-mass to be zero,
(ii) *internal symmetries* under groups like SU(2) or SU(3) to account for the known features of elementary particles,
(iii) *covariance* or the ability to be coupled to gravitation by working on curved spacetime."[34]

In this book, the instanton concept in electromagnetism is applied for the following two reasons: (1) in some sense, the instanton, or pseudoparticle, is a compactification of degrees of freedom due to the particle's boundary conditions; and (2) the instanton, or pseudoparticle, then exhibits the behavior (the transformation or symmetry rules) of a high-energy particle, *but without the presence of high energy*, i.e., the pseudoparticle shares certain behavioral characteristics in common (shares transformation rules, hence symmetry rules in common) with a particle of much higher energy.

[31]Cf. Atiyah, M.F. and Ward, R.S., Instantons and Algebraic Geometry, *Commun. Math. Phys.*, 55, 117–124, 1977.
[32]Atiyah, M.F., Hitchin, N.J., Drinfeld, V.G. and Manin, Yu.I., Construction of instantons. *Phys. Lett.*, 65A, 23–25, 1978.
[33]Atiyah, M., in *Michael Atiyah: Collected Works, Volume 5, Gauge Theories*, Clarendon Press, Oxford, p. 95, Instantons and Algebraic Geometry, 1988.
[34]Atiyah, M., in *Michael Atiyah: Collected Works, Volume 5, Gauge Theories*, Clarendon Press, Oxford, p. 96, Instantons and Algebraic Geometry, 1988.

Therefore, we have suggested[35] that the Mikhailov effect,[36] and the Ehrenhaft[37] effect, which address demonstrations exhibiting *magnetic charge-like* behavior, are examples of instanton or pseudoparticle behavior. Stated differently: (1) the instanton shows that there are ways, other than possession of high energy, to achieve high symmetry states; and (2) symmetry dictates behavior.

3.4. Polarization Modulation Over a Set Sampling Interval[38]

It is well-known that all static polarizations of a beam of radiation, as well as all static rotations of the axis of that beam can be represented on a Poincaré sphere[39] (Fig. 3.1(A)). A vector can be centered in the middle of the sphere and pointed to the underside of the surface of the sphere at a location on the surface which represents the *instantaneous* polarization and rotation angle of a beam. Causing that vector to trace a trajectory over time on the surface of the sphere represents a *polarization modulated* (and *rotation modulated*) beam (Fig. 3.1(B)). If, then, the beam is sampled by a device at a rate which is less than the rate of modulation, then the sampled output from the device will be a condensation of *two components* of the wave, which are continuously changing with respect to each other, into *one snapshot* of the wave, at *one location* on the surface of the sphere and *one instantaneous polarization and axis rotation.* Thus, from the viewpoint of a device sampling at a rate less than the

[35]Barrett, T.W., The Ehrenhaft–Mikhailov effect described as the behavior of a low-energy density magnetic monopole instanton. *Ann. Fondation Louis de Broglie*, 19, 291–301, 1994.
[36]A summary of the Mikhailov effect is: pp. 593-619 in Barrett, T.W. and Grimes, D.M. (eds) *Advanced Electromagnetism: Foundations, Theory and Applications*, World Scientific, Singapore, 1995.
[37]Felix Ehrenhaft (1879–1952).
[38]Based on: Barrett, T.W., On the distinction between fields and their metric: the fundamental difference between specifications concerning medium-independent fields and constitutive specifications concerning relations to the medium in which they exist, *Ann. Fondation Louis de Broglie*, 14, 37–75, 1989.
[39]Poincaré, H., *Théorie Mathématique de la Lumière*, Vol. 2, Georges Carré, Paris, Chapter 2, 1892.

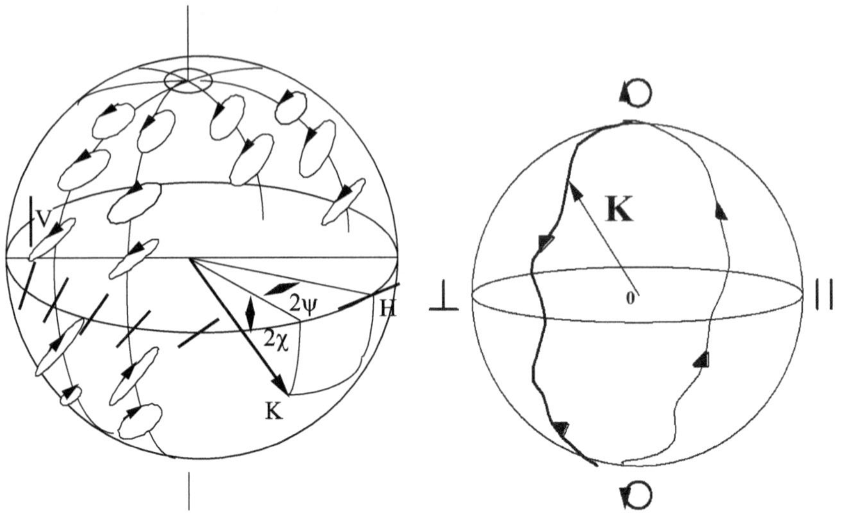

(a) (b)

Fig. 3.1. (a) Poincaré sphere representation of wave polarization and rotation. (b) A Poincaré sphere representation of signal polarization (longitudinal axis) and polarization rotation (latitudinal axis). A representational trajectory of polarization/rotation modulation is shown by changes in the vector centered at the center of the sphere and pointing at the surface. Waves of various polarization modulations $\partial\phi^n/\partial t^n$, can be represented as trajectories on the sphere. The case shown is an arbitrary trajectory repeating 2π. After Barrett (1997).[40]

modulation rate, a two-to-one mapping (over time) has occurred, which is the signature of an SU(2) field.

The modulations which result in trajectories on the sphere are infinite in number. Moreover, those modulations, at a rate of multiples of 2π greater than 1, which result in the return to a single location on the sphere at a frequency of exactly 2π, will all be detected by the device sampling at a rate of 2π as the same. In other words, the device cannot detect what kind of simple or complicated trajectory was performed between departure from, and arrival at, the same location on the sphere. To the relatively slowly sampling

[40]Barrett, T.W., Polarization–rotation modulated, spread polarization–rotation, wide-bandwidth radio-wave communications system. United States Patent 5,592,177 dated January 7, 1997.

device, the fast modulated beam can have "internal energies" quite unsuspected.

We can say that such a static device is *a* $U(1)$ *unipolar, set rotational axis, sampling device* and the fast polarization (and rotation) modulated beam is a *multipolar, multirotation axis,* SU(2) *beam.* The reader may ask: how many situations are there in which a sampling device, at set unvarying polarization, samples at a slower rate than the modulation rate of a radiated beam? The answer is that there is an infinite number, because nature is set up to be that way.[41] For example, the period of modulation can be faster than the electronic or vibrational or dipole relaxation times of any atom or molecule. In other words, *pulses or wave packets* (which, in temporal length, constitute the sampling of a continuous wave, continuously polarization and rotation modulated, but sampled only over a temporal length between arrival and departure time at the instantaneous polarization of the sampler of set polarization and rotation — in this case an electronic or vibrational state or dipole) have an internal modulation at a rate greater than that of the relaxation or absorption time of the electronic or vibrational state.

The representation of the sampling by a unipolar, single-rotation-axis, $U(1)$ sampler of an SU(2) continuous wave which is polarization and rotation modulated is shown in Fig. 3.2 which is the correspondence between the output space sphere and an Argand plane.[42] The Argand plane, Σ, is drawn in two-dimensions, x and y, with $z = 0$, and for a set snapshot in time. A point on the Poincaré sphere is represented as $P(t,\, x,\, y,\, z)$, and as in this representation $t = 1$ (or one step in the future), specifically as $P(1,\, x,\, y,\, z)$. The Poincaré sphere is also identified as a 3-sphere, $\mathbf{S}^{+}$, which is defined in Euclidean space as

$$x^2 + y^2 + z^2 = 1.$$

[41]See Barrett, T.W., Is quantum physics a branch of sampling theory? *Courants, amers, écueils en microphysique*, C. Cormier-Delanoue, G. Lochak, P. Lochak (eds), Fondation Louis de Broglie, 1993.

[42]After Penrose, R. and Rindler, W., *Spinors and Spacetime, Volume 1, Two-Spinor Calculus and Relativistic Fields*, Cambridge University Press, 1984.

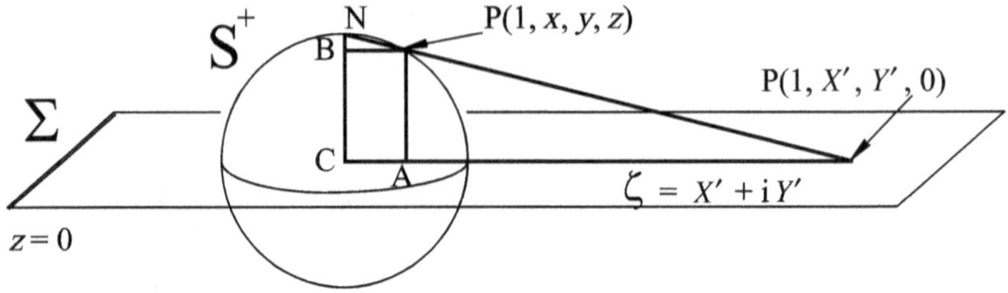

Fig. 3.2. Correspondence between the ouput space sphere and an Argand plane. After Penrose and Rindler (1984).[43]

The sampling described above is represented as a mapping of a point $P(1,\ x,\ y,\ z)$ in $\mathbf{S}^+$, and of SU(2) symmetry, to a point $P'(1,\ x', y',\ z')$ on Σ, and in $U(1)$ symmetry.

The point P' can then be labeled by a single complex parameter:

$$\zeta = X' + iY'.$$

Using the definition:

$$z = 1 - \frac{CA}{CP'} = 1 - \frac{NP}{NP'} = 1 - \frac{NB}{NC},$$

then:

$$\zeta = \frac{x + iy}{1 - z}.$$

A pair (ξ, η) of complex numbers can be defined:

$$\zeta = \frac{\xi}{\eta},$$

and Penrose and Rindler have shown[43], in another context, that what is identified here as the *pre-sampled* SU(2) polarization and rotation modulated wave can be represented in the units of:

$$W = \frac{1}{\sqrt{2}} (\xi\xi * + \eta\eta *),$$

$$X = \frac{1}{\sqrt{2}} (\xi\eta * + \eta\xi *),$$

[43] Penrose, R. and Rindler, W., *Spinors and Spacetime, Volume 1, Two-Spinor Calculus and Relativistic Fields*, Cambridge University Press, 1984.

$$Y = \frac{1}{i\sqrt{2}}\,(\xi\eta * -\eta\xi *),$$

$$Z = \frac{1}{\sqrt{2}}\,(\xi\xi * -\eta\iota *).$$

These definitions make explicit that a complex linear transformation of the $U(1)$ ξ and η results in a real linear transformation of the SU(2) (W, X, Y, Z).

Therefore, a complex linear transformation of ξ and η can be defined:

$$\xi \mapsto \xi' = \alpha\xi + \beta\eta, \tag{3.6a}$$

$$\eta \mapsto \eta' = \gamma\xi + \delta\eta,$$

$$\text{or}: \ \zeta \mapsto \zeta' = \frac{\alpha\zeta + \beta}{\gamma\zeta + \delta}, \tag{3.6b}$$

where α, β, γ, and δ are arbitrary nonsingular complex numbers.

Now the transformations, 3.6(A) and 3.6(B), are *spin transformations*, implying that

$$\zeta = \frac{X + iY}{T - Z} = \frac{W + Z}{X - iY},$$

and if a spin matrix, **A**, is defined:

$$A = \begin{pmatrix} \alpha & \beta \\ \gamma & \delta \end{pmatrix}, \ \det A = 1,$$

then the two transformations (3.6), are

$$A = \begin{pmatrix} \xi' & \xi \\ \eta' & \eta \end{pmatrix}, \tag{3.7}$$

which means that the spin matrix of a composition is given by the product of the spin matrix of the factors. Any transformation of the Fig. 3.2 form is linear and real and leaves the form $W^2 - X^2 - Y^2 - Z^2$ invariant.

Furthermore, there is a unimodular condition:

$$\alpha\delta - \beta\gamma = 1,$$

and the matrix **A** has the inverse:

$$A^{-1} = \begin{pmatrix} \delta & -\beta \\ -\gamma & \alpha \end{pmatrix},$$

which means that the spin matrix **A** and its inverse $\mathbf{A}^{-1}$ give rise to the same transformation of ζ even although they define different spin transformations. Due to the unimodular condition, the **A** spin-matrix is unitary or:

$$\mathbf{A}^{-1} = \mathbf{A}^*,$$

where $\mathbf{A}^*$ is the conjugate transpose of **A**.

The consequence of these relations is that *every proper* 2π *rotation on* $\mathbf{S}^+$ *— in the present instance the Poincaré sphere — corresponds to precisely two unitary spin rotations.* As every rotation on the Poincaré sphere corresponds to a polarization/rotation *modulation*, then *every proper* 2π *polarization-rotation modulation corresponds to precisely two unitary spin rotations.* The vector **K** in Fig. 3.1 corresponds to two vectorial components, one being the negative of the other. As every unitary spin transformation corresponds to a unique proper rotation of $\mathbf{S}^+$, then any *static* (unipolarized, e.g., linearly, circularly or elliptically polarized, as opposed to polarization modulated) representation on $\mathbf{S}^+$ (Poincaré sphere) corresponds to a tri-sphere representation (Fig. 3.3(A)). Therefore

$$A^{-1}A = \pm I,$$

where I is the identity matrix. *Thus, a spin transformation is defined uniquely up to sign by its effect on a static instantaneous snapshot representation on the* $\boldsymbol{S}^+$, *a Poincare sphere*:

$$\xi_1, \eta_1; \xi_2, \eta_2 \mapsto e^{i2\theta}\xi, e^{i2\theta}\eta;\ 0 < \theta < \pi.$$

Turning now to the case of polarization/rotation modulation, or *continuous rotation* of $\xi_1\eta_1; \xi_2\eta_2$: corresponding to a continuous rotation of $\xi_1\eta_1; \xi_2\eta_2$ through 2θ, there is a rotation of the resultant through θ. This correspondence is a consequence of the $A^{-1}A = \pm I$ relation, namely, that if the unitary transformation of **A** or $\mathbf{A}^{-1}$ is applied separately the identity matrix will *not* be obtained. However,

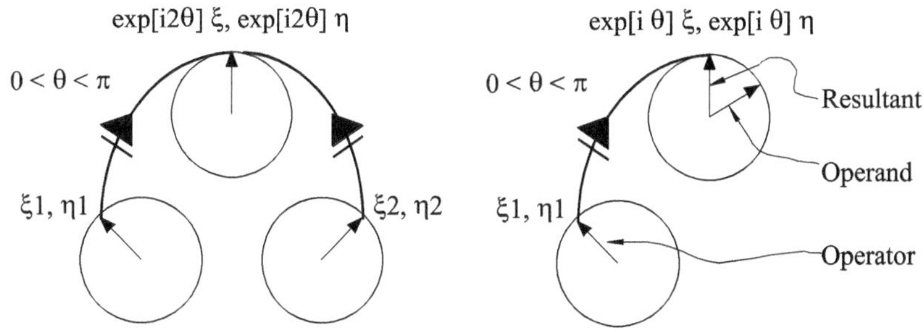

(a) (b)

Fig. 3.3. (a) Tri-sphere representation of *static polarization mapping*: $\xi_1\eta_1; \xi_2\eta_2 \mapsto e^{i2\theta}\xi; e^{i2\theta}\eta; 0 < \theta < \pi$. Note that a 360° excursion of $\xi_1\eta_1$ and $\xi_2\eta_2$ corresponds to a 360° excursion of $e^{i2\theta}\xi, e^{i2\theta}\eta$, i.e., this is a mapping for static polarization. (b) Bi-sphere representation of *polarization modulation mapping* (or $\xi_1\eta_1 \mapsto e^{i\theta}\xi, e^{i\theta}\eta\ 0 < \theta < \pi$) exhibiting the property of spinors that, corresponding to two unitary transformations of e.g., 2π, i.e., 4π, a null rotation of 2π is obtained. Notice that for a 360° rotation of the resultant (i.e., the final output wave), and with a stationary operand, the operator must be rotated through 720°.

if the unitary transformation is applied *twice*, then the identity matrix *is* obtained; and from this follows the remarkable properties of spinors that corresponding to two unitary transformations of, e.g., 2π, i.e., 4π, one null vector rotation of 2π is obtained. This is a bi-sphere correspondence and is shown in Fig. 3.3(B). This figure also represents the case of polarization/rotation modulation — as opposed to static polarization/rotation.

We now identify the vector, **K**, in Fig as a *null vector* defined:

$$\mathbf{K} = Ww + Xx + Yy + Zz,$$

the coordinates of which satisfy;

$$W^2 - X^2 - Y^2 - Z^2 = 0,$$

where W, X, Y and Z are functions of time: $W(t)$, $X(t)$, $Y(t)$ and $Z(t)$. The distinguishing feature of this null vector is that phase transformations $\xi \mapsto e^{i\theta}\xi, \eta \mapsto e^{i\theta}\eta$ leave **K** unchanged, i.e., **K** represents ξ and η only up to phase — which is the hallmark of a $U(1)$ representation.

K thus defines a static polarization/rotation — whether linear, circular or elliptical — on the Poincaré sphere. The ξ, η representation of the vector **K** *gives no indication of the future position of* **K**; i.e., the representation does not address the indicated hatched trajectory of the vector **K** around the Poincaré sphere. But it is precisely this trajectory which defines the particular polarization modulation for a specific wave. *Stated differently: a particular position of the vector* **K** *on the Poincaré sphere gives no indication of its next position at a later time*, because the vector can depart (be joined) in any direction from that position when only the static ξ, η coordinates are given.

In order to address *polarization/rotation modulation* — not *static* polarization or *static* rotation — an algebra is required which can reduce the ambiguity of a static representation. Such an algebra which is associated with ξ, η, and which reduces the ambiguity up to a sign ambiguity, is available in the twistor formalism.[44] In this formalism, polarization/rotation modulation can be accomodated, and a spinor, κ, can be represented not only by a null direction indicated by ξ, η or ζ, but also a real tangent vector **L** indicated in Fig. 3.4.

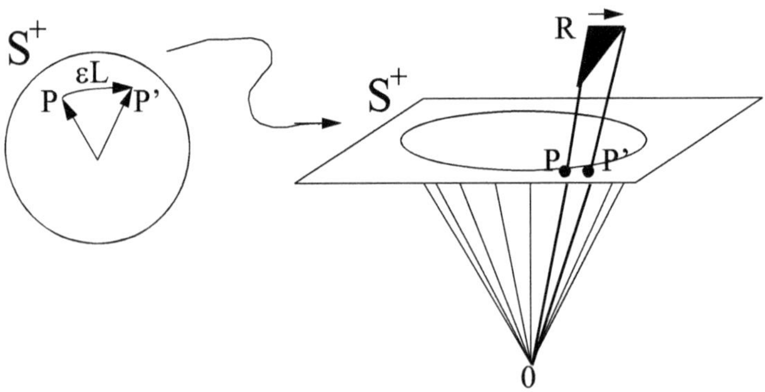

Fig. 3.4. Relation of a trajectory in a specific direction on an output sphere $\mathbf{S}^+$ and a null flag representation on the hyperplane, W, intersection with $\mathbf{S}^+$. After Penrose and Rindler (1984).[44]

[44]Penrose, R. and Rindler, W., *Spinors and Spacetime, Volume 1, Two-Spinor Calculus and Relativistic Fields*, Cambridge University Press, 1984.

Using this algebraic formalism, the Poincaré vector — *and its direction of change (up to sign ambiguity)* — can be represented. A real tangent vector **L** of $\mathbf{S}^+$ at P is defined:

$$L = \frac{\lambda\partial}{\partial\zeta} + \frac{\lambda * \partial}{\partial\zeta *},$$

where λ is some expression in ξ, η. With the choice $\lambda = -(\frac{1}{\sqrt{2}})\eta^{-2}$:

$$L = \left(\frac{1}{\sqrt{2}}\right)\left[\eta^{-2}\left(\frac{\partial}{\partial\zeta}\right) + \eta *^{-2}\left(\frac{\partial}{\partial\zeta *}\right)\right],$$

and thus knowing **L** at P (as an operator) means that the pair ξ, η is known completely up to sign, or, for any $f(\zeta, \zeta *)$:

$$\frac{1}{\varepsilon_{lim\ \varepsilon\to 0}}\left(fp' - fp\right) = Lf.$$

Succinctly: the tangent vector **L** in the abstract space $\mathbf{S}^+$ (Poincaré sphere) corresponds to a tangent vector **L** in the coordinate-dependent representation S^+ of $\mathbf{S}^+$. **L** is a unit vector if and only if, K, the null vector corresponding to ξ, η, defines a point actually on $\mathbf{S}^+$. Therefore, a plane of K and L can be defined by

$$aK + bL,$$

and if $b > 0$, then a half plane, Π, is defined bounded by K. K and L are both spacelike and orthogonal to each other. In the twistor formalism, Π and K are referred to as a *null flag* or a *flag*. The vector K is called the *flagpole*, its direction is the *flagpole direction* and the half-plane, Π, is the *flag plane*.

Our conclusions are that a polarization–rotation modulated wave can be represented as a periodic trajectory of polarization/rotation modulation on a Poincaré sphere, or a spinorial object. A defining characteristic of a spinorial object is that it is not returned to its original state when rotated through an angle 2π about some axis, but only when rotated through 4π. Referring to Fig. 3.5, we see that for the resultant to be rotated through 2π and returned to its original polarization state, the *operator* must be rotated through 4π. Thus a spinorial object (polarization/rotation modulated beams) exists in a different topological space from static polarized/rotated beams

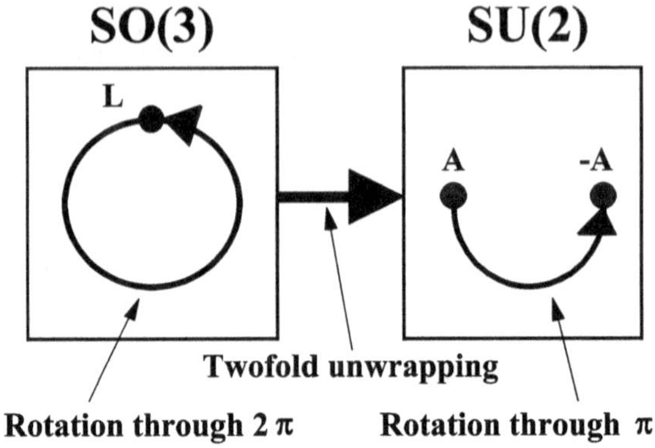

Fig. 3.5. The left side (SO(3)) describes the symmetry of the trajectory **K** on the Poincaré sphere; the left side describes the symmetry of the associated $Q_1(\psi, \chi)$ and $Q_2(\psi, \chi)$ which are functions of the ψ, χ angles on the Poincaré sphere. Adapted from Penrose and Rindler (1984).

due to the additional degree of freedom provided by the polarization bandwidth, which does not exist prior to modulation.

For example, let us consider constituent polarization vectors, $Q^i(\omega, \delta)$, and let **C** be the space orientations of $Q^i(\omega, \delta)$. A spinorized version of $Q^i(\omega, \delta)$ can be constructed provided the space is such that it possesses a twofold universal covering space $\mathbf{C}^*$, and provided the two different images, $Q_1(\psi, \chi)$ and $Q_2(\psi, \chi)$ existing in $\mathbf{C}^*$ of an element existing in **C** are interchanged after a continuous rotation through 2π is applied to a $Q^i(\omega, \delta)$. In the case we are considering, **C** has the topology of the SO(3) group. but $\mathbf{C}^*$ of the SU(2) group (which is the same as the space of unit quaternions). Thus, there is a $2 \to 1$ relation between the SU(2) object and the SO(3) object (Fig. 3.5).

We may take the $Q^i(\omega, \delta)$ to be polarization vectors (null flags) and **C** to be the space of null flags. The spinorized null flags, $Q_1(\psi, \chi)$ and $Q_2(\psi, \chi)$, are elements of $\mathbf{C}^*$, i.e., they are spin vectors. Referring to Figs. 3.3(B) and 3.5, we see that each null flag, $Q^i(\omega, \delta)$, defines two associated spin vectors, κ and $-\kappa$. A continuous rotation through 2π will carry κ into $-\kappa$ by acting on (ξ, η). On repeating the process, $-\kappa$ is carried back into κ:

$$-(-\kappa) = \kappa.$$

Furthermore, any spin vector, κ_1, can be represented as a linear combination of *two spin vectors* κ_2 and κ_3:

$$\{\kappa_2, \kappa_3\}\kappa_1 + \{\kappa_3, \kappa_1\}\kappa_2 + \{\kappa_1, \kappa_2\}\kappa_3 = 0,$$

where $\{\}$ indicates the antisymmetrical inner product. *Thus any arbitrary polarization can be represented as a linear combination of spin vectors.*

A generalized representation of spin vectors (and thus of polarization/rotation modulation) in terms of components is obtained using a normalized pair, a, b, as a spin frame:

$$\{a, b\} = -\{b.a\} = 1.$$

Therefore,

$$\kappa = \kappa^0 a + \kappa^1 b,$$

with

$$\kappa^0 = \{\kappa, b\},\ \kappa^1 = -\{\kappa, a\}.$$

The flagpole of a is $\frac{(t+z)}{\sqrt{2}}$ and of b is $\frac{(t-z)}{\sqrt{2}}$ and can be represented over time in Minkowski tetrad (t, x, y, z) form (t_1 representation) and for multiple time-frames or sampling intervals providing overall $(t_1, \ldots, t_n)$ a Cartan–Weyl form representation (Fig. 3.6) by using sampling intervals which "reset the clock" after every sampling of instantaneous polarization. Thus polarization modulation is represented by the continuous changes in a, b over time or the collection of samplings of a, b over time as depicted in Fig. 3.6.

The relation to the electromagnetic field is as follows. The (antisymmetrical) inner product of two spin vectors can represented as:

$$\{\kappa_1, \kappa_2\} = \varepsilon_{AB}\kappa^A\kappa^B = -\{\kappa_2, \kappa_1\},$$

where the ε (or the fundamental numerical metric spinors of second rank) are antisymmetrical:

$$\varepsilon_{AB}\varepsilon^{CB} = -\varepsilon_{AB}\varepsilon^{BC} = \varepsilon_{AB}\varepsilon^{BC} = -\varepsilon_{BA}\varepsilon^{CB} = \varepsilon_A^C = -\varepsilon_A^C,$$

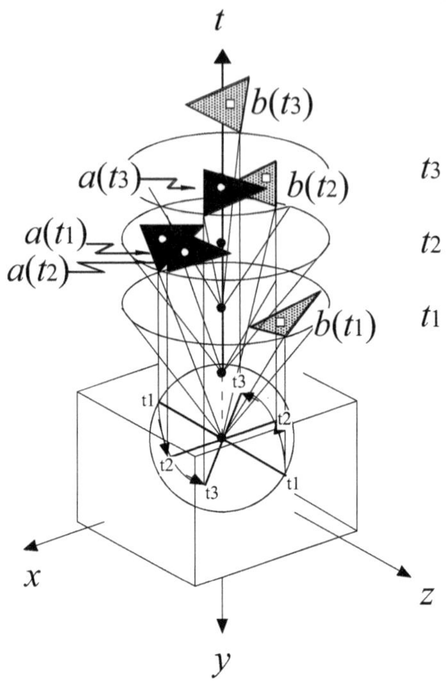

Fig. 3.6. Spin frame representation of a spin vector by flag pole normalized pair representation $\{a, b\}$ over the Poincaré sphere in Minkowski tetrad (t, x, y, z) form (t_1 representation) and for three time-frames or sampling intervals providing overall $(t_1, \ldots, t_n)$ a Cartan–Weyl form representation. The sampling intervals "reset the clock" after every sampling of instantaneous polarization. Thus polarization modulation is represented by the collection of samplings over time. Minkowski form after Penrose and Rindler (1984). This is an SU(2) $Q_i(\psi, \chi)$ in $\mathbf{C}^*$ over π representation, *not* an SO(3) $Q^i(\omega, \delta)$ in $\mathbf{C}$ representation over 2π. This can be seen by noting that $a \mapsto b$ or $b \mapsto a$ over π, not 2 π, while the polarization modulation in SO(3) repeats at a period of 2π. Adapted from Penrose and Rindler.[44]

with a canonical mapping (or isomorphism) between, e.g., $\kappa^B$and κ_B:

$$\kappa^B \mapsto \kappa_B = \kappa^A \varepsilon_{AB}.$$

A potential can be defined as

$$\Phi_A = i\,(\varepsilon\alpha)^{-1}\,\nabla_A \alpha,$$

where α is a gauge:

$$\alpha\alpha{*} = 1,$$

and ∇_A is a covariant derivative, $\partial/\partial x^A$, but without the commutation property. The covariant electromagnetic field is then:

$$F_{AB} = \nabla_A \Phi_B - \nabla_B \Phi_A + ig\,[\Phi_B, \Phi_A],$$

where g is generalized charge.

A physical representation of the polarization modulated (SU(2)) beam can be obtained using a Lissajous pattern[45] representation (Figs. 3.7–3.9) and controlling parameters shown in Table 3.3.

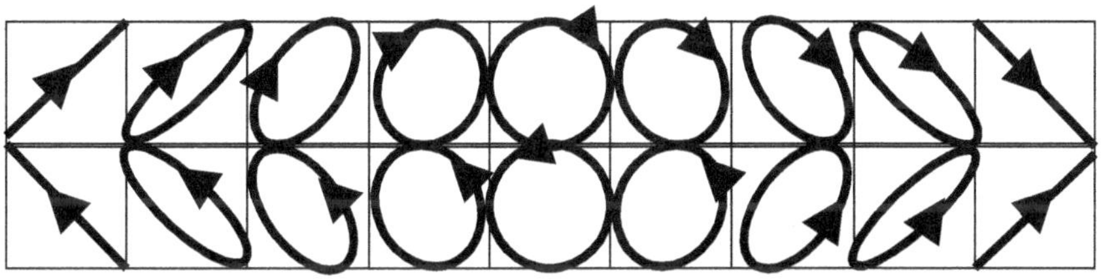

Fig. 3.7. Lissajous patterns representing a polarization modulated electric field over time, viewed in the plane of incidence, resulting from the two orthogonal s and p fields which are out of phase by the following degrees: 0, 21, 42, 64, 85, 106, 127, 148, 169 (top row); 191, 212, 233, 254, 275, 296, 318, 339, 360 (bottom row). In these Lissajous patterns, the plane polarizations are represented at 45° to the axes. In this example, there is a simple constant rate polarization with no rotation modulation. This is an SO(3) $Q^i(\omega, \delta)$ in $\mathbf{C}$ representation over 2π, *not* an SU(2) $Q_i(\psi, \chi)$ in $\mathbf{C}^*$ over π.

[45]Lissajous patterns are the locus of the resultant displacement of a point which is a function of two (or more) simple periodic motions. In the usual situation, the two periodic motions are orthogonal (i.e., at right angles) and are of the same frequency. The Lissajous Figures then represent the polarization of the resultant wave as a diagonal line, top left to bottom right in the case of linear perpendicular polarization; bottom left to top right, in the case of linear horizontal polarization; a series of ellipses, or a circle, in the case of circular corotating or contrarotating polarization, all of these corresponding to the possible differences in constant phase between the two simple periodic motions. If the phase is not constant, but is changing or modulated, as in the case of polarization modulation, then the pattern representing the phase is constantly changing over the time the Lissajous Figure is generated. Named after Jules Lissajous (1822–1880).

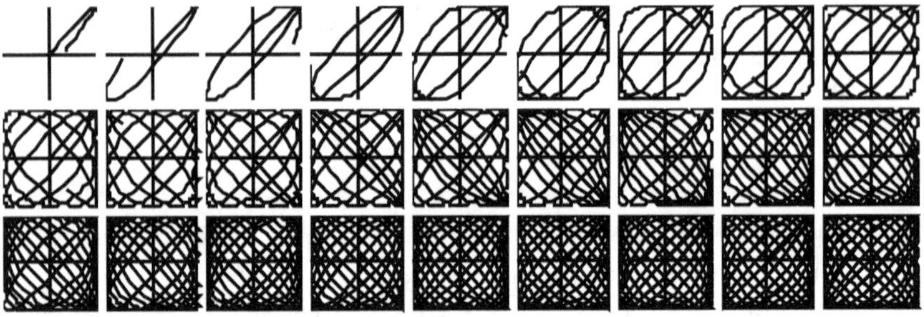

Fig. 3.8. Lissajous patterns representing the polarized electric field over time, viewed in the plane of incidence resulting from the two orthogonal s and p fields. The p field is phase modulated at a rate $d\phi/dt = 0.2\,t$. In these Lissajous patterns, the plane polarizations are represented at 45° to the axes. This is an SO(3) $Q^i(\omega, \delta)$ in **C** representation over 2π, *not* an SU(2) $Q_i(\psi, \chi)$ in $\mathbf{C}^*$ over π.

Fig. 3.9. A representation of a polarization modulated beam over 2p in the z-direction. These are SO(3) $Q^i(\omega, \delta)$ in **C** representations over 2π, *not* an SU(2) $Q_i(\psi, \chi)$ in $\mathbf{C}^*$ over π.

The controlling variables for polarization and rotation modulation are given in Table 3.3. We can note that the Stokes' parameters $(s_0,\ s_1,\ s_2,\ s_3)$ defined over the SU(2)-dimensional variables, ψ, χ, of $Q_i(\psi, \chi)$ are sufficient to describe *polarization/rotation modulation*, and relate those variables to the SO(3)-dimensional variables,

Table 3.3. Controlling variables.[46]

Field input variables (coordinate axes)	$E_x = a_1 \cos(\tau + \delta_1)$ $E_y = a_2 \cos(\tau + \delta_2)\ \tau = \omega t - \kappa z$
Field input variables (ellipse axes)	$E_\xi = a\cos(\tau + \delta) = E_x \cos\psi + E_y \sin\psi$ $E_\eta = \pm b\cos(\tau + \delta) = -E_x \sin\psi + E_y \cos\psi\tau$ $= \omega t - \kappa z$
Phase variables	$\delta = \delta_2 - \delta_1; \left(\frac{E_x}{a_1}\right)^2 + \left(\frac{E_y}{a_2}\right)^2 - 2\frac{\cos\delta}{a_1 a_2} = \sin^2\delta$
Auxiliary angle, α	$\frac{a_2}{a_1} = \tan(\alpha)$
Control variables	$a_1,\ a_2,\ \delta_1,\ \delta_2$
Resultant transmitted variables and relation of coordinate axes, a_1, a_2, to ellipse axes, a, b	$a^2 + b^2 = a_1^2 + a_2^2$
Rotation	$\tan(2\psi) = (\tan(2\alpha))\cos(\delta) = \frac{2a_1a_2}{a_1^2 - a_2^2}\cos\delta$
Ellipticity	$\sin(2\chi) = (\sin(2\alpha))\sin(\delta); \tan(\chi) = \pm\frac{b}{a}$
Rotation	ψ — resultant determined by a_1 and a_2 with δ constant
Ellipticity	χ — resultant determined by δ with a_1 and a_2 constant
Determinant of rotation ψ	a_1, a_2 with δ constant
Determinant of ellipticity χ	δ with a_1 and a_2 constant
Stokes parameters	$s_0 = a_1^2 + a_2^2$ $s_1 = a_1^2 - a_2^2 = s_0\cos(2\chi)\cos(2\psi)$ $s_2 = 2a_1a_2\cos(\delta) = s_0\cos(2\chi)\sin(2\psi) = s_1\tan(2\psi)$ $s_3 = 2a_1a_2\sin(\delta) = s_0\sin(2\chi)$
Linear polarization condition	$\delta = \delta_2 - \delta_1 - m\pi,\ m - 0, \pm1, \pm2, \ldots,$ $\frac{E_y}{E_x} = (-1)^m\frac{a_2}{a_1}$
Circular polarization condition	$a_1 = a_2 = a; \delta = \delta_2 - \delta_1 = \frac{m\pi}{2}\ m = \pm1, \pm3, \pm5, \ldots$ $E_x^2 + E_y^2 = a^2$
Right-hand polarization condition	$\sin\delta > 0$ $\delta = \frac{\pi}{2} + 2m\pi,\ m = 0, \pm1, \pm2, \ldots$ $E_x = a\cos(\tau + \delta_1)$ $E_y = a\cos\left(\tau + \delta_1 + \frac{\pi}{2}\right) = -a\sin(\tau + \delta_1)$ $\sin\delta < 0$ $\delta = \frac{\pi}{2} + 2m\pi,\ m = 0, \pm1, \pm2, \ldots$ $E_x = a\cos(\tau + \delta_1)$ $E_y = a\cos\left(\tau + \delta_1 + \frac{\pi}{2}\right) = -a\sin(\tau + \delta_1)$
Left-hand polarization condition	$\sin\delta < 0$ $E_y = a\cos\left(\tau + \delta_1 + \frac{\pi}{2}\right) = a\sin(\tau + \delta_1)$.

[46] After Born, M. and Wolf, E., *Principles of Optics*, 7th edition, Cambridge University Press, 1999.

$\omega(\tau, z), \delta$, of $Q^i(\omega, \delta)$, which are sufficient to describe the *static polarization/rotation conditions* of linear, circular, left and right-handed polarization/rotation.

We can also note the fundamental role that concepts of topology played in distinguishing static polarization–rotation from polarization–rotation *modulation.*

3.5. Aharonov–Bohm Effect

We consider again the Aharonov–Bohm effect as an example of a phenomenon understandable only from topological considerations. Summarizing the description in Chapter 1: beginning in 1959 Aharonov and Bohm[47] challenged the view that the classical vector potential produces no observable physical effects by proposing two experiments. The one which is most discussed is shown in Fig. 3.10, which is Fig. 1.1 of Chapter 1, provided here for convenience. A beam of mono-energetic electrons exists from a source at X and is diffracted into two beams by the slits in a wall at $Y1$ and $Y2$. The two beams produce an interference pattern at *III* which is measured. Behind the wall is a solenoid, the **B** field of which points out of the paper. The absence of a free local magnetic monopole postulate in conventional $U(1)$ electromagnetism ($\nabla \cdot \mathbf{B} = 0$) predicts that the magnetic field outside the solenoid is zero. Before the current is turned on in the solenoid, there should be the usual interference patterns observed at *III*, of course, due to the differences in the two path lengths.

Aharonov and Bohm made the important prediction that if the current is turned on, then due to the differently directed **A** fields along paths 1 and 2 indicated by the arrows in Fig. 3.10, additional phase shifts should be discernible at *III*. This prediction

[47] Aharonov, Y. and Bohm, D., Significance of the electromagnetic potentials in quantum theory. *Phys. Rev.*, 115, 485–491, 1959.

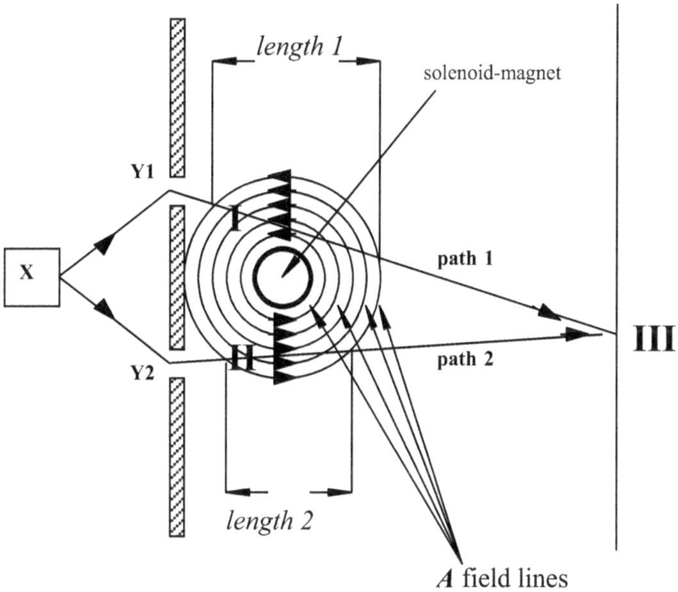

Fig. 3.10. Two-slit diffraction experiment of the Aharonov–Bohm effect. Electrons are produced by a source at X, pass through the slits of a mask at $Y1$ and $Y2$, interact with the **A** field at locations I and II over lengths l_1 and l_2, respectively, and their diffraction pattern is detected at III. The solenoid-magnet is between the slits and is directed out of the page. The different orientations of the external **A** field at the places of interaction I and II of the two paths 1 and 2 are indicated by arrows following the right-hand rule.

was confirmed experimentally[48] and the evidence for the effect has been extensively reviewed.[49]

[48] Chambers, R.G., Shift of an electron interference pattern by enclosed magnetic flux, *Phys. Rev. Lett.*, 5, 3–5, 1960.

Boersch, H., Hamisch, H., Wohlleben, D. and Grohmann, K., Antiparallele Weissche Bereiche als Biprisma für Elektroneninterferenzen, *Zeits. Phys.* 159, 397–404, 1960.

Möllenstedt, G and Bayh, W., Messung der kontinuierlichen Phasenschiebung von Elektronenwellen im kraftfeldfreien Raum durch das magnetische Vektorpotential einer Luftspule, *Die Naturwissenschaften* 49, 81–82, 1962.

Matteucci, G. and Pozzi, G., New diffraction experiment on the electrostatic Aharonov–Bohm effect. *Phys. Rev. Lett.*, 54, 2469–2472, 1985.

Tonomura, A., *et al*, Observation of Aharonov–Bohm effect by electron microscopy. *Phys. Rev. Lett.*, 48, 1443–1446, 1982.

It is our view that the topology of this situation is fundamental and dictates its explanation. Therefore, we must clearly note the topology of the physical layout of the design of the situation which exhibits the effect. The physical situation is that of an *interferometer.* That is, there are two paths around a central location — occupied by the solenoid — and a measurement is taken at a location, *III*, in Fig. 3.10, where there is overlap of the wave functions of the test waves which have traversed, separately, the two different paths. (The test waves or test particles are complex wave functions with phase.) It is important to note that the overlap area, at *III*, is the only place where a measurement can take place of the effects of the **A** field which occurred earlier and at other locations (*I* and *II*). The effects of the **A** field occur along the two different paths and at locations *I* and *II*, but they are *inferred*, and not measurable there. Of crucial importance in this special interferometer is the fact that the solenoid presents a *topological obstruction.* That is, if one were to consider the two joined paths of the interferometer as a raceway or a loop and one squeezed the loop tighter and tighter, then nevertheless one cannot in this situation — unlike as in most situations — reduce the interferometer's raceway of paths to a single point. (Another way of saying this is: not all closed curves in a region need have a vanishing line integral, because one exception is a loop with an obstruction.)

———, Is magnetic flux quantized in a toroidal ferromagnet? *Phys. Rev. Lett.*, 51, 331–334, 1983.

———, Evidence for Ahronov–Bohm effect with magnetic field completely shielded from electron wave, *Phys. Rev. Lett.*, 56, 792–795, 1986.

——— and Callen, E., Phase, electron holography and conclusive demonstration of the Aharonov–Bohm effect *ONRFE Sci. Bull.*, 12(3), 93–108, 1987.

[49]Berry, M.V., Exact Aharonov–Bohm wavefunction obtained by applying Dirac's magnetic phase factor, *Eur. J. Phys.*, 1, 240–244, 1980.

Peshkin, M., The Ahronov–Bohm effect: why it cannot be eliminated from quantum mechanics. *Phys. Rep.*, 80, 375–386, 1981.

Olariu, S. and Popescu, I.I., The quantum effects of electromagnetic fluxes, *Rev. Mod. Phys.*, 157, 349–436, 1985.

Horvathy, P.A., The Wu–Yang factor and the non-Abelian Aharonov–Bohm experiment, *Phys. Rev.*, D33, 407–414, 1986.

Peshkin, M and Tonomura, A., *The Ahaonov–Bohm Effect*, Springer–Verlag, New York, 1989.

The reason one cannot reduce the interferometer to a single point is because of the existence in its middle of the solenoid, which is a positive quantity, and acts as an obstruction.

Again, it is our view that the existence of the obstruction changes the situation entirely. *Without* the existence of the solenoid in the interferometer, the loop of the two paths *can be* reduced to a single point and the region occupied by the interferometer is then *simply-connected.* But *with* the existence of the solenoid, the loop of the two paths *cannot be* reduced to a single point and the region occupied by this special interferometer is *multiply-connected.* The Aharonov–Bohm effect only exists in the multiply-connected scenario. But we should note that the Aharonov–Bohm effect is a *physical* effect and simple and multiple connectedness are *mathematical descriptions* of physical situations.

The topology of the physical interferometric situation addressed by Aharonov and Bohm defines the physics of that situation and also the mathematical description of that physics. If that situation *were not* multiply-connected, but simply-connected, then there would be no interesting physical effects to describe. The situation would be described by $U(1)$ electromagnetics and the mapping from one region to another is conventionally one-to-one. However, as the Aharonov–Bohm situation is multiply-connected, there is a two-to-one mapping $(\mathrm{SU}(2)/\mathrm{Z}_2)$ of the two different regions of the two paths to the single region at III where a measurement is made. Essentially, at *III* a measurement is made of *the differential histories* of the *two* test waves which traversed the *two* different paths and experienced *two* different forces resulting in two different phase effects.

In conventional, i.e., normal $U(1)$ or simply-connected situations, the fact that a vector field, viewed axially, is pointing in one direction, if penetrated from one direction on one side, and is pointing in *the opposite direction*, if penetrated from the same direction, but *on the other side*, is of no consequence at all — because that field is of $U(1)$ symmetry and can be reduced to a single point. Therefore in most cases which are of $U(1)$ symmetry, we do not need to distinguish between the direction of the vectors of a field from one region to another of that field. However, the Aharonov–Bohm situation is not conventional or simply-connected, but special. (In other words, the physical situation associated with the Aharonov–Bohm effect has a nontrivial topology.) It is a multiply-connected situation and of $SU(2)/Z_2$ symmetry. Therefore, the direction of the **A** field on the separate paths is of crucial importance, because a test wave traveling along one path will experience an **A** vectorial component directed *against* its trajectory and thus be retarded, and another test wave traveling along another path will experience an **A** vectorial component directed *with* its trajectory and thus its speed is boosted. These "retardations" and "boostings" can be measured as phase changes, *but not at the time nor at the locations of, I and II, where their occurrence is separated along the two different paths*, but *later*, and at the *overlap location* of III. It is important to note that if measurements are attempted at locations *I* and *II* in Fig. 3.10, these effects will not be seen because there is no two-to-one mapping at either *I* and *II* and therefore no referents. The locations *I* and *II* are both simply-connected with the source and therefore only the conventional $U(1)$ electromagnetics applies at these locations (with respect to the source). It is only region *III* which is multiply-connected with the source and at which the histories of what happened to the test particles at *I* and *II* can be measured. In order to distinguish the "boosted" **A** field (because the test wave is traveling "with" its direction) from the "retarded" **A** field (because the test wave is traveling "against" its direction), we introduce the notation: $\mathbf{A}_+$ and $\mathbf{A}_-$.

Because of the distinction between the **A** oriented potential fields at positions *I* and *II* — which *are not* measurable and are *vectors or numbers* of $U(1)$ symmetry — and the A potential fields at *III* — which *are* measurable and are *tensors or matrix-valued functions* of (in the present instance) $SU(2)/Z_2 = SO(3)$ symmetry (or higher symmetry) — for reasons of clarity we might introduce a distinguishing notation. In the case of the potentials of $U(1)$ symmetry at *I* and *II* we might use the lower case, $\mathbf{a}_\mu$, $\mu = 0, 1, 2, 3$ and for the potentials of $SU(2)/Z_2 = SO(3)$ at *III* we might use the upper case $\mathbf{A}_\mu$, $\mu = 0, 1, 2, 3$. Similarly, for the electromagnetic field tensor at *I* and *II*, we might use the lower case, $\mathbf{f}_{\mu\nu}$, and for the electromagnetic field tensor at *III*, we might use the upper case, $\mathbf{F}_{\mu\nu}$. Then the following definitions for the electromagnetic field tensor are as follows:

At locations *I* and *II* the Abelian relationship is:

$$\boldsymbol{f}_{\mu\nu}(x) = \partial_\nu a_\mu(x) - \partial_\mu a_\nu(x), \tag{3.8}$$

where, as is well known, $\mathbf{f}_{\mu\nu}$ is Abelian and gauge *invariant*; But at location III the non-Abelian relationship is:

$$F_{\mu\nu} = \partial_\nu \mathbf{A}_\mu(x) - \partial_\mu \mathbf{A}_\nu(x) - ig_m\,[\mathbf{A}_\mu(x), \mathbf{A}_\nu(x)], \tag{3.9}$$

where $\mathbf{F}_{\mu\nu}$ is gauge *covariant*, g_m is the magnetic charge density and the brackets are commutation brackets. We remark that in the case of non-Abelian groups, such as SU(2), the potential field *can carry charge*. It is important to note that if the physical situation changes from SU(2) symmetry back to $U(1)$, then $\mathbf{F}_{\mu\nu} \to \mathbf{f}_{\mu\nu}$.

Despite the clarification offered by this notation, the notation can also cause confusion, because in the present literature, the electromagnetic field tensor is *always* referred to as $\mathbf{F}$, whether $\mathbf{F}$ is defined with respect to $U(1)$ or SU(2) or other symmetry situations. Therefore, although we prefer this notation, we shall not proceed with it. However, it is important to note that the $\mathbf{A}$ field in the $U(1)$ situation is a *vector or a number*, but in the SU(2) or non-Abelian situation, it is a *tensor or a matrix-valued function.*

We referred to the physical situation of the Aharonov–Bohm effect as an interferometer around an obstruction and it is two-dimensional. It is important to note that the situation is not provided by a toroid, although a toroid is also a physical situation with an obstruction and the fields existing on a toroid are also of SU(2) symmetry. However, the toroid provides a two-to-one mapping of fields in not only the x and y dimensions but also in the z dimension, and *without* the need of an electromagnetic field pointing in two directions + and −. The physical situation of the Aharonov–Bohm effect is defined only in the x and y dimensions (there is no z dimension) and in order to be of $\mathrm{SU}(2)/Z_2$ symmetry *requires* a field to be oriented differentially on the separate paths. If the differential field is removed from the Aharonov–Bohm situation, then that situation reverts to a simple interferometric raceway which can be reduced to a single point and with no interesting physics.

How does the topology of the situation affect the explanation of an effect? A typical previous explanation[50] of the Aharonov–Bohm effect commences with the Lorentz force law:

$$F = e\mathbf{E} + ev \times \mathbf{B}. \tag{3.10}$$

The electric field, **E**, and the magnetic flux density, **B**, are essentially confined to the inside of the solenoid and therefore cannot interact with the test electrons. An argument is developed by defining the **E** and **B** fields in terms of the **A** and φ potentials:

$$\mathbf{E} = -\frac{\partial \mathbf{A}}{\partial t} - \nabla\varphi, \ \mathbf{B} = \nabla \times \mathbf{A}. \tag{3.11}$$

Now we can note that these conventional $U(1)$ definitions of **E** and **B** can be expanded to SU(2) forms:

$$\mathbf{E} = -(\nabla \times \mathbf{A}) - \frac{\partial \mathbf{A}}{\partial t} - \nabla\varphi, \ \mathbf{B} = (\nabla \times \mathbf{A}) - \frac{\partial \mathbf{A}}{\partial t} - \nabla\varphi. \tag{3.12}$$

Furthermore, the $U(1)$ Lorentz force law, (3.10) can hardly apply in this situation because the solenoid is electrically neutral to the test electrons and therefore $\mathbf{E} = 0$ along the two paths. Using the definition of **B** in Eq. (3.12), the force law in this SU(2) situation is

$$F = e\mathbf{E} + ev \times \mathbf{B} = e\left(-(\nabla \times \mathbf{A}) - \frac{\partial \mathbf{A}}{\partial t} - \nabla\varphi\right) + ev$$
$$\times \left((\nabla \times \mathbf{A}) - \frac{\partial \mathbf{A}}{\partial t} - \nabla\varphi\right), \tag{3.13}$$

but we should note that Eqs. (3.10) and (3.11) are *still valid* for the conventional theory of electromagnetism based on the $U(1)$ symmetry Maxwell's equations provided in Table 3.1 and associated with the group $U(1)$ algebra. They are *invalid* for the theory based on the modified SU(2) symmetry equations, also provided in Table 3.2 and associated with the group SU(2) algebra.

[50]Ryder, L.H., *Quantum Field Theory*, 2nd edition, Cambridge University Press, 1996.

The typical explanation of the Aharonov–Bohm effect continues with the observation that a phase difference, δ, between the two test electrons is caused by the presence of the solenoid:

$$\Delta\delta = \Delta\alpha_1 - \Delta\alpha_2 = \frac{e}{\hbar}\left(\int_{l_2} \mathbf{A}\cdot dl_2 - \int_{l_1} \mathbf{A}\cdot dl_1\right)$$
$$= \frac{e}{\hbar}\int_{l_2-l_1} \nabla\times\mathbf{A}\cdot dS = \frac{e}{\hbar}\int \mathbf{B}\cdot dS = \frac{e}{\hbar}\phi_M, \qquad (3.14)$$

where $\Delta\alpha_1$ and $\Delta\alpha_2$ are the changes in the wave function for the electrons over paths 1 and 2, $\mathbf{S}$ is the surface area and φ_M is the *magnetic flux* defined:

$$\phi_M = \mathbf{A}_\mu(x)dx^\mu = F_{\mu\nu}d\sigma^{\mu\nu}. \qquad (3.15)$$

Now, we can extend this explanation further, by observing that the local phase change at *III* of the wave function of a test wave or particle is given by

$$\mathbf{\Phi} = \exp\left[ig_m\mathbf{A}_\mu(x)dx^\mu\right] = \exp\left[ig_m\phi_M\right]. \qquad (3.16)$$

$\mathbf{\Phi}$, which is proportional to the magnetic flux, ϕ_M, is known as the *phase factor* and is *gauge covariant.* Furthermore, $\mathbf{\Phi}$, this phase factor measured at position *III* is the *holonomy* of the *connection*, $\mathbf{A}_\mu$; and g_m is the SU(2) *magnetic charge density.*

We next observe that φ_M is in units of volt-seconds (V.s) or kg.m^2/(A.s^2) = J/A. From Eq. (3.14) it can be seen that $\Delta\delta$ and the phase factor, $\mathbf{\Phi}$, are dimensionless. Therefore, we can make the prediction that if the magnetic flux, φ_M, is known and the phase factor, $\mathbf{\Phi}$, is measured (as in the Aharonov–Bohm situation), *the magnetic charge density, g_m, can be found by the relation:*

$$g_m = \ln(\mathbf{\Phi})/(i\phi_M). \qquad (3.17)$$

Continuing the explanation: as was noted above, $\nabla\times\mathbf{A} = 0$ outside the solenoid and the situation must be redefined in the following way. An electron on path 1 will interact with the $\mathbf{A}$ field oriented in the positive direction. Conversely, an electron on path 2 will interact with the $\mathbf{A}$ field oriented in the negative direction. Furthermore, the $\mathbf{B}$ field can be defined with respect to a local stationary component $\mathbf{B}_1$ which is confined to the solenoid and a component $\mathbf{B}_2$ which is either a standing wave or propagates:

$$\begin{aligned} \mathbf{B} &= \mathbf{B}_1 + \mathbf{B}_2, \\ \mathbf{B}_1 &= \nabla \times \mathbf{A}, \\ \mathbf{B}_2 &= -\tfrac{\partial \mathbf{A}}{\partial t} - \nabla\varphi. \end{aligned} \tag{3.18}$$

The magnetic flux density, $\mathbf{B}_1$, is the confined component associated with $U(1)\times\mathrm{SU}(2)$ symmetry and $\mathbf{B}_2$ is the propagating or standing wave component associated *only* with SU(2) symmetry. In a $U(1)$ symmetry situation, $\mathbf{B}_1$ = components of the field associated with $U(1)$ symmetry, and $\mathbf{B}_2 = 0$.

The electrons traveling on paths 1 and 2 require different times to reach *III* from X, due to the different distances and the opposing directions of the potential $\mathbf{A}$ along the paths l_1 and l_2. Here we only address the effect of the opposing directions of the potential $\mathbf{A}$, i.e., the distances traveled are identical over the two paths. The change in the phase difference due to the presence of the $\mathbf{A}$ potential is then:

$$\begin{aligned} \Delta\delta = \Delta\alpha_1 - \Delta\alpha_2 = \frac{e}{\hbar}\left[\int_{l_2}\left(-\frac{\partial \mathbf{A}_+}{\partial t} - \nabla\varphi_+\right)\cdot dl_2\right. \\ \left. - \int_{l_1}\left(-\frac{\partial \mathbf{A}_-}{\partial t} - \nabla\varphi_-\right) dl_1\right]\cdot dS = \frac{e}{\hbar}\int \mathbf{B}_2 \cdot dS = \frac{e}{\hbar}\phi_M. \end{aligned} \tag{3.19}$$

There is no flux density $\mathbf{B}_1$ in this equation since this equation describes events outside the solenoid, but only the flux density $\mathbf{B}_2$ associated with group SU(2) symmetry; and the "+" and "−" indicate the direction of the $\mathbf{A}$ field encountered by the test electrons — as discussed above.

We note that the phase effect is dependent on $\mathbf{B}_2$ and $\mathbf{B}_1$, but not on $\mathbf{B}_1$ alone. Previous treatments found no convincing argument around the fact that whereas the Aharonov–Bohm effect depends on an interaction with the $\mathbf{A}$ field outside the solenoid, $\mathbf{B}$, defined in $U(1)$ electromagnetism as $\mathbf{B} = \nabla \times \mathbf{A}$, is zero at that point of interaction. However, when $\mathbf{A}$ is defined in terms associated with an SU(2) situation, that is not the case as we have seen.

We depart from former treatments in other ways. Commencing with a *correct* observation that the Aharonov–Bohm effect depends on the topology of the experimental situation and that the situation is not simply-connected, a former treatment then erroneously seeks an explanation of the effect in the connectedness of the $U(1)$ gauge symmetry of conventional electromagnetism, but for which (1) the potentials are ambiguously defined, (the $U(1)$ $\mathbf{A}$ field is gauge invariant) and (2) in $U(1)$ symmetry $\nabla \times \mathbf{A} = 0$ outside the solenoid.

Furthermore, whereas a former treatment again makes a *correct* observation that the non-Abelian group, SU(2), is simply-connected and that the situation is governed by a multiply-connected topology, there is the failure to observe that the non-Abelian group SU(2) defined over the integers modulo 2, $\mathrm{SU}(2)/\mathrm{Z}_2$, is, in fact, multiply-connected. Because of the two paths around the solenoid it is this group which describes the topology underlying the Aharonov–Bohm effect.[51] $\mathrm{SU}(2)/Z_2 \approx \mathrm{SO}(3)$ is obtained from the group SU(2) by identifying pairs of elements with opposite signs. The $\Delta\delta$ measured at location *III* in Fig. 3.10 is derived from a *single* path in SO(3)[52] because the *two* paths through locations *I* and *II* in SU(2) are

[51] Barrett, T.W., Electromagnetic phenomena not explained by Maxwell's equations, in Lakhtakia, A. (ed.) *Essays on the Formal Aspects of Maxwell's Theory*, World Scientific, Singapore, pp. 6–86, 1993.

——, Sagnac effect, in Barrett, T.W. and Grimes, D.M., (eds) *Advanced Electromagnetism: Foundations, Theory, Applications*, World Scientific, Singapore, pp. 278–313, 1995.

——, The toroid antenna as a conditioner of electromagnetic fields into (low energy) gauge fields, *Speculations Sci. Technol.*, 21(4), 291–320, 1998.

[52] ***O*(*n*) Group Algebra:** The orthogonal group, $O(n)$, is the group of transformation (including inversion) in an n-dimensional Euclidean space. The elements of $O(n)$ are represented by $n \times n$ real orthogonal matrices with $n(n-1)/2$ real parameters satisfying $\mathrm{AA}^{\mathrm{t}} = 1$.

***O*(3) Group Algebra:** The orthogonal group, $O(3)$, is the well known and familiar group of transformations (including inversions) in three-dimensional space with three parameters, those parameters being the rotation or Euler angles (α, β, γ). $O(3)$ leaves the distance squared, $x^2 + y^2 + z^2$, invariant.

***SO*(3) Group Algebra:** The collection of matrices in Euclidean three-dimensional space which are orthogonal and moreover for which the determinant is $+1$ is a subgroup of $O(3)$. SO(3) is the special orthogonal group in three variables and defines rotations in three-dimensional space.

regarded as a *single* path in SO(3). This path in $SU(2)/Z_2 \approx SO(3)$ cannot be shrunk to a single point by any continuous deformation and therefore adequately describes the multiple-connectedness of the Aharonov–Bohm situation. Because a former treatment failed to note the multiple connectedness of the $SU(2)/Z_2$ description of the Aharonov–Bohm situation, it *incorrectly* fell back on a $U(1)$ symmetry description.

Now back to the main point of this excursion to the Aharonov–Bohm effect: the reader will note that the author appealed to topological arguments to support the main points of his argument.

Rotation of the Riemann sphere is a rotation in $\Re^3$ or $\xi-\eta-\zeta$ space, for which

$$\xi^2+\eta^2+\zeta^2=1,\quad \xi=\frac{2x}{|z|^2+1},\eta=\frac{2y}{|z|^2+1},\zeta=\frac{|z|^2-1}{|z|^2+1},z=x+iy=\frac{\xi+i\eta}{1-\zeta}.$$

$$U_\xi(\alpha)=\frac{1}{\sqrt{2}}\begin{pmatrix}1&-1\\1&1\end{pmatrix}\begin{pmatrix}e^{i\alpha/2}&0\\0&e^{-i\alpha/2}\end{pmatrix}\frac{1}{\sqrt{2}}\begin{pmatrix}1&1\\-1&1\end{pmatrix}$$

$$=\begin{pmatrix}\cos\alpha/2&i\sin\alpha/2\\i\sin\alpha/2&\cos\alpha/2\end{pmatrix}\text{ or }\pm U_\xi(\alpha)\to R_1(\alpha),$$

$$U_\eta(\beta)=\frac{1}{\sqrt{2}}\begin{pmatrix}1&-i\\-i&1\end{pmatrix}\begin{pmatrix}e^{i\beta/2}&0\\0&e^{-i\beta/2}\end{pmatrix}\frac{1}{\sqrt{2}}\begin{pmatrix}1&i\\i&1\end{pmatrix}$$

$$=\begin{pmatrix}\cos\beta/2&-\sin\beta/2\\\sin\beta/2&\cos\beta/2\end{pmatrix}\text{ or }\pm U_\eta(\beta)\to R_2(\beta),$$

$$U_\zeta(\gamma)=\frac{1}{\sqrt{2}}\begin{pmatrix}1&0\\0&1\end{pmatrix}\begin{pmatrix}e^{i\gamma/2}&0\\0&e^{-i\gamma/2}\end{pmatrix}\frac{1}{\sqrt{2}}\begin{pmatrix}1&0\\0&1\end{pmatrix}$$

$$=\begin{pmatrix}\cos\gamma/2&-\sin\gamma/2\\\sin\gamma/2&\cos\gamma/2\end{pmatrix}\text{ or }\pm U_\zeta(\gamma)\to R_3(\gamma).$$

which are mappings from SL(2,C) to SO(3). However, as the SL(2,C) are all unitary with determinant equal to +1, they are of the SU(2) group. Therefore SU(2) is the covering group of SO(3). Furthermore, SU(2) is simply connected and SO(3) is multiply connected.

A simplification of the above is

$$U_\xi(\alpha)=e^{i(\alpha/2)\sigma_1},\ U_\eta(\beta)=e^{-i(\beta/2)\sigma_2},\ U_\zeta(\gamma)=e^{i(\gamma/2)\sigma_3},$$
$$\textit{where}\ \sigma_1=\begin{pmatrix}0&1\\1&0\end{pmatrix},\ \sigma_2=\begin{pmatrix}0&-i\\i&0\end{pmatrix},\ \sigma_3=\begin{pmatrix}1&0\\0&-1\end{pmatrix}.$$

$\sigma_1,\sigma_2,\sigma_3$ are the Pauli matrices.

Underpinning the $U(1)$ Maxwell theory is an Abelian algebra; underpinning the SU(2) theory is a non-Abelian algebra. The algebras specify the form of the equations of motion. However, whether one or the other algebra can be (validly) used can only be determined by topological considerations.

3.6. Discussion

We shown the fundamental explanatory nature of the topological description of solitons, instantons and the Aharonov–Bohm effect – and hence electromagnetism. In the case of electromagnetism, we have shown in previous chapters that, given a Yang–Mills description, electromagnetism can, and should be extended, in accordance with the topology with which the electromagnetic fields are associated.

This approach has further implications. If the conventional theory of electromagnetism, i.e., "Maxwell's theory", which is of $U(1)$ symmetry form, is the simplest *local* theory of electromagnetism, then those pursuing a unified field theory may wish to consider as a candidate for that unification, not only the simple local theory, but other electromagnetic fields of group symmetries lower than $U(1)$. Other such forms include symplectic gauge fields of lower group symmetry, e.g., SU(2), etc.

Chapter 4

Orthogonal Signal Spectrum Overlay

4.1. Overview

The cross referring disciplines enabling a multidisciplinary description addressed in this chapter are information theory, communication theory and thermodynamic laws and relations. These relations are captured in a new efficient — in terms of bandwidth efficiency and power efficiency — communication system modulation technique: Orthogonal Signal Spectrum Overlay (OSSO). The variables involved are: (1) the thermodynamic concepts of energy/power required to define (1) power efficiency under a bandwidth constraint; (2) available entropy defining the possible signal receptions for a specific communication receiver and known in engineering as a "constellation"; and (3) negative entropy known as Shannon information in terms of binary signals or bits. The transmissions can be described in group theory terms providing the link to topology (cf. Naber, 1997).

We commence a signal and system description with a short overview defining bandwidth and power efficiencies and in the following sections describe the concepts more fully. This new signal/symbol modulation technique is described in terms in which these relations appear.

OSSO exploits the eigenvector expansion of a given symbol envelope waveform associated with a given time bandwidth product (TBP) (Barrett, 2009). It is important to note that, for purposes of use of the spectrum commercially, a communications channel is always constrained in bandwidth and transmitted power, so comparison of efficiencies provided by various modulation devices,

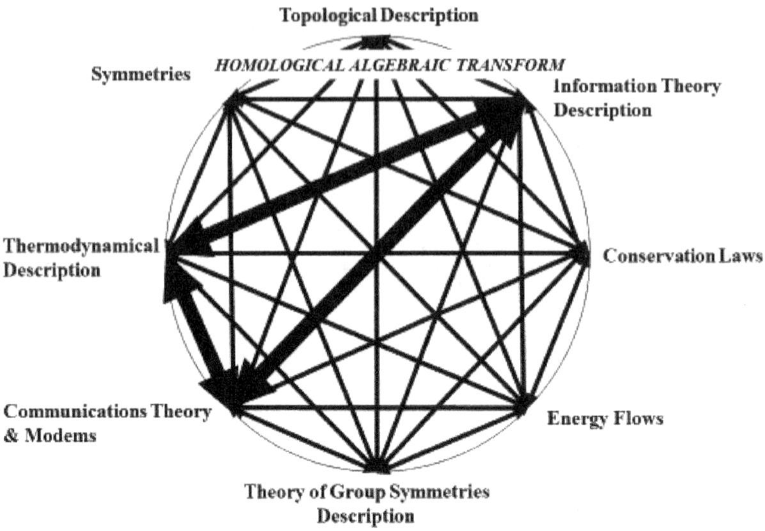

Fig. 4.1. A novel communication modulation technique, Orthogonal Signal Spectrum Overlay (OSSO), provides links between the descriptions of communication theory, information theory (or negentropy) and the thermodynamic concepts of entropy and energy. Optimum signals confined in both time and bandwidth — as described referring to the energy confinement problem — are addressed. By the variational principle the topological entropy is the supremum of the entropies of invariant measures. As shown by Addabbo & Blackmore (2019), there is a hierarchy of entropies: topological → Kolmogorov–Sinai (metric) → Shannon → Boltzmann → Gibbs → Clausius. Aspects of this hierarchy, bridging the continuous measurable phase space of Kolmogorov-Sinai or metric theory have been addressed by others.[1] The transmissions can be described in group theory terms providing the link to topology.

or modems, are always with reference to these constraints. Due to the orthogonality of the eigenvectors in a transmitted OSSO symbol waveform of a given TBP, it is possible to overlay by this modulation technique as many subchannels through a physical channel of given limited bandwidth as there are eigenvectors of the given symbol envelope associated with the TBP — however, with a penalty cost in SNR. The technical aim where channel bandwidth is limited is optimum spectral efficiency at set power efficiency; or alternatively, optimum power efficiency at set bandwidth efficiency. The concept

[1] Adler *et al.* (1965); Frigg (2004); and Delvenne (2019).

of power efficiency introduces thermodynamic (energy) constraints into communications.

The OSSO symbol is typically longer in duration than conventional symbols resulting in a greater number of sampled points per bit resulting in a "spread time" (the dual to spread spectrum) processing gain. This processing gain results in greater stability against multipath and colored and transient noise environments (see Fig 4.14, below). OSSO also provides covert communications — or 'hidden' channels (steganography), and capabilities regarding low probability of intercept and low probability of exploitation if intercepted capabilities, as well robustness against jamming.

4.2. Weber–Hermite Polynomials as Signals and their Overlay Forming Symbols

As in some other modulation techniques, an input binary stream, for example a digital output from a computer, is first serial-to-parallel converted to a vector and each vectorial component is assigned to a quaternary amplitude modulation (QAM) *constellation*[2] of signal/symbol possibilities shared by the transmitter (TX) and the receiver (RX), the in-phase (I) and quadrature (Q) components of which amplitude modulate separate I and Q Weber–Hermite (WH) wavelet signals (see Fig. 4.2). The number, n, of orthogonal channels is the number of I and Q pairs of WH(x) wavelets,

[2]A constellation is a two-dimensional table of all the defined and constrained possibilities of transmission and reception over a communications system channel. It defines the entropy or uncertainty, pretransmission, of the communications channel. Both transmitter and receiver have an agreed set of possible signals. The transmitter chooses which of the possible signals to encode a bit stream prior to transmission. Reception of a signal over the channel, resolves the receiver's uncertainty concerning which of the number of possible outcomes the transmitter chose. On reception and identification of the transmitted, analog/continuous, signals, the number of *possible* signals, by correlation "collapses" to the one received in digital/discontinuous form. Thus the received and identified signals, on resolving which of the possible outcomes represented in the preassigned constellation, may be considered to provide negative entropy or "negentropy" — equivalent to information — collapsing the entropic possibilities of the constellation and reducing the entropy of the constellation.

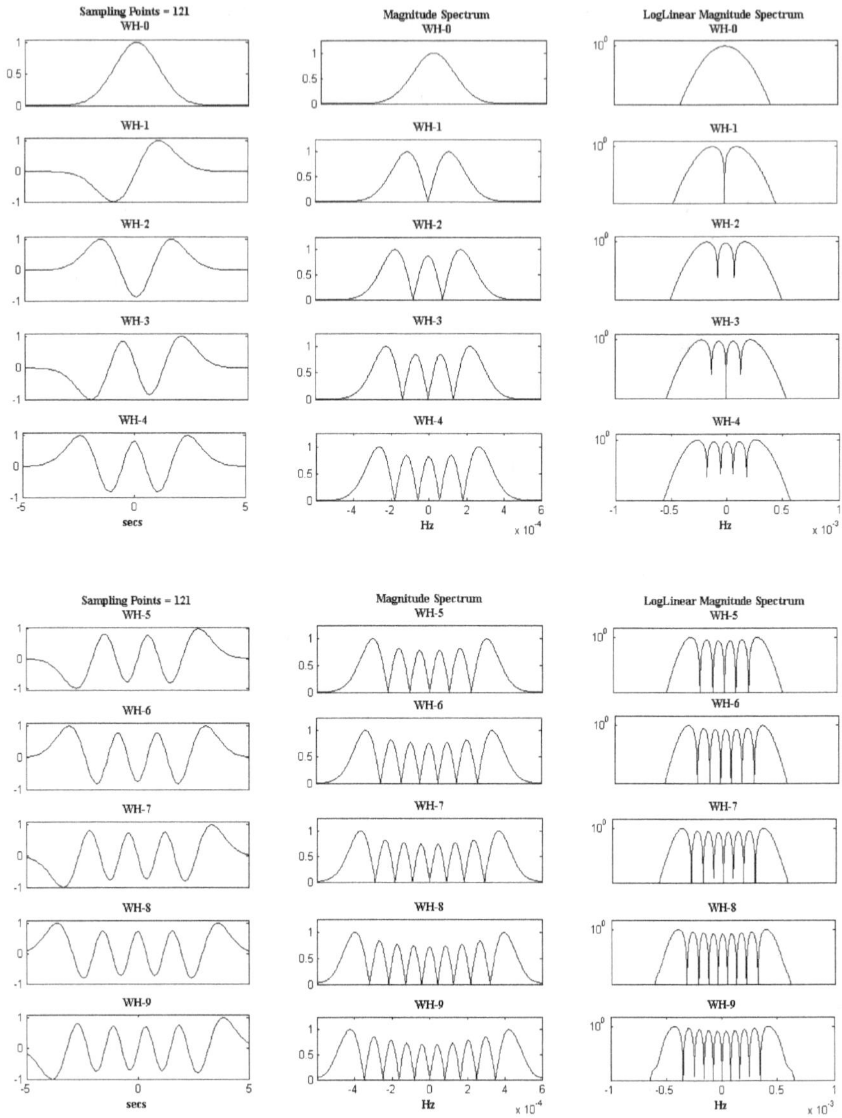

Fig. 4.2. Ten individual orthogonal WH wavelets (WH 0–9) shown from left column to right column: (a) in the time domain; (b) the magnitude spectra; and (c) the log-linear magnitude spectra.

$x = 0, 1, 2, \ldots, n$. The n separate, now amplitude modulated, WH wavelets, defined as *signals*, are linearly summed to provide two I and Q envelopes defined here as a *symbol*. These OSSO I and Q symbol envelopes then modulate an arbitrary carrier (see Figs. 4.3(a–e)). The OSSO encoding of QAM data coefficients on n signals/channels on *transmittance*, is homologous to a WH transform and the OSSO decoding on *reception* is homologous to a WH inverse transform — see Fig. 4.4 and Appendix 4.1. The use of OSSO WH wavelets is a solution to the "energy confinement problem", or optimum time-bandwidth product symbol limiting, as addressed in Appendix 4.2.

4.3. Information Encoding

The previous figures described the *unmodulated* OSSO signals and symbols. Moving now to the amplitude modulation, 16 QAM modulation for 6 OSSO signal/channels (or 6-OSO-16), together with the constellations representing reception possibilities and uncertainty in future reception, are shown in Fig. 4.4. In this example, each individual WH signal/channel transmits 4 bits of information. The 4×4 constellations require four amplitudes in both I and Q components.

The signals or channels, which, when summed, constitute a transmitted symbol, are orthogonal parabolic cylinder or Weber–Hermite wave functions (WHWFs) or Weber–Hermite wavelets (Barrett, 1972–1973, 2009). Here a short introduction is given to the signals and a fuller analysis in Appendix 4.1.

The WHWFs are related to *Weber's equation* (1869):

$$\partial^2\psi/\partial z + \left(n + \frac{1}{2} - \frac{1}{4}x^2\right)\psi(x) = 0. \tag{4.1}$$

for which the *general Weber equation*, or *parabolic cylinder differential equation*, is (Abramowitz and Stegun, 1972):

$$\partial^2\psi/\partial x(ax^2 + bx + c)\psi(x) = 0. \tag{4.2}$$

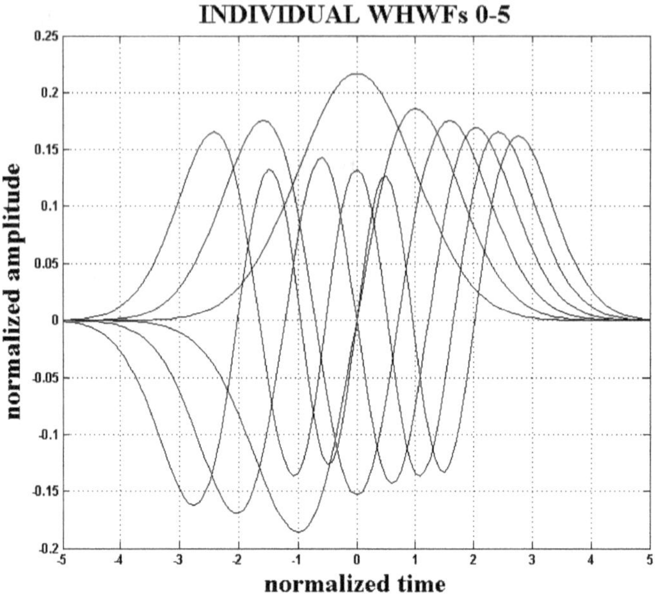

(a)

Fig. 4.3. (a) Six Weber–Hermite polynomial wave functions, WHWF 0–5, or WH wavelet signals 0–5, are overlaid in the time domain. Here they are shown (i) before QAM amplitude modulated, i.e., before information encoding; and (ii) before summation — occurring in an information-carrying transmitted symbol. It is noteworthy that these signals increase in time length the higher the wave function 0–5 with bandwidth constrained. (b) WHWF 0–5, spectra overlaid in the frequency domain of the signals above. Also as above, the description is before QAM amplitude modulation and summation, as the signals are in a transmitted information carrying symbol. With a time constraint these signals increase in bandwidth the higher the wave function. The higher the wave function, the larger the time-bandwidth required as indicated in this figure and (a). (c) It must be emphasized that to provide the envelope of a transmitted symbol, the individual signals are *overlaid* and summed — as shown in the upper figure. A summation resulting in a symbol is shown in figure (d). The signals are *not mid-frequency displaced with offsets* as in the OFDM modulation technique — and as shown figure (d). (d) Six WHWFs (WHWF 0–5) after summation providing the envelopes in either I or Q for a *symbol* shown here in the time domain. As six wave functions, i.e., six channels are summed, this symbol is designated 6-OSSO. Here, the WHWFs are not QAM modulated in amplitude, which is required for information encoding. When the WHWFs are modulated, the symbol shape can change separately for I and Q symbol forms. (e) The unmodulated 6-OSSO symbol envelope of figure (d), representing the envelope for either I or Q, in the frequency domain. Here again, the WHWFs are not QAM modulated in amplitude, which is require when information is encoded by QAM modulation of each channel.

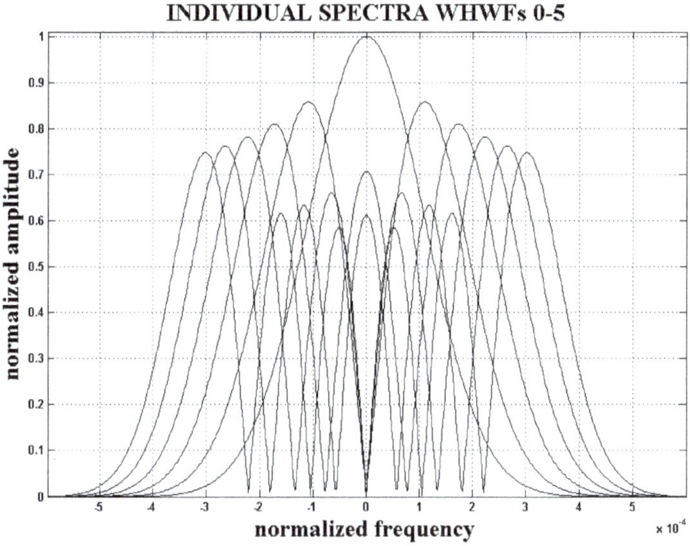

(b)

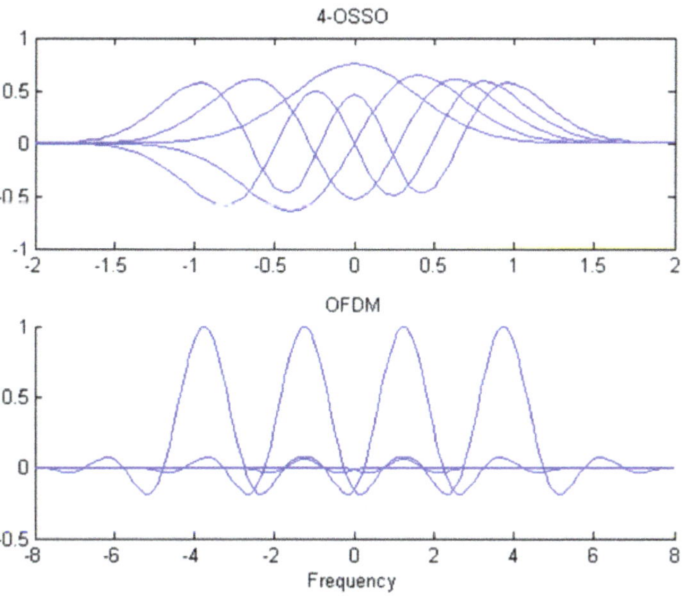

(c)

Fig. 4.3. (*Continued*)

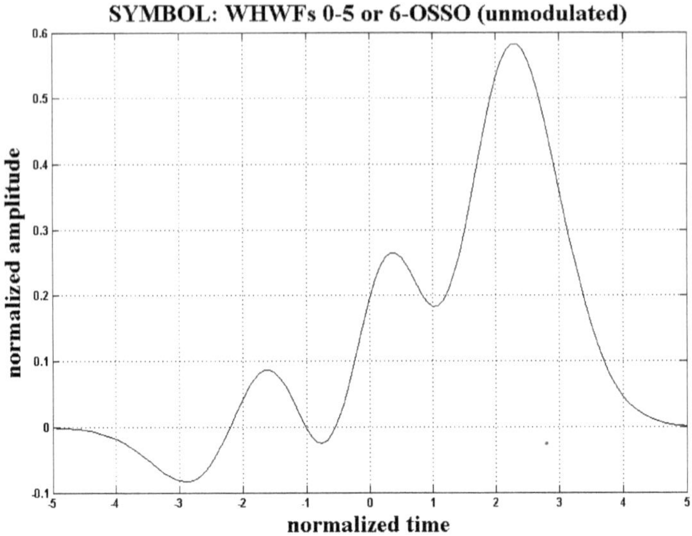

(d)

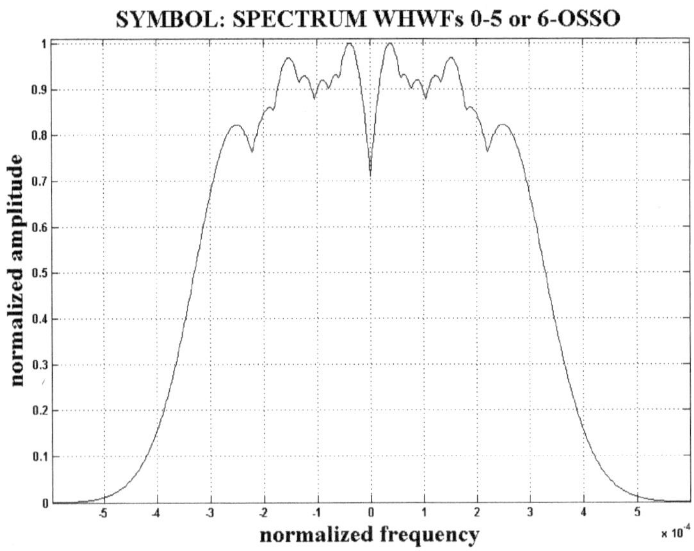

(e)

Fig. 4.3. (*Continued*)

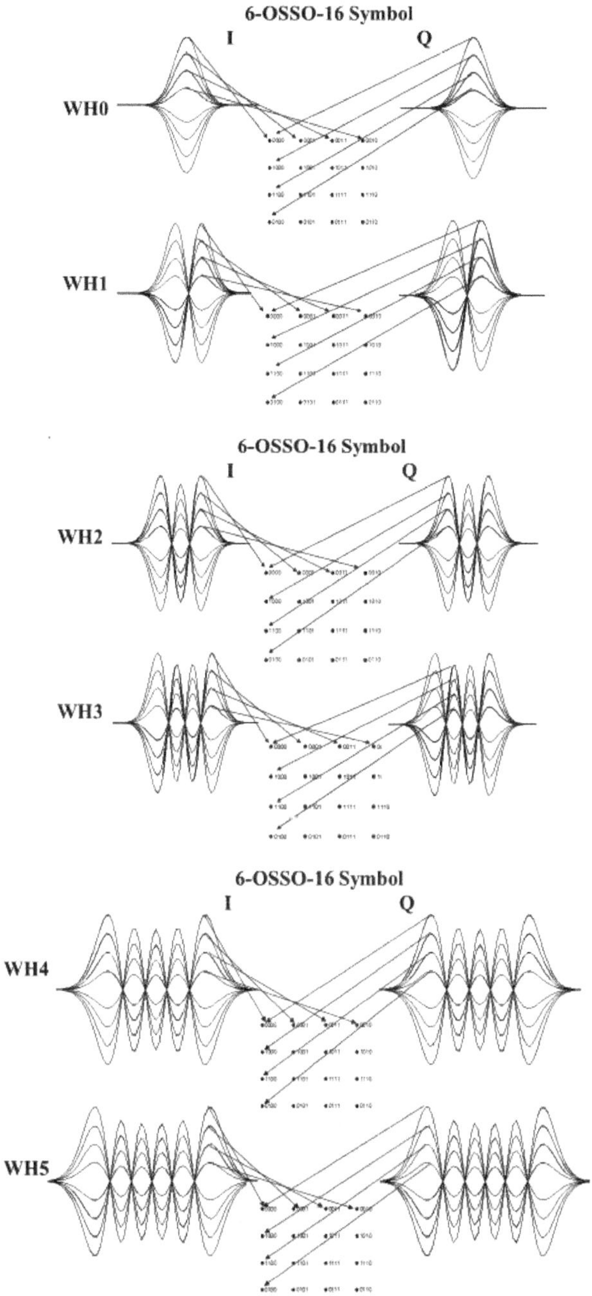

Fig. 4.4. (*Continued*)

Fig. 4.4. The relation of six WH wavelet (or WHWF) signals to 4 × 4 constellations in each of six channels (i.e., 16-QAM) for implementation of QAM coding. Each wave function/signal/channel is modulated to provide four amplitudes in both I (in-phase) and Q (in quadrature) providing OSSO-16. With WHs 0–5 overlaid and summed separately in I and Q, in a single symbol, six channels are provided for that symbol. Such a symbol is designated 6-OSSO-16.

The solutions of this equation are the orthogonal parabolic cylinder or Weber–Hermite wave functions (WHWFs):

$$\psi_n(x) = 2^{-\frac{n}{2}} e^{-\frac{x^2}{4}} H_n(x/\sqrt{2}), \quad n = 0, 1, 2, \ldots, \tag{4.3}$$

where the H_n are Hermite polynomials.

4.4. Transmitter and Receiver

The transmitted symbol is the linear summation of n WHWFs after QAM modulation in I and Q, creating envelopes which modulate an arbitrary carrier (Fig. 4.5(a)). The symbol reception follows a similar set of operations in reverse (Fig. 4.5(b)). The carrier is removed and the OSSO symbol envelope is received on I and Q channels. Inner products of unmodulated template WHWFs and the symbol envelope in both I and Q recover the data coefficients and hence the QAM constellations for each channel. Parallel to serial conversion recovers the data stream. If, for example, the first 6 WHWFs or 6 channels are used (Fig. 4.4), and on each there is 16-QAM modulation, we label the system 6-OSSO-16. The OSSO reception decoding to QAM data coefficients is homologous to the inverse WH transform — see Appendix 4.1.

The OSSO transceiver is an instantiation of this new unitary transform method that "tiles", "paves" or "discretizes" a communications channel in a unique way that reveals the maximum information carrying capacity of a channel.

This approach to waveform diversity employs solutions to the energy containment problem for maximum time-bandwidth product use (Slepian and Pollak, 1961; Slepian, 1964–1983). As the OSSO approach is based on developed mathematical and physical proofs

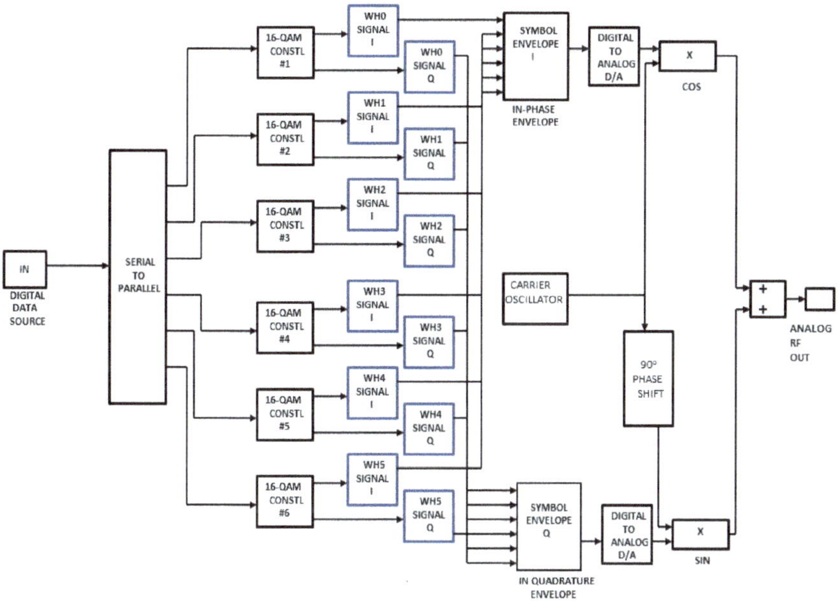

(a)

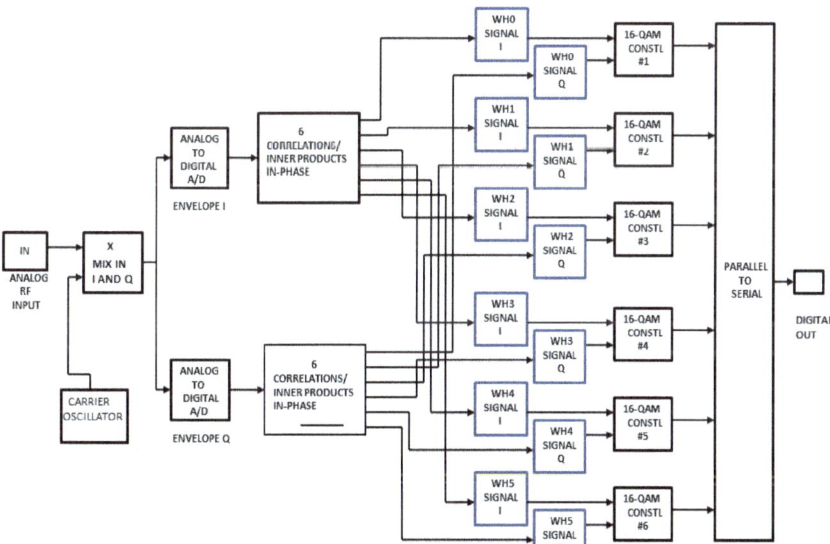

(b)

Fig. 4.5. (*Continued*)

Fig. 4.5. (a) 6-OSSO-16 Transmitter: The input binary stream is serial-to-parallel converted to a vector and each vectorial component is assigned to a QAM constellation, the in-phase (I) and quadrature (Q) components of which amplitude modulate a WHWF basis function separately in I and Q. The number of orthogonal channels is the number of WHWFs used and the OSSO symbol envelopes are the linear summation, separately in I and Q, of the amplitude modulated WHWFs, which then modulate an arbitrary carrier. (b) 6-OSSO-16 Receiver: The reception follows a similar set of operations in reverse. The carrier is discarded and the OSSO symbol envelope is received on I and Q channels. Inner products of template WHSFs and the symbol envelope in both I and Q recover the data coefficients and hence the QAM constellation assignments for each channel. Parallel to serial conversion recovers the data stream.

with regard to energy containment, it provides superior bandwidth and power efficiency.

A transmitter and receiver schematic for a 6 channel OSSO modem are shown in Figs. 4.5(a) and 4.5(b). The input binary stream is serial-to-parallel converted to vector form and each vectorial component is assigned to a QAM constellation, the in-phase (I) and quadrature (Q) components of which amplitude modulate a WHWF basis function separately in I and Q (Fig. 4.4). The number of orthogonal channels is the number of WHWF functions used and the OSSO symbol envelopes are the linear summation, separately in I and Q, of the amplitude modulated WHWFs, which then modulate an arbitrary carrier (Fig. 4.5(a)).

The reception (RX) follows a similar set of operations in reverse (Fig. 4.5(b)). The carrier is discarded and the OSSO symbol envelope is received on separate I and Q channels. Inner products of template WHWFs and the symbol envelope in both I and Q recover the data coefficients and hence the QAM constellation assignments for each channel — 6 in this case. Parallel-to-serial conversion recovers the data stream. If, for example, the symbol consists of the first 6 WHWFs or 6 channels, and on each there is 16-QAM modulation, the system is referenced as 6-OSSO-16. The OSSO reception decoding to QAM data coefficients is homologous to the inverse WH transform — see Appendix 4.1.

A full duplex wireless 6-OSSO-16 modem operating with a 2.144 GHz carrier was constructed (Barrett, 2007). The system design, included Tomlinson–Harashima precoding (THP) at the transmitter[3] and a decision feedback equalizer at the receiver.[4]

4.5. Time-Bandwidth Products and Power-Bandwidth-Data Rate Tradeoffs

The symbols can be partially overlaid in time to achieve a higher data rate and the size of the time-bandwidth product is set by the highest WHWF overlaid in the symbol. The greater the overlap, the smaller the symbol time-bandwidth product (TBP) and therefore, the higher the date rate at set bandwidth occupancy. The TBP describes the tradeoff between the symbol bandwidth occupancy and the symbol rate and hence the data rate for a set bandwidth occupancy and set modulation.

The overlapping introduces inter-symbol interference (ISI) and inter-channel interference (ICI), which is removed by equalization/cancellation up to a maximum overlap — 33%. If, for example the previously described system were overlapped by 33% of the symbol temporal length, the system is referenced as 6-OSSO-16-33. Table 4.1 shows the relation between symbol overlap and the TBP for 1–10 WHWFs (WH0–9) and the relation between the TBP to the number of WH signals per symbol as a function of percentage overlap of those signals is shown in Figs. 4.6(a) and 4.6(b).

4.6. Trade-offs Between Transmission Errors and SNR

The reliability of the system is given by a bit-error rate (*BER*) measure versus the energy per bit to noise power spectral density ratio, i.e., the normalized signal-to-noise ratio (*SNR*) — stated as $E_b/N_0(dB)$. The *BER* is the number of bit errors divided by the

[3]Tomlinson (1971); Harashima and Miyakawa (1972); Fischer (2002).

[4]The system design is moe fully described in Barrett (2007) and Barrett *et al.* (2008).

Table 4.1. OSSO time-bandwidth products.

Nonoverlapped condition		33% overlapped condition	
	TBP		TBP
WH0	4.21	WH0	2.78
WH1	6.54	WH1	4.32
WH2	8.30	WH2	5.48
WH3	9.80	WH3	6.47
WH4	11.00	WH4	7.26
WH5	12.11	WH5	7.99
WH6	13.18	WH6	8.70
WH7	14.12	WH7	9.32
WH8	15.10	WH8	9.97
WH9	16.19	WH9	10.68

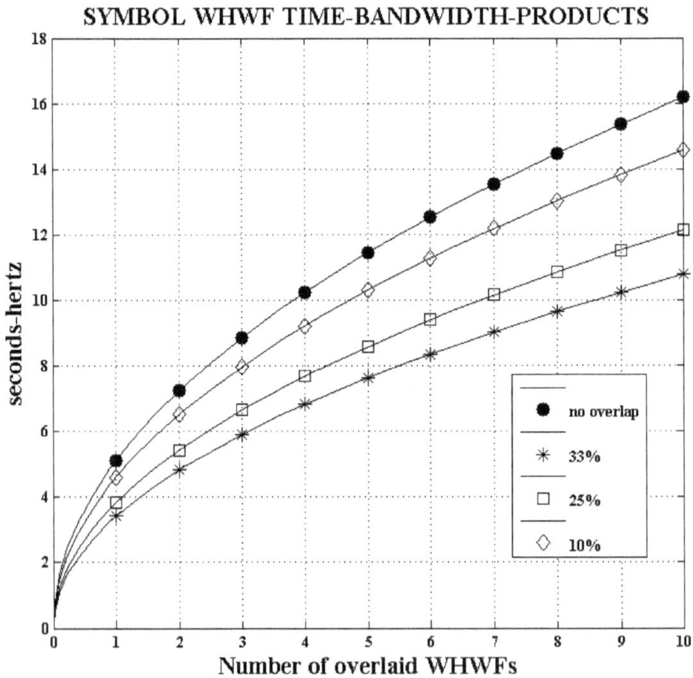

(a)

Fig. 4.6. (a) Time-bandwidth products (ordinate) for OSSO symbols with respect to: (i) The number of WHWFs per OSSO symbol, e.g., $1, 2, \ldots, 10$, providing $1, 2, \ldots, 10$ channels (or OSSO-1, OSSO-2, ..., OSSO-10 (abscissa)). (ii) The percentage overlap of these symbols (insignia). (b) The data of figure (a) in a log-log plot.

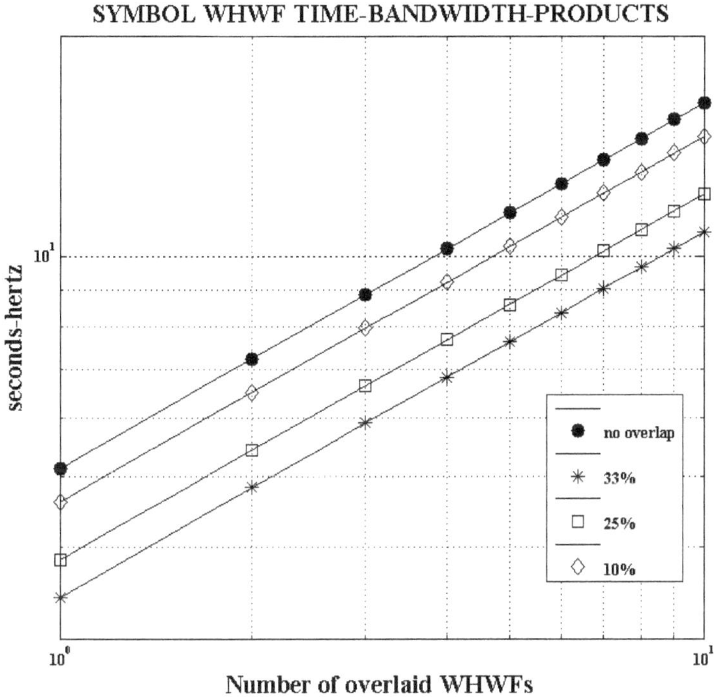

(b)

Fig. 4.6. (*Continued*)

total number of transferred bits during a specific time interval. The *BER* versus E_b/N_0 calculated relation for 6-OSSO with 16-QAM modulation and with *no* precoder and *DFE* is shown in Fig. 4.7(a). In contrast Fig. 4.7(b) shows the same relations ($E_{b/}N_0$ versus *BER*) and *with* precoder and *DFE* for (a) no symbol overlap in the time domain, i.e., 6-OSSO-16-00; (b) with symbols overlapped 16% i.e., 6-OSSO-16-16 or quarter symbol overlap; (c) 6-OSSO-16-33 (half symbol overlap).

As an example of an arbitrary coding assignment with one symbol being transmitted composed of six signal/channels, and referring back to Fig. 4.4 and the constellations shown there, the QAM possible modulations of each of the 6 WH signals related to the constellations is shown in Table 4.2.

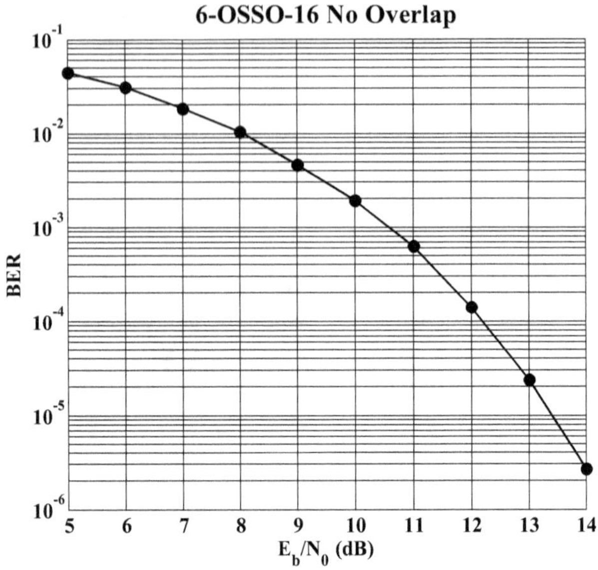

(a)

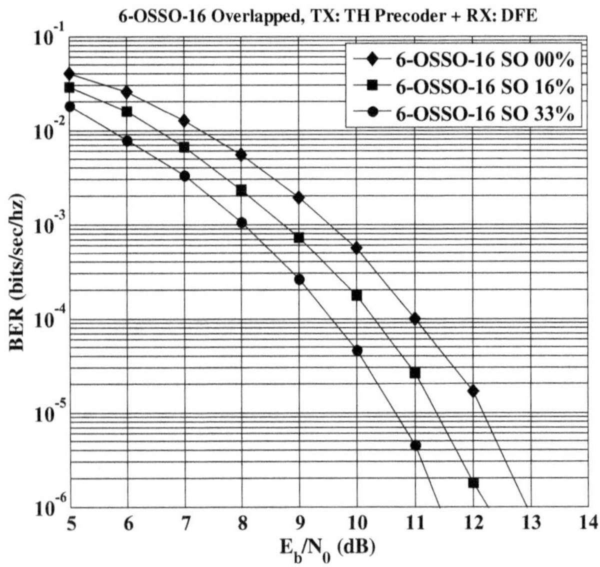

(b)

Fig. 4.7. (*Continued*)

Fig. 4.7. (a) Plot of E_b/N_0 (dB) (shorthand for the energy per bit to noise power spectral density ratio, or the normalized signal-to-noise ratio (SNR); i.e., the SNR per bit (abscissa)) versus BER (bit error rate is the number of bit errors divided by the total number of transferred bits during a specific time interval. BER is a unitless performance measure, — ordinate) at set bandwidth occupancy for modulations 16 QAM. There is no overlap of symbols and no precoder or equalizer was used. Compare with figure (b) with symbol overlap, precoder and equalizer. (b) E_b/N_0 (dB) versus BER at set bandwidth occupancy and number of channels (WHWFs) per symbol — here, 6, for modulation 16 QAM, i.e., 6-OSSO-16 with symbol overlap in the time domain, providing a higher symbol rate and hence a higher data rate. A Tomlinson–Harashima (TH) precoder and decision feedback equalizer (DFE) were used. Compare with figure (a) in which no precoder, DFE and symbol overlap were used.

Table 4.2. An example.

4-Digit packets	In-phase signal amplitude	In-quadrature signal amplitude	Signal/channel
1000	1.00	0.75	WH0
0101	0.75	0.25	WH1
1110	0.50	0.50	WH2
1111	0.25	0.25	WH3
1010	1.00	0.75	WH4
1001	0.75	0.75	WH5

The symbol, composed of this assignment, is transmitted, received and with the inner product of 6 WH analog signal templates with the analog symbol in I and Q, the analog signals are identified permitting their relative amplitudes to be recovered.

Using Dirac notation[5] then for the OSSO symbol *before* QAM modulation:

- $|\psi_i^I\rangle = WH_i^I$ and $|\psi_i^Q\rangle = WH_i^Q$ are WH *signals* for either the I and Q channels,

[5]Bra-ket or Dirac notation: a "ket" vector is a column vector denoted, e.g., $|\psi\rangle$ and a "bra" is a vector denoted, e.g., $\langle\phi|$. $\langle\phi|\psi\rangle$ provides an inner product and $|\psi\rangle\langle\phi|$, an outer product. Furthermore, the Dirac notation, is defined as noncommutative *by definition.* Writing the transmitted analog signal or symbol as TX $= |y\rangle$ and the analog RX templates as $= \langle x|$, we then recover the QAM coding by the inner product $\langle x|y\rangle$, i.e., the a's and the b's, above.

- $|\Psi_i^I\rangle = \sum_{i=0}^{i=n-1} |\psi_i^I\rangle$ is the modulating envelope for the I *symbol* channel,
- $|\Psi_i^Q\rangle = \sum_{i=0}^{i=n-1} |\psi_i^Q\rangle$ is the modulating envelope for the Q *symbol* channel.[6]

The complete *symbol* modulating envelope in I and Q *after* QAM modulation is:

$$|\Psi^{I+Q}\rangle = \sum_{i=0}^{i=n-1} a_i|\psi_i^l\rangle + ib_i||\psi_i^Q\rangle, \tag{4.4}$$

where a_i and b_i are the *chosen* QAM modulating amplitudes referring to specific 4-bit code words. For example, referring to a *full* QAM modulating assignment scheme (a constellation) shown in Fig. 4.4:

	A	1.00	0.75	0.50	0.25
B					
1.00		0000	0001	0011	0010
0.75		1000	1001	1011	1010
0.50		1100	1101	1111	1110
0.25		0100	0101	0111	0110

So if an example bit stream to be transmitted is 1000 0101 1110 1111 1000 1001, the QAM modulation with chosen modulating amplitudes, a and b, is as follows:

- $\psi_0^I, \psi_0^Q = 1.00, 0.75$ which encodes the 4 bits: 1000,
- $\psi_1^I, \psi_1^Q = 0.75, 0.25$ which encodes the 4 bits: 0101,
- $\psi_2^I, \psi_2^Q = 0.50, 0.50$ which encodes the 4 bits: 1110,
- $\psi_3^I, \psi_3^Q = 0.25, 0.25$ which encodes the 4 bits: 1111,
- $\psi_4^I, \psi_4^Q = 1.00, 0.75$ which encodes the 4 bits: 1000,
- $\psi_5^I, \psi_5^Q = 0.75, 0.75$ which encodes the 4 bits: 1001.

[6]Thus, if the transmitted and received analog symbol is, e.g., $|y\rangle$ and the analog RX templates as $= \langle x|$, we then recover the QAM coding by the inner product $\langle x|y\rangle$, i.e., the a's and the b's, above, are recovered.

A question arises concerning whether the time-bandwidth product (TBP) commutes. In the analysis until now, the TBP was shown as if applying to only one space dimension plus time. This can be expressed as

$$t = x = 1; \quad y = z = 0,$$

and so, for example, in one space dimension:

$$\Delta f \Delta t = 1/2$$
$$\Delta f \Delta t - \Delta t \Delta f = 0,$$

indicating a commutation relation associated with an Abelian algebra.

However, the transmitted field, according to field theory, requires more than one dimension. Therefore in two space dimensions:

$$\Delta f_{x,y} = \begin{bmatrix} f_{11} & f_{12} \\ f_{21} & f_{22} \end{bmatrix},$$
$$\Delta t_{x,y} = \begin{bmatrix} t_{11} & t_{12} \\ t_{21} & t_{22} \end{bmatrix}$$

and then

$$\Delta f_{x,y} \Delta t_{x,y} - \Delta t_{x,y} \Delta f_{x,y} = \begin{bmatrix} f_{12}t_{21} - f_{21}t_{12} & f_{11}t_{12} - f_{12}t_{11} + f_{12}t_{22} - f_{22}t_{12} \\ f_{21}t_{11} - f_{11}t_{21} - f_{21}t_{22} + f_{22}t_{21} & f_{21}t_{12} - f_{12}t_{21} \end{bmatrix}.$$

indicating noncommutation and non-Abelian algebra.

With three dimensions:

$$\Delta f_{x,y,z} = \begin{bmatrix} f_{11} & f_{12} & f_{13} \\ f_{21} & f_{22} & f_{23} \\ f_{31} & f_{32} & f_{33} \end{bmatrix}$$

$$\Delta t_{x,y,z} = \begin{bmatrix} t_{11} & t_{12} & t_{13} \\ t_{21} & t_{22} & t_{23} \\ t_{31} & t_{32} & t_{33} \end{bmatrix}$$

and with

$$\Delta f_{x,y,z}\Delta t_{x,y,z} - \Delta t_{x,y,z}\Delta f_{x,y,z} = \begin{bmatrix} A & B & C \\ D & E & F \\ G & H & I \end{bmatrix},$$

where

$$\begin{aligned}
A &= f_{12}t_{21} - f_{21}t_{12} + f_{13}t_{31} - f_{31}t_{13},\\
B &= f_{11}t_{12} - f_{12}t_{11} + f_{12}t_{22} - f_{22}t_{12} + f_{13}t_{32} - f_{32}t_{13},\\
C &= f_{11}t_{13} - f_{13}t_{11} + f_{12}t_{23} - f_{23}t_{12} + f_{13}t_{33} - f_{33}t_{13},\\
D &= f_{21}t_{11} - f_{11}t_{21} - f_{21}t_{22} + f_{22}t_{21} + f_{23}t_{31} - f_{31}t_{23},\\
E &= f_{21}t_{12} - f_{12}t_{21} + f_{23}t_{32} - f_{32}t_{23},\\
F &= f_{21}t_{13} - f_{13}t_{21} + f_{22}t_{23} - f_{23}t_{22} + f_{23}t_{33} - f_{33}t_{23},\\
G &= f_{31}t_{11} - f_{11}t_{31} - f_{21}t_{32} + f_{32}t_{21} - f_{31}t_{33} + f_{33}t_{31},\\
H &= f_{31}t_{12} - f_{12}t_{31} - f_{22}t_{32} + f_{32}t_{22} - f_{32}t_{33} + f_{33}t_{32},\\
I &= f_{31}t_{13} - f_{13}t_{31} - f_{23}t_{32} + f_{32}t_{23},
\end{aligned}$$

also indicating noncommutation — and so on depending on the number of spatial dimensions.

Furthermore, and in a nutshell, the relation of the OSSO communication system to topology is as follows.

1. Since the time of Maxwell[7] a distinction has been made between the fields **E** and **H** and **D** and **B**. Maxwell considered **E** and **H** as quantities defined with respect to a line, and **D** and **B** as quantities defined with respect to a plane, indicating that **E** and **H** are polar vectors and **D** and **B** are axial vectors. A polar vector is transformed to its negative under inversion of its coordinate axes and only two space dimensions are required. *However, a pseudo vector, or axial vector, is invariant under inversion of its coordinate axes and requires a volume, or three space dimensions, to be described.*

[7]Maxwell (1873, 1881, 1891).

2. Referring now to quaternion algebra, the description of three space dimensional continuous wave transmission in group theory terms is described:

$$\mathrm{SU}(2) = \{x\mathbf{i} + y\mathbf{j} + z\mathbf{k} + t\mathbf{I} : x^2 + y^2 + z^2 + t^2 = 1\}. \qquad (4.5)$$

   Furthermore, as QAM modulation involves both an in-phase (I) and an in quadrature (Q) transmission to define a symbol/signal, in a complete symbol/signal transmission there are two describable SU(2)[8] group transmissions.
3. At the transmitter there is a constellation in SO(3) defined by these $2 \times \mathrm{SU}(2)$ continuous analog waves — i.e., the I and the Q waves. This is a $1 \to 2$ mapping, or $\mathrm{SO}(3) \to 2 \times \mathrm{SU}(2)$ I and Q waves.
4. At the receiver, there is a reverse $2 \to 1$ mapping of $2 \times \mathrm{SU}(2)$ I and Q waves back to an SO(3) group constellation.
5. In topological terms, each SU(2) group — in both I and in Q — describes a torus.

The question also arises concerning whether OSSO signals are entangled.[9] A method to detect entanglement is Schmidt decomposition, which is a way of expressing a vector as the tensor product of two inner product spaces, equivalent to a restatement of the singular value decomposition in a different context. Applying Schmidt decomposition to an unmodulated 6-OSSO symbol (Fig. 4.8(a)) and an arbitrary modulated 6-OSSO symbol (Fig. 4.8(b)) we obtain the

[8]The group U is unitary because $U * U$ = an identity matrix; S = special, because the determinant of $U = 1$; 2 because 2×2 matrices are described. As a manifold, this algebra is equivalent to a three-dimensional sphere, S^3, sitting inside $\mathbf{R}^4$. But there is a 2-to-1 homomorphism of SU(2) onto SO(3), also called a double covering.

[9]A state, or signal, is entangled if it is not a mixture of product states or signals. Alternatively, the circumstances for entanglement are global states of a composite system which cannot be written as the product of the states of individual systems (Horodecki *et al.* 2009). Entangled states are superpositions of vectors from different vector spaces, whenever the superpositions cannot be rearranged into a single product from two vector spaces (Qian *et al.* 2015). It should be emphasized that entanglement, per se, does not imply nonlocality (Ghose and Mukherjee, 2013; Berg-Johansen *et al.* 2015).

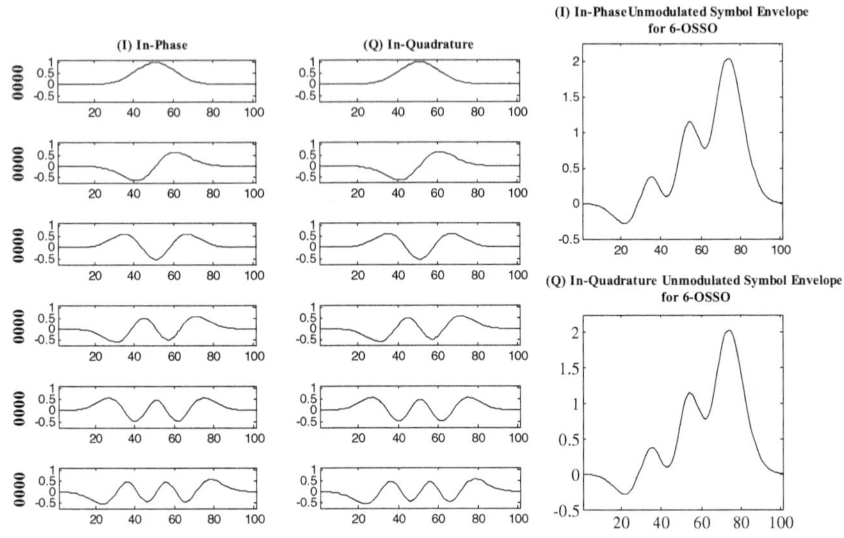

(a)

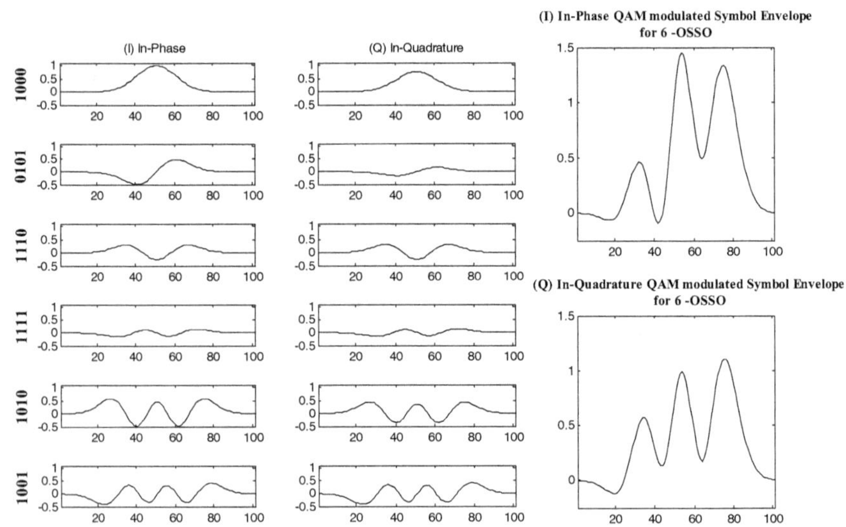

(b)

Fig. 4.8. (a) The premodulated symbol envelopes in I and Q for 6-OSSO (right), composed of 6 channels of WH signals (left). (b) The modulated (with arbitrary modulation) symbol envelopes in I and Q for 6-OSSO (right), composed of 6 channels of WH signals (left).

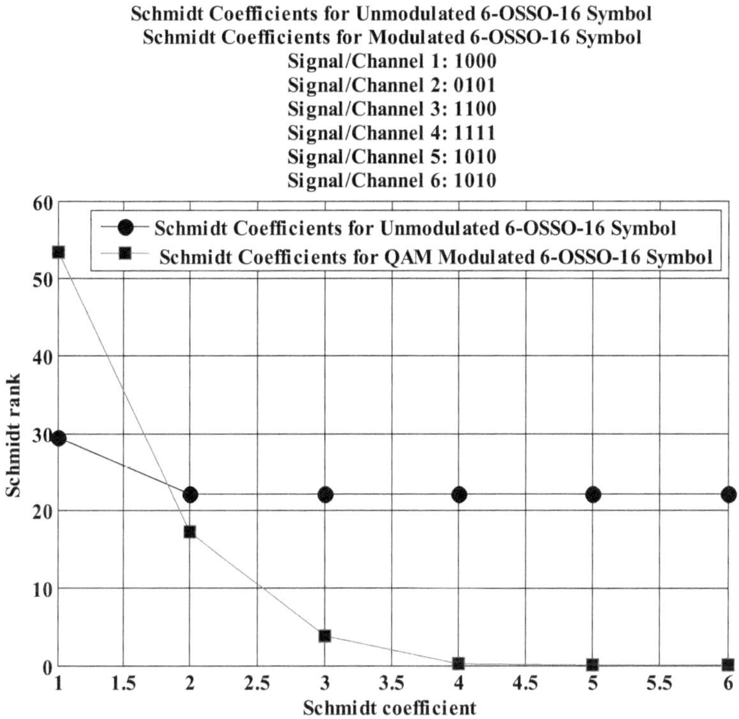

Fig. 4.9. Schmidt coefficients for unmodulated (Fig. 4.8(a)) and QAM modulated (Fig. 4.8(b)) 6-OSSO-16, and indicating a high Schmidt rank or degree of entanglement (without nonlocality). In the case of nonentangled systems there would be only one coefficient. As indicated in the figure, and as above, the arbitrary example QAM coding for this modulated channel set was:

$$1.00 \times \psi_0^I + 0.75 \times \psi_0^Q; \quad \text{or } 1000,$$
$$0.75 \times \psi_1^I + 0.25 \times \psi_1^Q; \quad \text{or } 0101,$$
$$0.50 \times \psi_2^I + 0.50 \times \psi_2^Q; \quad \text{or } 1100,$$
$$0.25 \times \psi_3^I + 0.25 \times \psi_3^Q; \quad \text{or } 1111,$$
$$1.00 \times \psi_4^I + 0.75 \times \psi_4^Q; \quad \text{or } 1010,$$
$$0.75 \times \psi_5^I + 0.75 \times \psi_5^Q; \quad \text{or } 1010.$$

Schmidt coefficients shown in Fig. 4.9. A symbol (or state) is said to be entangled if its Schmidt rank (number of singular values) is greater than 1, and not entangled other wise. Figure 4.9 indicates that the WH signals/channels are locally entangled.

In the case of the constellation of all signal/symbol possibilities of reception or the channel's entropy, in the case of 6-OSSO and for the signals and symbol defined before:

- $|\psi_i^I\rangle = WH_i^I$ and $|\psi_i^Q\rangle = WH_i^Q$ are uncoded WH *signals* in the I and Q channels,
- $|\Psi_i^I\rangle = \sum_{i=0}^{i=n-1} |\psi_i^I\rangle$ is the modulating envelope for the I *symbol* channel,
- $|\Psi_i^Q\rangle = \sum_{i=0}^{i=n-1} |\psi_i^Q\rangle$ is the modulating envelope for the Q *symbol* channel,

and for 16-QAM:

$|A\rangle = |B\rangle = [1.0, 0.75, 0.5\ 0.25]$, which are *possible* amplitude modulations. So the complete set of possible symbol receptions, i.e., the constellation of *possibilities*, is: 4.6a

$$|\Psi^{I+Q}\rangle = [|A\rangle \otimes |\Psi^I\rangle + |B\rangle \otimes |\Psi^Q\rangle], \tag{4.6a}$$

where $\otimes$ represents the tensor product. Equation (4.5), describing the chosen encoding for this example, can be compared with Eq. (4.6) of constellation possibilities.

In both I and Q, the reception *possibilities* are as follows:

$$|A\rangle \otimes |\Psi^I\rangle = |B\rangle \otimes |\Psi^Q\rangle =$$

$1.00 * \psi_0$	$0.75 * \psi_0$	$0.50 * \psi_0$	$0.25 * \psi_0$
$1.00 * \psi_1$	$0.75 * \psi_1$	$0.50 * \psi_1$	$0.25 * \psi_1$
$1.00 * \psi_2$	$0.75 * \psi_2$	$0.50 * \psi_2$	$0.25 * \psi_2$
$1.00 * \psi_3$	$0.75 * \psi_3$	$0.50 * \psi_3$	$0.25 * \psi_3$
$1.00 * \psi_4$	$0.75 * \psi_4$	$0.50 * \psi_4$	$0.25 * \psi_4$
$1.00 * \psi_5$	$0.75 * \psi_5$	$0.50 * \psi_5$	$0.25 * \psi_5$

Modulation and information encoding then appears as one choice for each $|a\rangle$ (chosen from $|A\rangle$) and $|b\rangle$ (chosen from $|B\rangle$) for each ψ separately in I and Q. For example for 6-OSSO-16 modulation, if in I are chosen:

			$0.25 * \psi_0$
		$0.50 * \psi_1$	
$1.00 * \psi_2$			
			$0.25 * \psi_3$
		$0.50 * \psi_4$	
			$0.25 * \psi_5$

and if in Q are chosen:

		$0.50 * \psi_0$	
	$0.75 * \psi_1$		
			$0.25 * \psi_2$
$1.00 * \psi_3$			
	$0.75 * \psi_4$		
		$0.50 * \psi_5$	

then these arbitrary choices are:

	ψ_I	ψ_Q
ψ_0	0.25	0.50
ψ_1	0.50	0.75
ψ_2	1.00	0.25
ψ_3	0.25	1.00
ψ_4	0.50	0.75
ψ_5	0.25	0.50

Referring to the 16-QAM modulation constellation of *possibilities* (see Fig. 4.4):

$\psi_Q \downarrow \psi_I \rightarrow$	1.00	0.75	0.50	0.25
1.00	0000	0001	0011	0010
0.75	1000	1001	1011	1010
0.50	1100	1101	1111	1110
0.25	0100	0101	0111	0110

and this particular example QAM assignment identifies:

$$1110\ 1011\ 0100\ 0010\ 1011\ 1110$$

Thus, a single symbol of 6 channels each separately modulated by 16-QAM, i.e., 6-OSSO-16, is worth, or carries, $6 \times 4 = 24$ bits. Also, the entropy of the full constellation of possibilities is reduced by the reception of negentropy or information of the coding assignment.

The OSSO system is adaptable to a model of correlated events suggested by Barut and Meystre (1984, 1986) for quantum events. Barut and Meystre consider a system which splits into two parts with opposite spins in two directions[10] — spins taking on the role of the two polarizations in OSSO. It is shown that such as system can produce the results of quantum mechanics. As in the OSSO system's analog-to-digital (A-to-D) conversion, Barut and Myer "discretize" continuous measurements providing the equivalent of quantum mechanical "yes" or "no" experiments. However, Aspect (1986) rejected the Barut and Meystre model, on the basis of there being no analog-to-digital converter fast enough at that time. Since time has marched on, A-to-D converters have become considerably faster (see, e.g., Erdogan *et al.* 2017); and one might observe that "quantum jumps", once considered instantaneous, have shown to be not so (Minev *et al.* 2019). The quantum jumps between a ground and excited state are not instantaneous but take place gradually over time.

The so-called wave function reduction also takes place over time, and there is an analogy to correlation of the continuous received symbol/signals with templates of the signals to recover the a's and b's. To an observer confined to be situated *behind* a receiver — that is: post detection and, especially, post analog-to-digital conversion — the

[10]A beam splitter performs a Hadamard transformation:

$$\begin{pmatrix} |0'\rangle \\ |1'\rangle \end{pmatrix} = \frac{1}{\sqrt{2}} \begin{pmatrix} 1 & 1 \\ 1 & -1 \end{pmatrix} \begin{pmatrix} |0\rangle \\ |1\rangle \end{pmatrix}$$

Splitting transforms require identifications — a capability that only a Maxwell demon can supply — with concomitant requirement for entropy/energy (Bennett, 1987, 2003; Tribus and McIrvine 1971; Plesch *et al.* 2014; Canales, 2020).

analog or continuous transmissions do not exist. Only the digital or discretely sampled reduction of the continuous wave exists for this observer. However, this observer's "belief" in only the existence of the discrete is biased by his ignorance of the complete transmission process from transmitter by means of continuous wave symbols to the receiver, where, as stated, there is an analog-to-digital conversion, or continuous-to-discrete conversion. This hypothetical observer, who views the world as completely digital or discrete, may deny the existence of the continuous analog world of the channel.

The delusion of this fictitious observer is countered by the position taken by the position of counter-factual definiteness (CFD). CFD is the assumption of objects, properties of objects and transmissions, even when they have not been measured, but which are indicated by theoretical consistency (Kupczynski, 2020). CFD thus points to sub-phenomena. For the present interest, an OSSO system, the observer does not measure the analog, continuous waves traveling through a channel, but nevertheless can assume they exist.

The channel and the specific noise in the channel, as well as multipath, must also be assumed, as differences impact signal identification and measurements. Thus system performance measures are *contextual*,[11] with each performance measurement requiring its own probability space, which dictates the statistics used.

An OSSO system transmission commences with discontinuous, digital values that are transmitted, and also finishes with discontinuous, digital values. To an observer situated *before* the transmitter and an observer situated *after* the finish of reception, the inbetween (the channel) is unknown but postulated. These continuous e.m. fields are treated as unknown but in principal given defined values. Therefore, OSSO transmission exemplifies counterfactual definiteness (CFD), as the definiteness of these discontinuous or digital results of measurements [of analog or continuous (e.m. fields)] are allowed, which may not be retained on a given individual system, while the channel and e.m. fields exist *sub-phenomena*. Furthermore, the outcomes of a communication are not predetermined but are created

[11]Kupczynski, (2017a,b, 2020); Nieuwenhuizen (2016).

in the interaction of continuous analog transmissions with receivers containing A-to-D converters. RF transmission reception also does not provide predictions for a unique single outcome of a transmission but predictions consistent with specific channel noise, absorption and multipath. Each situation has a different (Kolmogov) probability model and probability space and is therefore contextual. Thus, there are major similarities of these OSSO classical requirements with quantum measurements.[12]

We turn now to describe the instantiation and testing of an OSSO system.

4.7. OSSO Modem System and Test Arrangement[13]

In Chapter 3, theoretical reasons were stated for a new form of communications modulation, OSSO. In this appendix, instantiations of an OSSO communications system are described which conform well with theoretical predictions. These prototypes to be described are far from being optimized and the detailed necessary testing under a variety of conditions has yet to be realized. Yet these systems serve the purpose as proof of a theoretical concept that has promise in applications to a number of fields.

Two systems, one transmitting horizontally polarized, the other vertically, each consisting of six WH channels, each channel 16-QAM modulated, i.e., a two orthogonally polarized system with combined modulation: 2 × 6-OSSO-16, were constructed conforming with the theoretical system description in this chapter (see Fig. 4.4). Figure 4.10 shows a digital board and a digital and RF board.

Using 2 orthogonally polarized channels with 6-OSSO-16 on each, enabled overlay of transmissions. Figure 4.11(a) and 4.11(b) show an overlay of the received and recovered constellations of the two channels over a series of transmissions.

The bandwidth occupancy and spectral characteristics for the system are shown in Fig. 4.12 and Table 4.3 in comparison with

---

[12]Kupczynski (2017a,b, 2020).

[13]This section describes work performed under Defense Advanced Research Projects Agency Contract No. FA8750-05-C-0259, 2007.

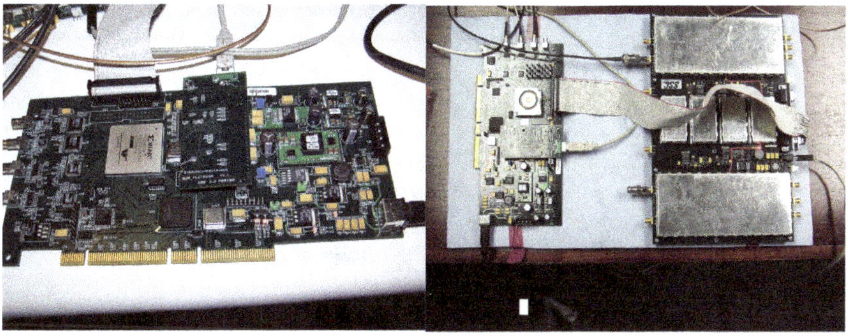

Fig. 4.10. Left: An OSSO digital board. Right: OSSO digital and RF boards. This system was one of four constructed.

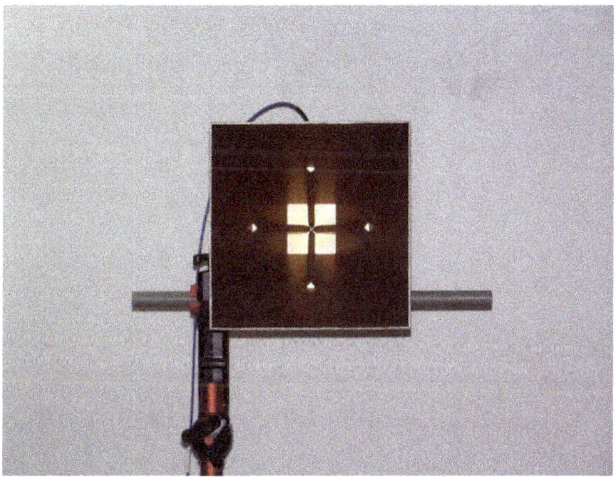

(a)

Fig. 4.11. (a) Two orthogonal beam antenna one used for transmittance (TX) and another for reception (RX) and providing a two orthogonal beam channel. (b) The reception (RX) and decoding into constellation form of two orthogonal beam channels each of six channel OSSO with 16-QAM constellations overlaid (6-OSSO-16 or 6×4 symbol). The symbols were transmitted, received, over two orthogonal channels and the decode constellations represented on an oscilloscope. With two orthogonal channels, this RX decoded constellation representation shows a $2 \times 6 \times 4 = 48$ bit *received and reconstructed from multiple transmissions*, constellation overlay for a sequence of transmissions.

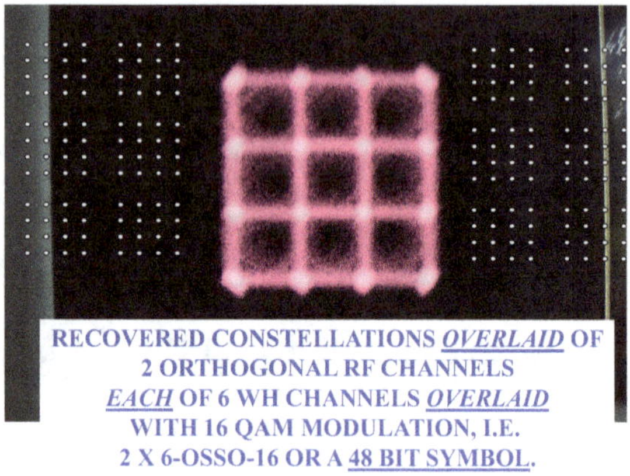

(b)

Fig. 4.11. (*Continued*)

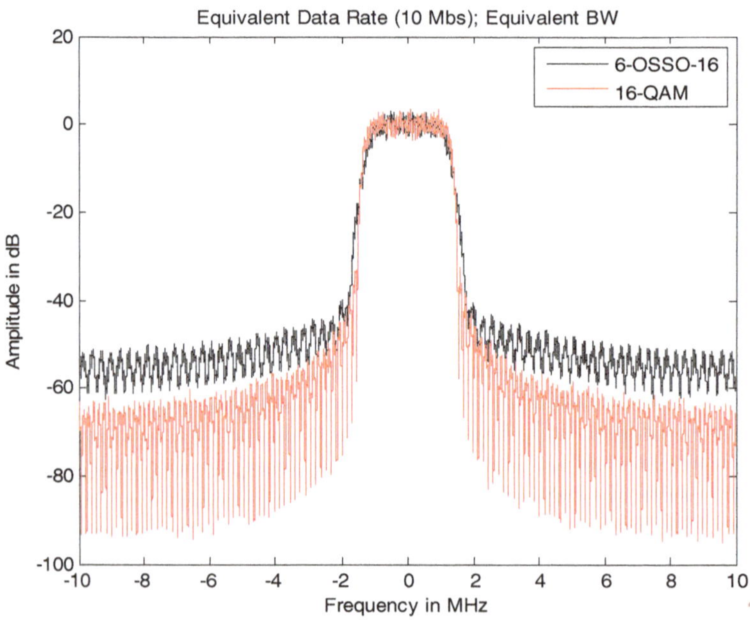

Fig. 4.12. For all tests of the wireless system components shown in Fig. 4.10, the transmission bandwidth was equal at the -50 dB level for both the 6-OSSO-16 system and a 16-QAM system -3.5 MHz.

Table 4.3. Bandwidth use comparison: at equal data rate: 10 Mbs E_b/N_0 at BER $= 10^{-6}$ for OSSO with Tomlinson precoder at TX and linear equalizer at RX.

	OSSO (MHz)	QAM (MHz)
3 dB Spectrum	2.3035	2.3962
6 dB Spectrum	2.4976	2.5854
30 dB Spectrum	3.3472	3.4192
90% of energy	2.1045	2.2362
95% of energy	2.2998	2.4170

Table 4.4. Bandwidth an power efficiency performance comparisons bandwidth occupancy: 100 MHz. sampling rate: 1 GSS X-OSSO-Y-Z, where X = # channels; Y = QAM per channel; Z = % symbol overlap.

Modulation	Symbol duration (Ns)	Time-bandwidth product	Symbol rate (Mss)	Bits per symbol	Data rate (Mbs)	Band-width efficiency	E_b/N_0 at BER $= 10^{-6}$ (dB)
6-OSO-16-00	121	12.1	8.3	24	200.0	2.00	13.10
6-OSSO-16-16	101	10.1	9.9	24	237.6	2.38	12.40
16-OSSO-16-33	81	8.1	12.3	24	296.2	2.96	11.50
6-OSS0-16-50	61	6.1	16.4	24	393.4	3.93	24.00
6-OFDM-16	141	14.1	7.1	24	170.2	1.70	15.10
16-QAM	12	1.2	83.3	4	333.3	3.33	14.50

a conventional 16-QAM system. These initial tests were conducted without optimizing TX energy partitioning among the WH subchannels (Fig. 4.13). Predicted power and power efficiencies are shown in Table 4.4. The time-bandwidth products for the nonoverlapped and overlapped conditions are shown in Table 4.5.

It can be seen in Table 4.4 that the best power efficiency is achieved by 6-OSSO-16-33 (or 6 channels, each 16-QAM modulated amd 33% overlapped) at an *acceptable* bandwidth efficiency (2.96 bits/sec/Hz). The best bandwidth efficiency is achieved by 6-OSSO-16-50 (or 6 channels, each 16 QAM modulated, but at an *unacceptable* bandwidth efficiency ($E_b/N_0 = 24$).

Table 4.5. OSSO time-bandwidth products.

Nonoverlapped condition		33% overlapped condition	
WH0	4.21	WH0	2.78
WH1	6.54	WH1	4.32
WH2	8.30	WH2	5.48
WH3	9.80	WH3	6.47
WH4	11.00	WH4	7.26
WH5	12.11	WH5	7.99
WH6	13.18	WH6	8.70
WH7	14.12	WH7	9.32
WH8	15.10	WH8	9.97
WH9	16.19	WH9	10.68

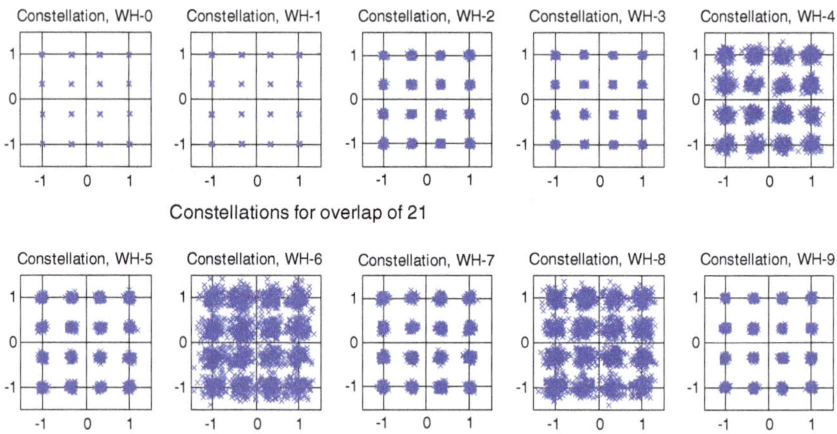

Fig. 4.13. Testing 10-OSSO-16 constellations for the subchannels WH0:WH9 indicating nonoptimum energy partitioning. Notice that (i) each even WH number signal has lower signal-to-noise than the paired odd WH number signal; (ii) The higher the WH number, the worse the signal-to-noise. These differences can be equalized by optimum transmit energy partitioning. Future prototypes will optimize energy partitioning between the WH0 channels prior to transmittance.

Figure 4.13 shows transmitted test symbols after reception and the build of received full constellations over time for 10-OSSO-16 channels. There is an obvious indication that optimum energy partitioning must be addressed in future prototypes.

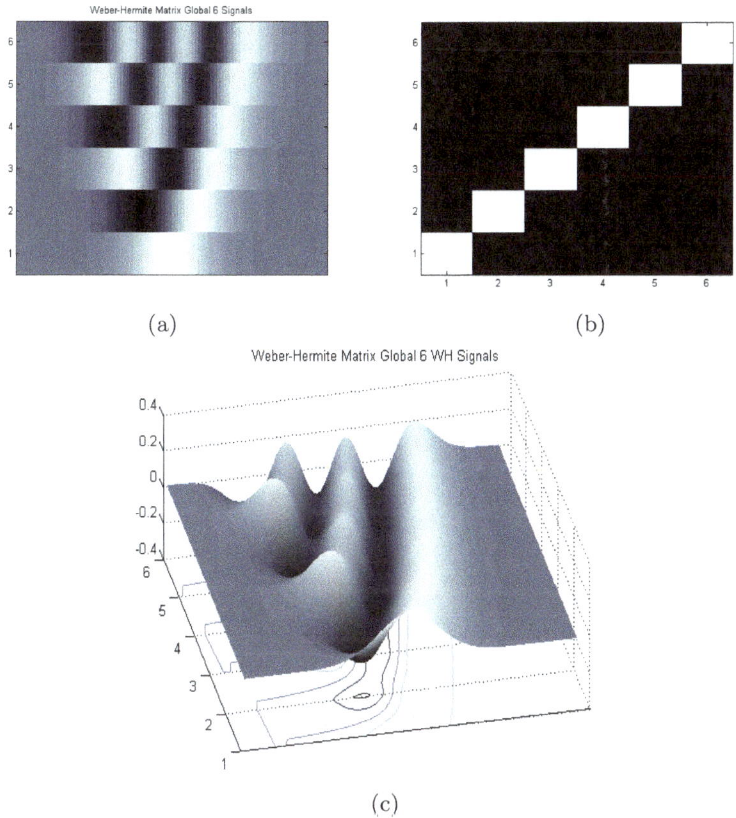

Fig. 4.14. (a) The unitary WH Matrix, M, composed of just six WH Signals, $n = 0, 1, 2, \ldots 5$; (b) The identity matrix $I = MM^*$; and (c) matrix M viewed at an angle. The transmittance and recovery of data in 6-OSSO, or six channel OSSO (or 6-OSSO addressed in these simulations) is homologous to these operations.

Figure 4.14 shows the unitary WH matrix at the receiver identifying each of the 6 WH signals in a symbol and their amplitudes — in this case all equal — permitting reconstruction for the digital transmission from this analog form.

A square root raised cosine (SQRC) 16-QAM pulse is represented by fewer samples than any OSSO channel carrying the same number of bits at equal sampling rate. This indicates superior OSSO performance in the presence of transient noise environments (Fig. 4.15).

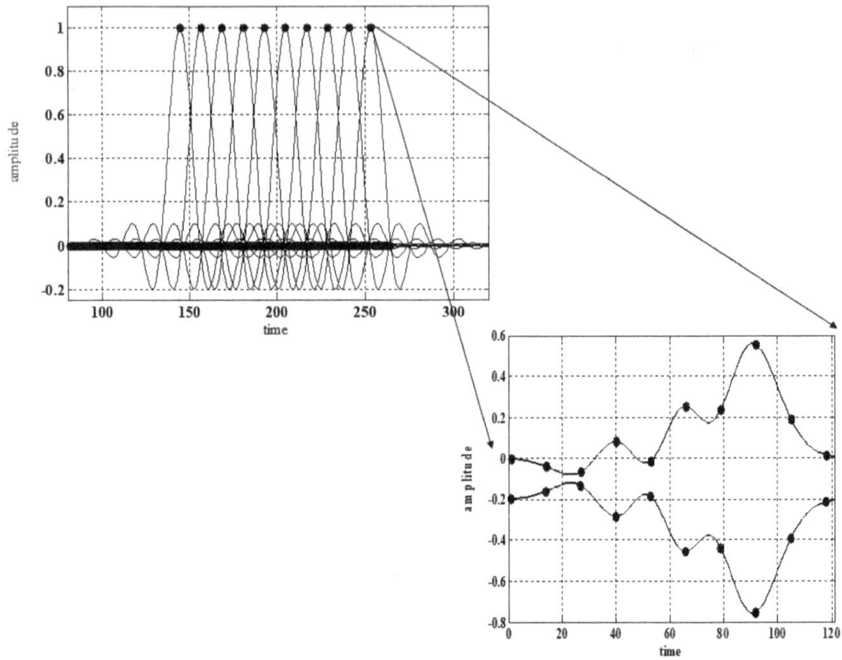

Fig. 4.15. Comparison of OSSO and QAM in colored and pulsed noise. Upper: 16-QAM – 4 bits per symbol. Lower: 6-OSSO-16 – 4 bits per channel/signal; 24 bits per symbol. Both occupying 100 MHz bandwidth. Both sampled at 1 GHz Hz. Time-Bandwidth Product overlapped 16-QAM: 1.2. Time-Bandwidth Product 6-OSSO-16: 12.1 With the time-bandwidth products and sampling rates indicated, each SQRC 16-QAM pulse carrying 4 bits of information is represented by one digital sample. In contrast, each 16-OSSO-16 symbol, consisting of six channels/signals, or, 6×4 bits of information is represented by 10 digital samples spread over time. If pulsed noise eliminates one SQRC 16-QAM pulse, the 4 bits are lost. In comparison, in the case of 16-OSSO-16 each of the 24 bits of information is spread over 10 samples. Therefore the 6-OSSO-16 channels in the symbol will be degraded, but not lost.

The OSSO system indicates the importance of sampling, A-to-D and D-to-A conversion in the preservation of information. In analogy to quantum mechanics, the OSSO symbols and signals exist as (a) continuous waves (in the channel) and (b) discrete digital particles (before transmission and after reception). However, it should be emphasized that while there is a wave-particle duality (of information) in an OSSO transmission system, *the waves and the particles do not coexist.*

Due to the importance of the time-bandwidth product in defining the translation between waves and particles and vice versa, we examine a number of derivations of the product in the following Appendix 4.1.

Appendix 4.1 The Energy Confinement Problem: Maximum Concentration of Signal Energy in a Fixed Time-Bandwidth Area

The problem of maximally concentrating signal energy in a fixed time and frequency region has been approached in two ways. The first is that of Heisenberg and Gabor, and is referred to, appropriately, as the Heisenberg–Gabor approach. The second is that of Slepian, Pollack and Landau and is referred to, also appropriately, as the Slepian–Pollack–Landau approach and the third is the Weber–Hermite confinement. We consider each in turn.

A.4.1.1. *The Heisenberg–Gabor approach*

The Heisenberg–Gabor approach addresses the impossibility of a signal having arbitrarily small support in both time and frequency (Heisenberg, 1927; Gabor, 1946). Couched in terms of energy confinement, energy concentrations can be formulated in terms of signal truncation operators (Flandrin, 1999). As absolute concentration in time and frequency is precluded by sampling considerations, the problem is considered in terms of signals that are *practically* confined in time and frequency.

This operational formalism when applied to signal theory (cf. Bonnet, 1968; Flandrin, 1982, 1999) time plays a role analogous to the position variable in quantum mechanics, and can be associated with the operator:

$$(\hat{t}x)(t) \equiv tx(t). \tag{A.4.1}$$

Similarly, frequency, analogous to momentum in quantum mechanics, can be associated with the operator:

$$(\hat{f}x)(t) = \frac{1}{i2\pi}\frac{dx}{dt}(t). \tag{A.4.2}$$

The product of these two operators depends on the order, and the operators satisfy the non-Abelian commutation relation:

$$[\hat{t}, \hat{f}] = \hat{t}\hat{f} - \hat{f}\hat{t} = \frac{1}{i2\pi}\hat{I}, \tag{A.4.3}$$

where $\hat{I}$ is the identity operator. This relation expresses the fact that the time and frequency variables are canonically conjugate.

The expectation value of an operator is expressed with the inner product, $\langle , \rangle$. The squared standard deviations can then be defined as

$$\Delta t^2 = \frac{1}{E_x}\langle \hat{t}^2 \rangle_x \quad \Delta f^2 = \frac{1}{E_x}\langle \hat{f}^2 \rangle_x. \tag{A.4.4}$$

Introducing the operator (Flandrin, 1999): $\hat{t} + i\lambda\hat{f}$, with λ being an arbitrary real number and using the positivity of the inner product, gives:

$$0 \leq \langle (\hat{t} + i\lambda\hat{f})x, (\hat{t} + i\lambda\hat{f})x \rangle = \langle (\hat{t} - i\lambda\hat{f})(\hat{t} + i\lambda\hat{f}) \rangle_x, \tag{A.4.5}$$

which simplified is

$$\langle \hat{t}^2 \rangle - \frac{E_x}{2\pi}\lambda + \langle \hat{f}^2 \rangle \lambda^2 \geq 0. \tag{A.4.6}$$

This inequality is satisfied for all λ if and only if the discriminant of the polynomial in λ is negative (Flandrin, 1999), which provides the relation:

$$\Delta t \cdot \Delta f \geq \frac{1}{4\pi}, \quad \text{or} \quad \Delta t \cdot \Delta \omega \geq \frac{1}{2} \tag{A.4.7}$$

which is the minimum (one standard deviation) signal time-bandwidth product. This demonstrates that there cannot be an unambiguous joint distribution based on two canonically conjugate variables, if their associated operators do not commute.

A.4.1.2. *The Slepian–Pollak–Landau approach*

Major contributions were made to the energy concentration problem by Slepian and collaborators,[14] the approach being similar to that used to address the Sturm–Liouville problem. The solutions that were found, prolate spheroidal wave functions (PSWFs), provide an orthonormal basis for the space of σ-bandlimited functions (support in $[-\sigma, +\sigma]$) and are maximally concentrated on an interval $[-\tau, +\tau]$. The PSWFs can be characterized as (Walter and Soleski, 2005):

(1) eigenfunctions of the Helmholtz equation on a prolate spheroid.
(2) providing the maximum energy concentration of a σ-bandlimited functions on an interval $[-\tau, +\tau]$, i.e., compact support in time and frequency.
(3) eigenfunctions of an integral or operator with kernel arising from the sync function, or of a differential operator.
(4) eigenvalues of a matrix operator equivalent to an integral operator (Walter and Soleski, 2005).

The PSWFs are eigenfunctions of the finite Fourier transform and of the sinc-kernel.[15]

However, as indicated by Khare and George (2003), when the energy concentration problem is couched in terms of an inversion problem, there is no obvious connection to the prolate spheroidal differential equation and wave functions. Their observation, translated from the spatial optical to the one-dimensional time signal domain, is that when a signal, $s(t)$, defined over $[-\sigma, +\sigma]$ and $[-T, +T]$ is sampled using the Whittaker-Shannon sampling theorem

[14]Slepian (1964, 1978, 1983); Slepian and Pollak (1961); Landau and Pollak (1961, 1962).

[15]Stratton *et al.* (1956); Flammer (1957); Walter (2005); Walter and Shen (2004); Walter and Soleski (2005); Walter and Shen (2003); Khare and George (2003).

(Whittaker, 1915; Shannon, 1949), the resulting signal is

$$\begin{aligned}\psi(t) &= \int_{-\sigma}^{+\sigma} df \exp[i2\pi ft] \int_{-T}^{+T} d\tau \exp[-i2\pi f\tau]s(\tau) \\ &= 2\sigma \int_{-T}^{+T} d\tau\, \mathbf{sinc}[2\sigma(t-\tau)]s(\tau). \end{aligned} \tag{A.4.8}$$

and the inversion problem is in terms of eigenfunctions of the sinc-kernel.

Khare and George (2003) considered the homogeneous Fredholm integral equation of the second kind with $\psi_n(t)$ as eigenfunctions and λ_n the associated eigenvalues:

$$\begin{aligned}\lambda_n\psi_n(t) &= \int_{-T}^{+T} d\tau\, \mathbf{sinc}[2\sigma(t-\tau)]\psi_n(\tau) \\ &= \sum\nolimits_{m=-\infty}^{+\infty} \lambda_n\psi_n\left(\frac{m}{2\sigma}\right) \mathbf{sinc}(2\sigma t - m). \end{aligned} \tag{A.4.9}$$

Using a substitution:

$$\lambda_n\psi_n\left(\frac{m}{2\sigma}\right) = \int_{-T}^{+T} d\tau\, \mathbf{sinc}\left[2\sigma\left(\frac{m}{2\sigma}-\tau\right)\right]\psi_n(\tau), \tag{A.4.10}$$

the following matrix representation is obtained:

$$\lambda_n\psi_n\left(\frac{m}{2\sigma}\right) = \sum_{k=-\infty}^{+\infty} A_{mk}\psi_n\left(\frac{k}{2\sigma}\right), \tag{A.4.11}$$

where

$$A_{mk} = \int_{-T}^{+T} d\tau\, \mathbf{sinc}(2\sigma\tau - m)\mathbf{sinc}(2\sigma\tau - k), \tag{A.4.12}$$

which is a real symmetric matrix.

In the discrete case, the prolate spheroidal wave functions are referred to as discrete prolate spheroidal sequences (Slepian, 1978). In the case of a sequence $s[n]$:

$$s[n] = 0 \quad \text{for } n < N_0 \quad \text{and for } n > N_0 + N - 1, \tag{A.4.13}$$

the critical measure of the extent to which the discrete Fourier transform of $s[n]$ is concentrated in the frequency interval $[-\sigma, +\sigma]$

is the ratio:

$$\mu = \frac{\int_{-\sigma}^{+\sigma} |H(f)|^2 df}{\int_{-1/2}^{+1/2} |H(f)|^2 df}. \tag{A.4.14}$$

The concentration ratio can be expressed as

$$\begin{aligned}\mu &= \frac{\sum_{n=N_0}^{N_0+N-1} \sum_{m=N_0}^{N_0+N-1} \frac{\sin(2\pi\sigma(n-m))}{\pi(n-m)} s[n]s^*[n]}{\sum_{n=N_0}^{N_0+N-1} |s[n]|^2} \\ &= \frac{\sum_{n=0}^{N-1} \sum_{m=0}^{N-1} \frac{\sin(2\pi\sigma(n-m))}{\pi(n-m)} s[n+N_0]s^*[n+N_0]}{\sum_{n=N_0}^{N_0+N-1} |s[n+N_0]|^2}. \end{aligned} \tag{A.4.15}$$

The sequence that maximizes this ratio for the bandwidth σ is the sequence:

$$s[n-N_0] = c\vartheta[n]^{(0)}(N,\sigma) \quad \text{for } n = 0, 1, \ldots, N-1, \\ \text{and the scalar, } c, \tag{A.4.16}$$

and for interval $[-\sigma, +\sigma]$ is the sequence:

$$s[n] = \vartheta[n]_{n-N0}^{(0)}(N,\sigma) \quad \text{for } N_0 \leq n \leq N_0 + N - 1, \tag{A.4.17}$$

with spectrum:

$$\Sigma(f) = d\Omega_0(N,\sigma)\exp[i\pi(2N_0 + N - 1)]f, \tag{A.4.18}$$

where Ω_0 is the discrete prolate spheroidal wave function of zeroth order corresponding to the discrete prolate spheroidal series of order k, length N and frequency interval $[-\sigma, +\sigma]$:

$$\begin{aligned}&\sum_{m=0}^{N-1} \frac{\sin(2\pi\sigma(n-m))}{\pi(n-m)} \vartheta[n]^{(k)}(N,\sigma) \\ &\quad = \lambda^{(k)}(N,\sigma)\vartheta[n]^{(k)}(N,\sigma), \quad n = 0, \pm 1, \pm 2, \ldots. \end{aligned} \tag{A.4.19}$$

for $k = 0, 1, 2, \ldots, N-1$, and the $\lambda^{(k)}(N,\sigma)$ are the eigenvalues of the system:

$$0 < \lambda_0^{(k)} < \lambda_1^{(k)} < \cdots < \lambda_{N-1}^{(k)} < 1. \tag{A.4.20}$$

These eigenvalues are clustered near zero and one, indicating that the eigenvectors are ill conditioned numerically.

A more physical interpretation is as follows. The wave equation, the Helmholtz equation, in prolate spheroidal coordinates is (Morse and Feshbach, 1953, p. 642):

$$(z^2) - 1)\frac{\partial^2\psi}{\partial z^2} + 2(a+1)z\frac{\partial\psi}{\partial z} + (h^2z^2 - b)\psi = 0, \quad z = x + iy, \tag{A.4.21}$$

and separable (Morse and Feshbach, 1953, pp. 514, 661), with regular singular points at $z = \pm 1$ and an irregular singular point at $z = \infty$. The methods of solution are as indicated above, and there are no analytic methods. New methods of evaluating the eigenvalues of prolate spheroidal harmonics have recently been proposed (Li *et al.* 1998) and earlier methods have been criticized as inaccurate. The new methods are:

Let D = time-bandwidth product, be an allocated tile, in the time-frequency plane representing the time-bandwidth product of an optimum symbol. Then the energy of that symbol is defined as:

$$E = \iint_D C(t, \nu; f) dt d\nu, \tag{A.4.22}$$

where

$$C(t, \nu; f) = \iiint_R e^{i2\pi\xi(s-t)} f(\xi, \tau) x\left(s + \frac{\tau}{2}\right) x * \left(s - \frac{\tau}{2}\right) e^{-i2\pi\nu\tau} d\xi ds d\tau \tag{A.4.23}$$

is a Cohen's class distribution (Cohen, 1989, 1995), for an arbitrary parameter function f. Computing the maximal energy is equivalent to finding the signal that belongs to the largest eigenvalue of the projection onto the time-bandwidth product, D. Given a function $G(t, \nu)$ and the arbitrary parameter function, f, then the operator associated with the function and the arbitrary parameter function is

$$(\hat{G}_f)x(t) = \int_{-\infty}^{+\infty} \gamma_f(t, s) x(s) ds \tag{A.4.24}$$

where

$$\gamma_f(s,t) = \int_{-\infty}^{+\infty} F\left(\frac{t+s}{2} - \theta, t-s\right)\gamma(\theta, t-s)d\theta, \tag{A.4.25}$$

$$\gamma(t,\tau) = \int_{-\infty}^{+\infty} G(t,\nu)e^{-i2\pi\nu\tau}d\nu, \tag{A.4.26}$$

and F is the Fourier transform of f.

The eigenvalue equation is then:

$$\int_{-\infty}^{+\infty} \gamma_f(t,s)x(s)ds = \lambda x(t). \tag{A.4.27}$$

As

$$\frac{E(D,f)}{E} \leq \lambda_{\max}, \tag{A.4.28}$$

every signal that constitutes an eigenfunction to the eigenlargest eigenvalue yields the best possible energy concentration of the distribution $C(t.\nu, f)$ to the time-bandwidth product D. There are no general solutions to this problem presently available, but there are special solutions.

Maximizing the symbol energy is obtained if the original arbitrary signal is an eigenfunction of an integral equation equal to the ratio of the energy of the twice truncated signal and the original signal. The eigenvalue equation has a spectrum of positive eigenvalues:

$$\mathbf{0} < \cdots < \lambda_n < \cdots < \lambda_1 < \lambda_0 = \lambda_{\mathbf{max}}, \tag{A.4.29}$$

and each eigenvalue is a function of the time-bandwidth product. The time-bandwidth products of the dominant eigenvalues are close to 1, while the others are close to 0. This development is in contrast to that for OSSO, as described above, where each eigenvalue is a function of a series of *increasing* products:

$$\lambda = 1/2(2n+1), \quad n = 0,1,2,\ldots \tag{A.4.30}$$

associated with a sequence of increasing energy levels.

There is one and only one eigenfunction associated with each eigenvalue. Properly normalized, this collection of eigenfunctions are the prolate spheroidal wave functions (PSWFs). A bandlimited signal which maximizes the energy concentration in a set time interval is an eigenfunction relative to a maximal eigenvalue. The eigenfunctions form an orthonormal system, so that:

$$x(t) = \sum_{n=0}^{+\infty} x_n \phi_n(t), \quad \text{where } x_n = \int_{-\infty}^{+\infty} x(t)\phi_n(t)dt. \tag{A.4.31}$$

As

$$\frac{\sin \pi B(t-s)}{\pi(t-s)} = \sum_{n=0}^{\infty} \phi_n(t)\phi_n(s), \tag{A.4.32}$$

we have

$$\sum_{n=0}^{\infty} \lambda_n = \int_{-T/2}^{+T/2} \left[\frac{\sin \pi B(t-s)}{\pi(t-s)}\right]_{s=t} dt = \textit{time-bandwidth product}. \tag{A.4.33}$$

An example of the first prolate spheroid is shown in Fig. 4.16.

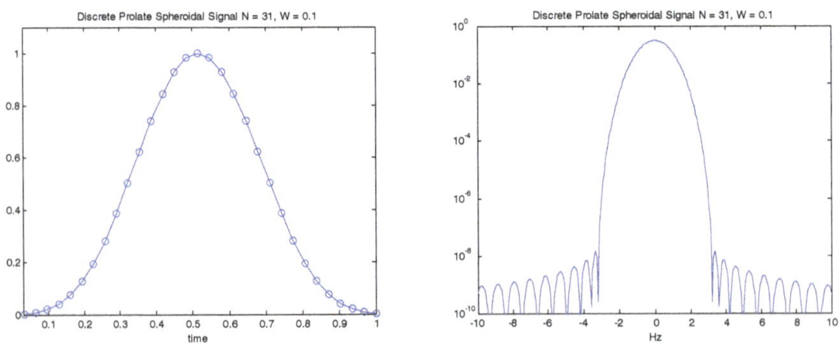

Fig. 4.16. Discrete first prolate spheroid for window length N = 31, and bandwidth concentration $W = 0.1$. Time domain representation on the left, frequency domain representation on the right.

A.4.1.3. *The Weber–Hermite transform and signal energy confinement*

The parabolic cylinder functions, or Weber–Hermite functions, are solutions to Weber's equation (Weber, 1869):

$$\frac{d^2\Psi}{dx^2} + \left(n + \frac{1}{2} - \frac{1}{4}x^2\right)\Psi(x) = 0, \tag{A.4.34}$$

for which the general Weber equation, or parabolic cylinder differential equation, is (Abramowitz and Stegun, 1972, p. 686):

$$\frac{d^2\Psi}{dx^2} + (ax^2 + bx + c)\Psi = 0. \tag{A.4.35}$$

and for which, point $x = \infty$ is strongly singular.

A.4.1.3.1. *Whittaker and Watson's derivation*

Equation (A.4.34) permits two solutions (Whittaker and Watson, p. 347). The substitutions $\Psi = x^{-1/2}W, z = x^2/2$, convert the Weber equation to the Whittaker equation, which is a special case of the confluent hypergeometric equation.[16] In particular, taking the solution for which $R(z) > 0$, the solution is

$$\begin{aligned}\psi_n &= 2^{n/2+1/4}z^{-1/2}W_{n/2+1/4,-1/4}(1/2z^2)\\ &= \frac{1}{\sqrt{z}}2^{n/2}e^{-z^2/4}(-iz)^{1/4}(iz)^{1/4}{}_1F_1\left(-\frac{1}{2}n, \frac{1}{2}, \frac{1}{2}z^2\right),\end{aligned} \tag{A.4.36}$$

where $W_{k,m}$ is the Whittaker function (Whittaker and Watson, p. 347) and ${}_1F_1(a,b,c)$ is a confluent hypergeometric function (Gauss,

[16]The hypergeometric differential equation is a second-order linear ordinary differential equation whose solutions are given by the hypergeometric series. The hypergeometric series have the form:

$$\left(\sum_{n=0}^{\infty} a_n\right)\left(\sum_{n=0}^{\infty} b_n\right) = \sum_{n=0}^{\infty} c_n, \quad \text{where } c_n = a_k b_{n-k}.$$

The confluent hypergeometric equation is a degenerate form of the hypergeometric equation.

1812). If

$$w = z^{-1/2} W_{k,-1/4}\left(\frac{1}{2}z^2\right), \tag{A.4.37}$$

then

$$\frac{\partial^2 w}{\partial z^2} + \left(2k - \frac{1}{4}z^2\right) w = 0, \tag{A.4.38}$$

The Weber equation (A.4.34) can be separated into:

$$\frac{d^2 U}{du} - (c + k^2 u^2)U = 0, \tag{A.4.39}$$

$$\frac{d^2 V}{du^2} - (c - k^2 u^2)V = 0. \tag{A.4.40}$$

For nonnegative n, and after renormalization, the solutions of Weber's equation (A.4.35) reduce to:

$$U_n(x) = 2^{-n/2} e^{-x^2/4} H_n(x/\sqrt{2}), \quad n = 0, 1, 2, \ldots, \tag{A.4.41}$$

which are parabolic cylinder functions or Weber–Hermite functions, and where H_n is a Hermite polynomial.

A.4.1.3.2. *Weber's derivation*

Completing the square, the Weber's second equation (A.4.35) can be rewritten as

$$\frac{d^2\Psi}{dx^2}\left[a\left(x + \frac{b}{2a}\right)^2 - \frac{b^2}{4a} + c\right]\Psi = 0. \tag{A.4.42}$$

Defining $u = x + \frac{b}{a}$; $du = dx$, and substituting, gives

$$\frac{d^2\Psi}{du^2} + (au^2 + d)\Psi = 0, \tag{A.4.43}$$

where $d = \frac{b^2}{4a} + c$.

Again, this equation admits of two solutions: an even and an odd function. Continuing with the even solution, the solution is

$$\Psi(x) = e^{-x^2/4} {}_1F_1\left(\frac{1}{2}a + \frac{1}{4}; \frac{1}{2}; \frac{1}{2}x^2\right), \tag{A.4.44}$$

where ${}_1F_1(a; b; z)$, as in Eq. (A.4.36), is the confluent hypergeometric function. The solutions of this equation are the parabolic cylinder or Weber–Hermite functions:

$$\psi_n(x) = 2^{-n/2} e^{-x^2/4} H_n(x/\sqrt{2}), \quad n = 0, 1, 2, \ldots, \tag{A.4.45}$$

A.4.1.3.3. *Morse and Feshbach's derivation*

A parallel derivation is as follows. The one-dimensional wave equation is

$$-\frac{1}{2m}\frac{\partial^2 \Psi}{\partial x^2} + V(x)\Psi = E\Psi, \tag{A.4.46}$$

with spring potential:

$$V(x) = \frac{1}{2}kx^2 = \frac{1}{2}m\omega^2 x^2, \tag{A.4.47}$$

where $\omega = \sqrt{\frac{k}{m}}$ = angular frequency, k is the stiffness constant, m is the mass, and x is the field deflection of the oscillator, or

$$-\frac{1}{2m}\frac{\partial^2 \Psi}{\partial x^2} + kx^2\Psi = E\Psi. \tag{A.4.48}$$

This wave equation can be written in dimensionless form by defining the independent variables $\xi = \alpha x$ and an eigenvalue, λ, and requiring:

$$\alpha^4 = mk; \quad \lambda = 2E\left(\frac{m}{k}\right)^{1/2} = \frac{2E}{\omega}. \tag{A.4.49}$$

The dimensionless form is

$$\frac{\partial^2 u}{\partial \xi^2} + (\lambda - \xi^2)u = 0. \quad \text{(Weber's equation)} \tag{A.4.50}$$

The wave equation for a vibrating string is associated with the difference between the total kinetic energy of the string and its potential energy being as small as possible. If a string vibrates with simple harmonic motion, then the time dependence is expressed as $\Psi = \psi(x)e^{-i\varepsilon\alpha^2 t}$ and the function ψ must satisfy Helmholtz's equation:

$$\frac{\partial^2\psi}{\partial x^2} + k^2\psi = 0, \tag{A.4.51}$$

with k a real constant. When $k = 0$, the equation is a one-dimensional Laplace equation; and when k^2 is a function of the coordinates and $\varepsilon = (2M/\hbar^2)E$, the equation is Schrödinger's equation for a particle with constant E (Morse and Feshbach, 1953, p. 494).

When ψ is space-dependent, the wave equation is

$$\frac{\partial^2\psi}{\partial x^2} + (\varepsilon - \alpha^2 x^2)\psi = 0 \quad \text{(Weber's equation)}, \tag{A.4.52}$$

where $\alpha = M\omega/\hbar$. This equation permits solutions as a function of $n = \frac{\varepsilon}{2\beta} - \frac{1}{2} = \frac{E}{\hbar\omega} - \frac{1}{2}$. In order for the solutions to be quadratically integrable, it is necessary that n take on integer values: $n = 0, 1, 2, \ldots$ (Morse and Feshbach, 1953, p. 1641). With normalization factors, the solutions are the Weber–Hermite or parabolic cylinder functions (Morse and Feshbach, 1953, p. 1642):

$$\psi_n(t) = \frac{1}{\sqrt{2^n n!}}\left(\frac{\alpha}{\pi}\right)^{1/4} \exp\left(-\frac{\alpha t^2}{2}\right) H_n(t\sqrt{\alpha}), \tag{A.4.53}$$

where $\alpha = M\omega/\hbar$. For the classical result, we substitute $\hbar \to 1$, and α is a time-frequency trade parameter/variable. This is the general form of the Weber–Hermite functions, and for the particular Weber–Hermite functions (Eqs. (A.4.41) and (A.4.45)), $\alpha = 1/2$, and there is alternative normalization.

If for a function, $f(x)$, an expansion of the form

$$f(x) = a_0\psi_0(x) + a_1\psi_1(x) + \cdots + a_n\psi_n(x) + \cdots \tag{A.4.54}$$

exists, and if it is legitimate to integrate term-by-term between the limits $-\infty$ and ∞, then:

$$a_n = \frac{1}{(2\pi)^{1/2} n!}\int_{-\infty}^{\infty} \psi_n(t) f(t)dt. \tag{A.4.55}$$

Thus by considering the envelope of a transmitted symbol as a function, that function can be expanded in terms of n Weber–Hermite functions with coefficients a_n.

Another picture, arriving at the same result, is as follows. In parabolic cylinder coordinates, the Helmholtz equation is

$$\frac{1}{u^2+\nu^2}\left(\frac{\partial^2\Psi}{\partial u^2}+\frac{\partial^2\Psi}{\partial \nu^2}\right)+\frac{\partial^2\Psi}{\partial z^2}+k^2\Psi=0. \tag{A.4.56}$$

This equation can be separated with:

$$\Psi(u,\nu,z)=U(u)V(v)Z(z), \tag{A.4.57}$$

resulting in:

$$\frac{1}{u^2+\nu^2}\left(VZ\frac{\partial^2 U}{\partial u^2}+UZ\frac{\partial^2 V}{\partial \nu^2}\right)+UV\frac{\partial^2 Z}{\partial z^2}+k^2UVZ=0. \tag{A.4.58}$$

Dividing by UVZ and separating out the Z part gives

$$\frac{\partial^2 Z}{\partial z^2}=-(k^2+m^2)Z, \quad \text{(Weber's equation 1)} \tag{A.4.59}$$

which can be solved, permitting the derivation:

$$\left(\frac{1}{U}\frac{\partial^2 U}{\partial u^2}-k^2u^2\right)+\left(\frac{1}{V}\frac{\partial^2 V}{\partial v^2}-k^2v^2\right)=0. \tag{A.4.60}$$

With

$$\left(\frac{1}{U}\frac{\partial^2 U}{\partial u^2}-k^2u^2\right)=c, \tag{A.4.61}$$

$$\left(\frac{1}{V}\frac{\partial^2 V}{\partial v^2}-k^2v^2\right)=-c, \tag{A.4.62}$$

we have

$$\frac{\partial^2 U}{\partial u^2}=-(c+k^2u^2)U, \quad \text{(Weber's equation 2)}, \tag{A.4.63}$$

$$\frac{\partial^2 V}{\partial v^2}=+(c-k^2v^2)V, \quad \text{(Weber's equation 3)}. \tag{A.4.64}$$

The solutions of Weber's equation 2 reduce to:

$$U_n(x) = 2^{-n/2}e^{-x^2/4}H_n(x/\sqrt{2}), \quad n = 0, 1, 2, \ldots, \tag{A.4.65}$$

which are parabolic cylinder functions or Weber–Hermite functions, where H_n is a Hermite polynomial – as for Eqs. (A.4.41) and (A.4.45).

A.4.1.3.4. *The classical electromagnetic wave derivation*

Hertz showed that the electromagnetic field can be expressed in terms of a single vector function (Hertz, 1889). A vector $\boldsymbol{p}$ is defined:

$$\frac{\partial \mathbf{p}}{\partial t} = \mathbf{J}, \tag{A.4.66}$$

$$\nabla \mathbf{p} = -\rho, \tag{A.4.67}$$

where $\mathbf{J}$ is the current density, and ρ is the charge density. It can then be shown (Jones, 1964) that:

$$\begin{aligned} &\mathbf{A} = \mu\varepsilon\frac{\partial \mathbf{\Pi}}{\partial t}, \quad V = -\nabla \cdot \mathbf{\Pi}, \quad \nabla^2\mathbf{\Pi} - \mu\varepsilon\frac{\partial^2 \mathbf{\Pi}}{\partial t^2} = -\boldsymbol{p}/\varepsilon, \\ &\mathbf{E} = \nabla\nabla \cdot \mathbf{\Pi} - \mu\varepsilon\frac{\partial^2 \mathbf{\Pi}}{\partial t^2}, \quad \boldsymbol{B} = \mu\varepsilon\nabla \times \frac{\partial \mathbf{\Pi}}{\partial t}. \end{aligned} \tag{A.4.68}$$

where μ is the magnetic permeability, and ε is the dielectric constant. The vector $\mathbf{\Pi}$ is the electric Hertz vector.

In parabolic cylinder coordinates ξ, η, z, the electric Hertz vector, $\mathbf{\Pi}$ is (Jones, 1964, p. 85):

$$\mathbf{\Pi} = \Xi(\xi) \cdot \mathbf{H}(\eta) \cdot \boldsymbol{Z}(z), \tag{A.4.69}$$

and

$$\begin{aligned} &\boldsymbol{Z}'' = \mu^2\boldsymbol{Z}, \\ &\Xi'' + [(\mu^2 + k^2)\xi^2 - h]\Xi = 0, \\ &\mathbf{H}'' + [(\mu^2 + k^2)\eta^2 + h]\mathbf{H} = 0, \end{aligned} \tag{A.4.70}$$

where μ and h are separation constants (Jones, p. 85). Both the equations in Ξ and $\mathbf{H}$ can be treated similarly. A substitution of

$$\xi = (4(\mu^2 + k^2)^{-1/4})\mathrm{X}, \tag{A.4.71}$$

gives

$$\Xi'' + (1/4\mathbf{X}^2 - \boldsymbol{i}(n + 1/2))\Xi = 0. \tag{A.4.72}$$

A further substitution of $\mathrm{X} = x \exp[\boldsymbol{i}1/4\pi]$ gives

$$\Xi'' + (n + 1/2 - 1/4x^2)\Xi = 0, \tag{A.4.73}$$

which is the Weber's equation (Weber, 1869). A solution is

$$\psi_n(x) = 2^{-1/2n} e^{-1/4x^2} H_n(2^{-1/2}x), \quad n = 0, 1, 2, \ldots, \tag{A.4.74}$$

where the H_n are Hermite polynomial functions and the ψ_n are again Weber–Hermite functions.

A.4.1.3.5. *The parallel quantum mechanical wave derivation*

The Hamiltonian of a particle is

$$H = \frac{p^2}{2m} + \frac{1}{2} m\omega^2 x^2, \tag{A.4.75}$$

where x is the position operator, p is the momentum operator $p = -i\hbar\frac{d}{dx}$, where the first term is the kinetic energy of the particle, and the second term the potential energy.

The one dimensional Schrödinger wave equation is

$$-\frac{\hbar^2}{2m}\frac{d^2\Psi}{dx^2} + V(x)\Psi = E\Psi, \quad \text{or} \quad -\frac{\hbar^2}{2m}\frac{d^2\Psi}{dx^2} + \frac{kx^2\Psi}{2} = E\Psi. \tag{A.4.76}$$

Equation (A.4.72) is the quantum mechanical version of Eq. (A.4.46). This equation can be written in dimensionless form using the following substitutions:

$$\xi = \alpha x.$$

$$\alpha^4 = \frac{mk}{\hbar^2}, \tag{A.4.77}$$

$$\lambda = \frac{2E}{\hbar}\left(\frac{m}{k}\right)^{1/2} = \frac{2E}{\hbar\omega}.$$

The wave equation in dimensionless form is then:

$$\frac{d^2\Psi}{d\xi^2} + (\lambda - \xi^2)\Psi = 0, \quad \text{(Weber's equation)} \tag{A.4.78}$$

Alternatively, by coordinate transformation:

$$x = \left(\frac{\hbar}{m\omega}\right)^{1/2} \xi \tag{A.4.79}$$

the wave equation becomes:

$$\left(\frac{d^2}{d\xi^2} + \frac{2E}{\hbar\omega} - \xi^2\right) \Psi = 0. \quad \text{(Weber's equation)} \tag{A.4.80}$$

Except when $\xi = \infty$, this wave equation has asymptotic solutions:

$$\Psi = \exp\left(\pm\frac{1}{2}\xi^2\right)^{1/2}, \tag{A.4.81}$$

Simplification of Eq. (A.4.77) can be achieved if the asymptotic factors are excluded by introducing the variable, y:

$$\Psi = \exp\left(\pm\frac{1}{2}\xi^2\right)^{1/2} y, \tag{A.4.82}$$

So that the following equation for y results:

$$\left(\frac{d^2}{d\xi^2} - 2\xi\left(\frac{d}{d\xi}\right) + \frac{2E}{\hbar\omega} - 1\right) y = 0. \tag{A.4.83}$$

In the case of solutions of y for polynomials of degree n:

$$-2n + \frac{2E}{\hbar\omega} - 1 = 0, \tag{A.4.84}$$

and the eigenvalues are as follows:

$$E_n = (n + 1/2)\hbar\omega. \tag{A.4.85}$$

The solutions of Eq. (A.4.78) are as follows:

$$y_n = H_n(\xi) = (-1)^n e^{\xi^2} \frac{d^n}{d\xi^n} e^{-\xi^2}, \tag{A.4.86}$$

where $H_n(\xi)$, $n = 0, 1, 2, \ldots$ are Hermite polynomials, satisfying:

$$\left(\frac{d^2}{d\xi^2} - 2\xi\frac{d}{d\xi} + 2n\right) H_n(\xi) = 0, \tag{A.4.87}$$

for which $2n = \lambda - 1$, or $\lambda = 2n + 1$. Therefore, substituting, the solutions of Eq. (A.4.80) are:

$$\Psi = \exp\left(\pm\frac{1}{2}\xi^2\right)^{1/2} H_n(\xi), \quad n = 0, 1, 2, \ldots, \tag{A.4.88}$$

which, as solutions to Weber's equation involving Hermite functions, are Weber–Hermite functions.

An alternative view is to write the time-independent one-dimensional Schrödinger equation in the form:

$$H|\Psi\rangle = E|\Psi\rangle, \quad \text{or} \quad \frac{-\hbar^2}{2m}\frac{d^2\Psi(x)}{dx^2} + \frac{1}{2}m\omega^2x^2\Psi(x) = E\Psi(x)$$
$$\text{(Weber's equation)} \tag{A.4.89}$$

The general solution is

$$\langle x|\psi_n\rangle = \frac{1}{\sqrt{2^n n!}}\left(\frac{m\omega}{\pi\hbar}\right)^{1/4}\exp\left(-\frac{m\omega x^2}{2\hbar}\right) H_n\left(\sqrt{\frac{m\omega}{\hbar}}x\right),$$
$$n = 0, 1, 2, \ldots, \tag{A.4.90}$$

where $H_n(x) = (-1)^n e^{x^2}\frac{d^n}{dx^n}e^{-x^2}$ are Hermite polynomials; or

$$\psi_n(z) = \left(\frac{\alpha}{\pi}\right)^{1/4}\frac{1}{\sqrt{2^n n!}}H_n(z)\exp[-z^2/2], \quad n = 0, 1, 2, \ldots, \tag{A.4.91}$$

where $z = \sqrt{\alpha}x$ and $\alpha = \frac{m\omega}{\hbar}$. This is a normalized form of the Weber–Hermite functions — see Eq. (A.4.51).

The corresponding energy levels are as follows:

$$E_n = \hbar\omega\left(n + \frac{1}{2}\right), \tag{A.4.92}$$

and the expectation value for the potential energy is

$$\langle V_n \rangle = \frac{1}{2}E_n = \int_{-\infty}^{+\infty} (1/2)\psi_n^* kx^2 \psi_n(x)dx = (1/2)k\frac{2n+1}{2\alpha^2}$$
$$= (1/2)(n+1/2)\hbar\omega = 1/2E_n, \tag{A.4.93}$$

Therefore,

$$\Delta x \Delta p = \frac{1}{2}(2n+1)\hbar. \tag{A.4.94}$$

Substituting $x \to t$, $p \to f$ and $\hbar \to 1$ for the classical case gives the time-bandwidth products for the WH signals:

$$\Delta t \Delta f = \text{time-bandwidth product} = \frac{1}{2}(2n+1), \quad n = 0, 1, 2, \ldots \tag{A.4.95}$$

It should be noted that these time-bandwidth products refer to one standard deviation of the signal duration, and one standard deviation of the signal bandwidth, i.e., not the 90% or 99% support of signal duration and bandwidth.

Appendix 4.2: The WH Transform and WH Wavelets

As the appropriate analysis method for both linear time invariant (LTI) and linear time-varying (LTV) system signals is time-frequency analysis, a local wavelet decomposition of RX signals can be used to analyze test results. The specific wavelet analysis method used provides a solution to the well-known energy confinement problem discussed above, or time-bandwidth product limiting problem, addressed by Slepian and collaborators.[17] Here is a short introduction.

As discussed above, Slepian *et al.*'s solution to the energy confinement problem was the prolate spheroidal wave function (PSWF) series, that can be recursively generated. Above an alternative

[17]Landau and Pollak (1961, 1962); Slepian (1964, 1978); Slepian and Pollak (1961).

solution, the parabolic cylinder or Weber–Hermite wave function (WHWF) series (Barrett, 1972, 1973a,b), was used.

As previously noted, these series are related to *Weber's equation* (Weber, 1869):

$$\frac{d^2\psi_n(x)}{dx^2} + \left(n + \frac{1}{2} - \frac{1}{4}x^2\right)\psi_n(x) = 0, \qquad \text{(A.4.96)}$$

for which the *general Weber equation*, or *parabolic cylinder differential equation*, is (Abramowitz & Stegun, 1972):

$$\frac{d^2\psi_n(x)}{dx^2} + (ax^2 + bx + c)\psi_n(x) = 0. \qquad \text{(A.4.97)}$$

The solutions of this equation are the parabolic cylinder or Weber–Hermite wave functions (WHWFs):

$$\psi_n(x) = 2^{-n/2}\exp[-x^2/4]H_n(x/\sqrt{2}), \quad n = 0, 1, 2, \ldots, \qquad \text{(A.4.98)}$$

where the H_n are Hermite polynomials. When n is an integer, the Weber–Hermite functions become proportional to the Hermite polynomials. Other names are used for the Weber–Hermite wave functions, e.g., Hermite-Gaussian functions.[18]

The WHWFs are given a physical representation as follows. The one-dimensional wave equation is

$$-\frac{1}{2m}\frac{\partial^2\psi}{\partial x^2} + V(x)\psi = E\psi, \qquad \text{(A.4.99)}$$

with spring potential:

$$V(x) = \frac{1}{2}kx^2 = \frac{1}{2}m\omega^2x^2, \qquad \text{(A.4.100)}$$

[18]We prefer the designation Weber–Hermite wave function because (a) Hermite-Gaussian implicates Gaussian in all polynomials, $n = 0, 1, 2, \ldots$ but the Gaussian is only just the first, for the case $n = 0$; (b) Weber's equation is more general than Hermite's equation; (c) the name "Weber–Hermite" follows the *Mathematical Encyclopedia* (Hazewinkel, 2002), usage; and (d) other texts, e.g., (Morse and Feshback, 1953, vol. 2, p. 1642; Jones, 1964, p. 86), have also used the name"Weber–Hermite".

where $\omega = \sqrt{\frac{k}{m}}$ is the angular frequency, k is the stiffness constant, m is the mass, and x is the field deflection of the oscillator. The wave equation can be written in dimensionless form by defining the independent variables $\xi = \alpha x$ and an eigenvalue, λ, and requiring:

$$\alpha^4 = mk, \quad \lambda = 2E\left(\frac{m}{k}\right)^{1/2} = \frac{2E}{\omega}. \tag{A.4.101}$$

The dimensionless form is then

$$\frac{\partial^2\psi}{\partial\xi^2} + (\lambda - \xi^2)\psi = 0, \tag{A.4.102}$$

which is also a form of Weber's equation and permits solutions as a function of $n = \frac{E}{2\beta} - \frac{1}{2}$. In order for the solutions to be quadratically integrable, it is necessary that n takes on integer values: $n = 0, 1, 2, \ldots$ (Morse and Feshbach, 1953, p. 1641). With normalization factors, the solutions are the Weber–Hermite or parabolic cylinder functions:

$$\psi_n(t) = \frac{1}{\sqrt{2^n n!}}\left(\frac{\alpha}{\pi}\right)^{1/4} \exp\left(-\frac{\alpha t^2}{2}\right) H_n(t\sqrt{\alpha}), \tag{A.4.103}$$

where $\alpha = m\omega$ and the H_n are Hermite polynomials, and α is a time-frequency trade parameter/variable.

If in the case of a function, $f(x)$, an expansion of the form:

$$f(x) = a_0\psi_0(x) + a_1\psi_1(x) + \cdots + a_n\psi_n(x) + \cdots \tag{A.4.104}$$

exists, and if it is legitimate to integrate term-by-term between the limits $-\infty$ and $+\infty$, then:

$$a_n = \frac{1}{\sqrt{(2\pi)^{1/2} n!}} \int_{-\infty}^{+\infty} \psi_n(t) f(t) dt. \tag{A.4.105}$$

The first 6 WHWFs (WHWF 0–5) are shown in the time and frequency domains in Fig. 4.15. It is apparent that the WHWFs

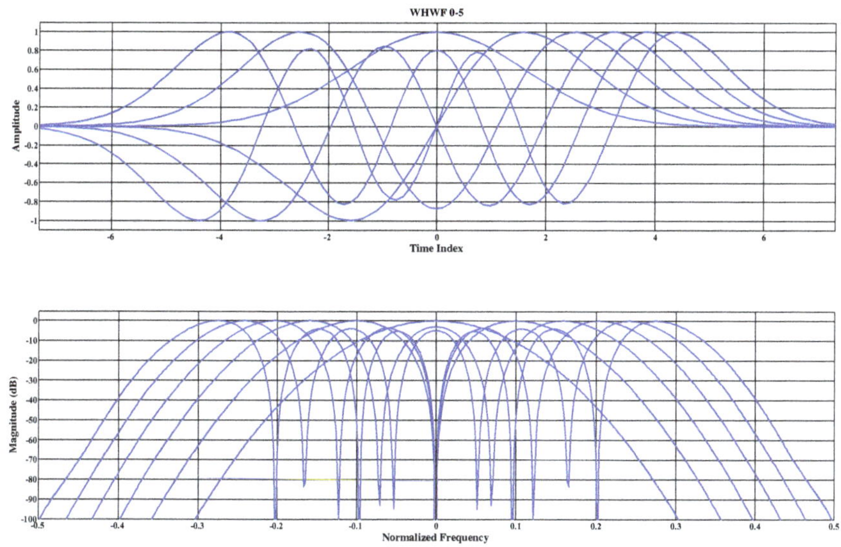

Fig. 4.17. The first 6 WHWFs (upper) and their Log Magnitude Spectra (lower) labeled according to an increasing n ($n = 0, 1, 2, 3, 4, 5$). As n increases, the temporal length and bandwidth increase, i.e., time-bandwidth product increases.

increase in (temporal length $\times$ bandwidth), i.e., in time-bandwidth product, as n increases (Fig. 4.17).

In a manner similar to that of the Fourier transform, a WH transform can be constructed by matrix methods. Figure 4.18(a) shows a 128×256 WH (magnitude) matrix. If the complex WH matrix is designated, $\mathbf{W}$, then, as $\mathbf{W}$ is a unitary matrix, $\mathbf{W}\mathbf{W}^\dagger = \mathbf{I}$, where $\mathbf{W}^\dagger$ is the conjugate transpose (Hermitian adjoint) of $\mathbf{W}$, and $\mathbf{I}$ is the identity matrix shown in Fig. 4.18(b). The inverse of $\mathbf{W}$, or $\mathbf{W}^{-1}$, is equal to the conjugate transpose: $\mathbf{W}^{-1} = \mathbf{W}^\dagger$.

Figure 4.19 — also Fig. 4.14 above — is a 6 WH section of Fig. 4.18.

A multiple–window time–frequency analysis (MWTFA) spectrum is a WHWF expression of Thomson's multiple window method (Thomson, 1982), that is a time–frequency distribution estimator for a random process and is a substitute for a Fourier transform

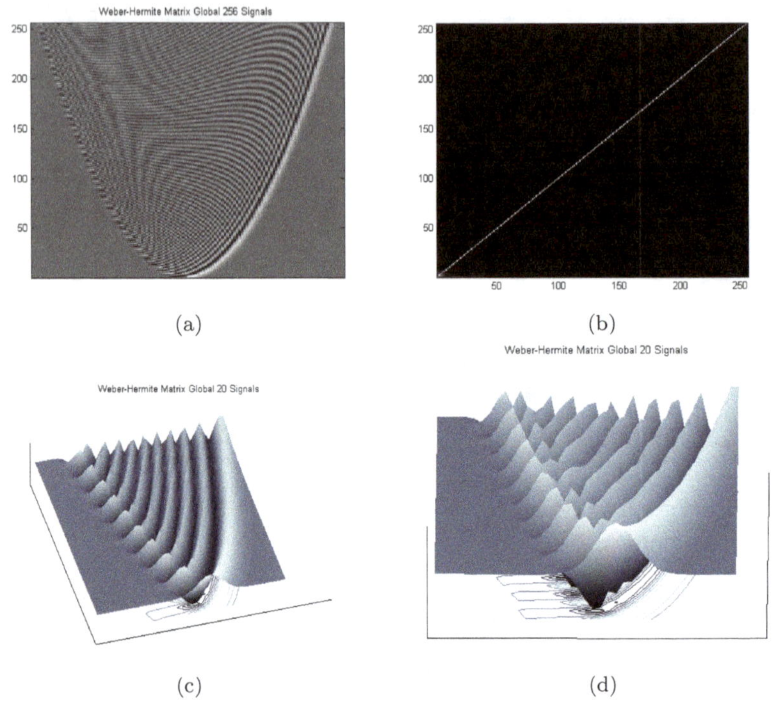

Fig. 4.18. (a) WH matrix composed of 256 WHWF Signals. All WH matrices are unitary. If the matrix is designated, M, then $MM^{\dagger} = I$, the identity matrix shown in (b), where $M^{\dagger}$ is the conjugate transpose (Hermitian adjoint) of M and I is the identity matrix. The inverse of M, or M^{-1}, is equal to the conjugate transpose: $M^{-1} = M^{\dagger}$. Images (c) and (d) are different views of the matrix M composed of 20 WHWF signals. The first, WHWF-0, a Gaussian, is at the bottom of the figure in both cases.

periodogram — an inappropriate estimator for short, time-limited signals. Thomson's approach to spectral estimation of such signals is to compute several periodograms using a set of orthogonal windows that are locally concentrated in frequency and then averaged (Xu *et al.* 1999).

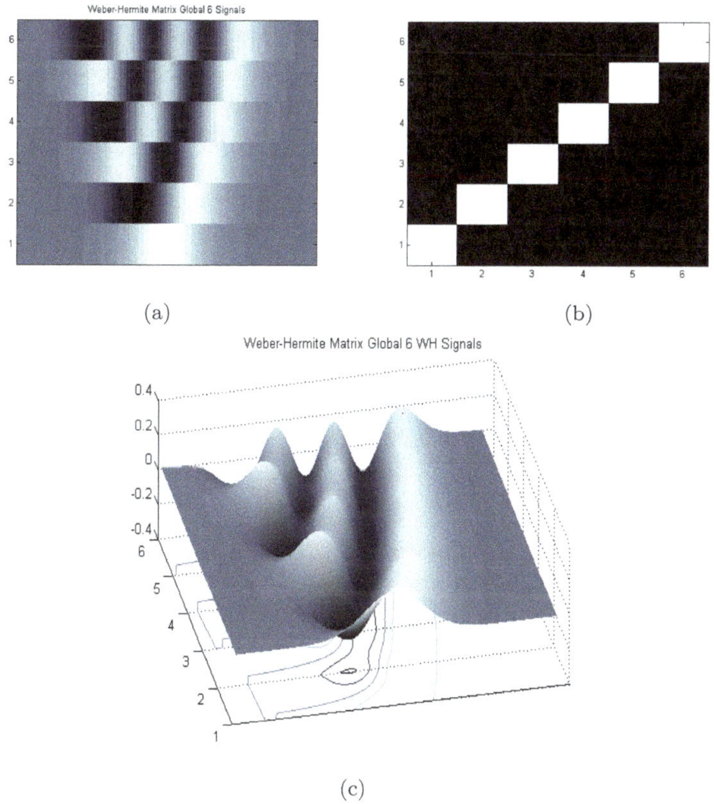

Fig. 4.19. (a) The unitary WH Matrix, M, composed of just six WH signals, $n = 0, 1, 2, \ldots, 5$. (b) the identity matrix $I = MM^*$; and (c) matrix M viewed at an angle. The transmittance and recovery of data in 6-OSSO, or 6 channel OSSO (or 6-OSSO addressed in these simulations), is homologous to these operations.

Bibliography

Abramowitz, M. and Stegun, C.A. (eds.) Parabolic cylinder functions, Chapter 19, in *Handbook of Mathematical Functions with Formulas, Graphs and Mathematical Tables*, New York, Dover, 685–700, 1972.

Addabbo R. and Blackmore, D., A dynamical systems-based hierarchy for Shannon, metric and topological entropy. *Entropy*, 21, 2019.

Adler, R.L., Konheim, A.G. and McAndrew, M.H., Topological entropy. *Trans. American Mathematical Society*, 114, 309–319, 1965.

Aspect, A., Comment on A classical model of EPR experiment with quantum mechanical correlations and Bell inequalities *Frontiers of Nonequilibrium Statistical Physics*, NATO ASI Series, Series B: Physics, Vol. 135, Plenum Press, New York, 185–189, 1986.

Barrett, T.W., The information content of an electromagnetic field with relevance to sensory processing of information. *T.I.T. J. Life Sci.*, 1, 129–135, 1971.

Barrett, T.W., On vibrating strings and information theory. *J. Sound Vibration*, 20, 407–412, 1972.

Barrett, T.W., Conservation of information. *Acustica*, 27, 44–47, 1972.

Barrett, T.W., Definition precedence of signal parameters: sequential versus simultaneous information. *Acustica*, 27, 90–93, 1972.

Barrett, T.W., The conceptual basis for two information theories — a reply to some criticisms. *J. Sound Vibration*, 25, 638–642, 1972.

Barrett, T.W., Analytical information theory. *Acustica*, 29, 65–67, 1973.

Barrett, T.W., Structural information theory. *J. Acoust. Soc. Am.*, 54, 1092–1098, 1973.

Barrett, T.W., Structural information theory based on electronic configurations. *T.I.T. J. Life Sci.*, 5, 29–42, 1975.

Barrett, T.W., Nonlinear analysis and structural information theory: a comparison of mathematical and physical derivations. *Acustica*, 33, 149–165, 1975.

Barrett, T.W., On linearizing nonlinear systems. *J. Sound Vibration*, 39, 265–268, 1975.

Barrett, T.W., Linearity in secular systems: four parameter superposition. *J. Sound Vibration*, 41, 259–261, 1975.

Barrett, T.W., Information measurement I. On maximum entropy conditions applied to elementary signals. *Acustica*, 35, 80–85, 1976.

Barrett, T.W., Information measurement II. On minimum conditions of energy order applied to elementary signals. *Acustica*, 36, 282–286, 1976.

Barrett, T.W., Structural information theory of sound. *Acustica*, 36, 272–281, 1976.

Barrett, T.W., Quantum statistical foundations for structural information theory and communication theory. In V. Lakshmikantham (ed.) *Nonlinear Systems & Applications: An International Conference*, Academic Press, New York, 391–409, 1977.

Barrett, T.W., Method and application of wavelet filter hierarchies as universal representations of N-dimensional signals and images. Serial No 10/353,096. U.S. Patent Pending, 2003.

Barrett, T.W., Final Report, Defense advanced research projects agency contract No. FA8750-05-C-0259, 2007.

Bayliss, W.E., *Electrodynamics: A Modern Geometric Approach*, Birkhäuser, 1999.

Barrett, T.W., Method and application of orthogonal signal spectrum overlay (OSSO) for communications, U.S. Patent No. 7,577,165, August 18, 2009.

Barrett, T.W., Harris, F., Vuletic, D. and W. Lowdermilk, W., Modulation with spectrally efficient prolate spheroidal wave functions overlaid on a conventional polyphase channelizer. *Software Defined Radio Conference*, Washington D.C., 2008.

Barut, A.O. and Meystre, P., A classical model of EPR experiment with quantum mechanical correlations and Bell inequalities. *Phys. Lett.*, 105A 458–462, 1984.

Barut, A.O., Three lectures on the foundations of quantum theory and quantum electrodynamics. *Frontiers of Nonequilibrium Statistical Physics*, NATO ASI Series, Series B: Physics Vol. 135, Plenum Press, New York, 119–128, 1986.

Bell, J.S., On the Einstein-Podolsky-Rosen paradox. *Physics*, 195, 1965.

Bell, J.S., *Speakable and Unspeakable in Quantum Mechanics*, Cambridge University Press, 2004.

Bennett, C.H., Demons, engines and the second law. *Sci. American*, 257, 104–114, 1987.

Bennett, C.H., Notes on Landauer's principle, reversible computation, and Maxwell's demon. *Studies History Philos. Modern Phys.*, 34, 501–510, 2003.

Berg-Johansen, S., Töpel, F., Stiller, B., Banzer, P., Ornigotti, M., Giacobino, E., Leuchs, G., Aiello, A. and Marquardt, C., Classically entangled optical beams for high-speed kinematic sensing. *Optica*, 2, 864–868, 2015.

Bonnet, G., Considérations sur la representation et l'analyse harmonique des signaux determinists ou aléatoires. *Ann. Télécom.*, 23, 62–86, 1968.

Brillouin, L., Maxwell demon cannot operate: information and entropy 1. *J. Appl. Phys.*, 22, 334, 1951.

Brillouin, L., Negentropy and information in telecommunications, writing, and reading. *J. Appl. Phys*, 25, 595, 1954.

Canales, J., *Bedeviled: A Shadow History of Demons in Science*, Princeton University Press, 2020.

Clauser, J.F., Hone, M.A., Shimony, A. and Holt, R.A., Proposed experiment to test local hidden-variable theories. *Phys. Rev. Lett.*, 1969.

Cohen, L., Time-frequency distributions — a review. *Proc. IEEE*, 77, 941–981, 1989.

Cohen, L., *Time-Frequency Analysis*, Prentice-Hall, 1995.

Delvenne, J-C., Category theory of autonomous and networked dynamical systems. *Entropy*, 21, 302, 2019.

Donoho, D.L. and Stark, P.B., Uncertainty principles and signal recovery. *SIAM J. Appl. Math.*, 49, 906–931,1989.

Dym, H. and McKean, H.P., *Fourier Series an Integrals*, Academic Press, New York, 1972.

Dyson, F., Why is Maxwell's Theory so hard to understand? Conference paper, December 2007, *Conference Antennas and Propagation (EuCAP2007) The Second European Conference*, Antennas 2, Propagation.

Erdogan, O.E., Shen, G., Anantharарman, R., Taparia, A., Javid, B. and Mahmud, S.T., Continuous time correlation architecture, *United States Patent* 9,625,507 B2, April 18, 2017.

Fancourt, C.L. and Principe, J.L., On the relationship between the Karhunen–Loeve transform and the prolate sphere. Acoustics, Speech and Signal Processing, ICASSP-88, International Conference, 1988.

Fischer, R.F.H., *Precoding and Signal Shaping for Digital Transmission*, Wiley, 2002.

Flammer, C., *Spheroidal Wave Functions*, Stanford University Press, Stanford, CA, 1957.

Flandrin, P., Représentations des signaux dans le plan temps-fréquence. Thèse Doct-Ing., INPG, Grenoble, 1982.

Flandrin, P., *Time-Frequency/Time-Scale Analysis*, Academic press, 1999.

Frigg, R., In what sense is the Kolmogorov-Sinai entropy a measure for chaotic behavior? — Bridging the gap between dynamical systems theory and communication theory. *British Soc. the Philos. Sci.*, 55, 411–434, 2004.

Fuchs, W.H.J., On the magnitude of Fourier transforms. In *Proc. Int. Congress of Math. Amsterdam*, 106–107, 1954.

Gabor, D., Theory of communication. *J.I.E.E.*, 93, 429–457, 1946.

Gauss, C. F. Disquisitiones Generales Circa Seriem Infinitam $\left[\frac{\alpha\beta}{1-\gamma}\right]\chi + \left[\frac{\alpha(\alpha+1)\beta(\beta+1)}{1-2-\gamma(\gamma+1)}\right]\chi^2 + \left[\frac{\alpha(\alpha+1)(\alpha+2)\beta(\beta+1)(\beta+2)}{1-2-3\gamma(\gamma+1)(\gamma+2)}\right]\chi^3 +$ etc. Pars Prior. *Commentationes Societiones Regiae Scientiarum Gottingensis Recentiores, Vol. II.* 1812. Reprinted in *Gesammelte Werke, Bd. 3*, 123–163 and 207–229, 1866.

Ghose, P. and Mukherjee, A., Entanglement in classical optics. *Phys. Optics* arXiv:1308.6154v2, 1–27,12 Sep, 2013.

Harashima, H. and Miyakawa, H., Matched-transmission technique for channels with intersymbol-interference. *IEEE Trans. Commun.*, COM-20, 774–780, 1972.

Heisenberg, W., Über den anschaulichen Inhalt der quantentheoretischen Kinematik un Mechanik. *Z. Phys.*, 43, 172–198, 1927.

Hertz, H., The forces of electric oscillations, treated according to Maxwell's theory. *Wiedemann's Ann.* 36, 1–23, 1889. Reprinted in H. Hertz, Electric Waves, Macmillan, 1893, reprinted Dover Publications, 1962.

Horodecki, R., Horodecki, P., Horodecki, M. and Horodecki, K. Quantum entanglement. *Rev. Mod. Phys.*, 81, 865, 2009.

Jancewicz, B., *Multivectors and Clifford Algebras in Electrodynamics*, World Scientific, 1988.

Jones, D.S., *The Theory of Electromagnetism*, Pergamon, 1964.

Joot, P.,*Geometric Algebra for Electrical Engineers*, Multivector Electromagnetism, 2021.

Kupczynski, M., Contextuality as the key to understanding quantum paradoxes, Preprint, May, 2020.

Kupczynski, M., Is the moon there when nobody looks. Bell inequalities and physical reality. *Found. Physics*, 23 Sep, 2020.

Kupczynski, M., Can Einstein with Bohr debate on quantum mechanics be closed? *Phil. Trans. Roy. Soc. A.*, 375, 2016039, 2017.

Kupczynski, M., Is Einstein non-signalling in Bell Tests? *Open Phys.* 15, 739–753, 2017.

Khare, K. and George, N., Sampling theory approach to prolate spheroidal wave functions. *J. Phys. A., Math. Gen.*, 36, 10011–10021, 2003.

Landau, H.J., Sampling, data transmission, and the Nyquist rate. *Proc. IEEE*, 55, 1701–1706, 1967.

Landau, H.J. and Pollak, H.O., Prolate spheroidal wave functions, Fourier analysis and uncertainty — II. *Bell Syst. Tech. J.*, 40, 65–84, 1961.

Landau, H.J. and Pollak, H.O., Prolate spheroidal wave functions, Fourier analysis and uncertainty — III: The dimension of the space of essentially time- and band-limited signals. *Bell Syst. Tech. J.*, 41, 1295–1336, 1962.

Landau, H.J. and Widom, H., The eigenvalue distribution of time and frequency limiting. *J. Math. Anal. Appl.*, 77, 469–481, 1980.

Li, L-W., Leong, M-S., Yeo, T-S., Kooi, P-S. and Tan, K-Y., Computation of spheroidal harmonics with complex arguments: a review with an algorithm. *Phys. Rev.*, E58, 6792–6806, 1998.

Maxwell, J.C., *A Treatise on Electricity and Magnetism*, Clarendon Press, 1873, 1st edition; 1881, 2nd edition; 1891, 3rd edition.

Minev, Z.K., Mundhada1, S.O., Shankar, S., Reinhold, P., Gutiérrez-Jáuregui, R., Schoelkopf, R.J., Mirrahimi, M., Carmichael, H.J. and Devoret, M.H., To catch and reverse a quantum jump mid-flight. *Nature*, doi.org/10.1038/s41586-019-1287-z, 2019.

Morse, P.M. and Feshbach, H., *Methods of Theoretical Physics*, 2 Volumes, McGraw-Hill, 1953.

Naber, G.L., *Topology, Geometry, and Gauge Fields: Foundations*, Springer, 1997.

Nieuwenhuizen, Th.M., Where Bell went wrong. arXiv:0812.3058v2, 30 Nov, 2016.

Papoulis, A., *Signal Analysis*, McGraw-Hill, New York, 1977.

Plesch, M., Dahlsten, O., Goold J. and Vedrai, V., Maxwell's Daemon: Information versus Particle Statistics. *Scientific Rep.*, 11(4), 6995. 2014.

Qian, X-F. and Eberly, J.H., Entanglement and classical polarization states. *Optics Lett.*, 36, 4110–4112, 2011.

Qian, X-F., Little, B. Howell, J.G. and Eberly, J.H., Shifting the quantum-classical boundary: theory and experiment for statistically classical optical fields. *Optica*, 2, 611–615, 2015.

Shannon, C.E., Communication in the presence of noise. *Proc. IRE*, 37, 10–21, 1949.

Shen, X.A., Guo, Y. and Walter, G.G., Slepian semi-wavelets and their use in modeling of fading envelope. *IEEE Topical Conference on Wireless Communication Technology (TCWCT)*, Honolulu, Hawaii, USA, October, 2003.

Slepian, D., Prolate spheroidal wave functions, Fourier analysis and uncertainty — IV: Extensions to many dimensions; generalized prolate spheroidal functions. *Bell System Tech. J.*, 43, 3009–3057, 1964.

Slepian, D., On bandwidth. *Proc. IEEE*, 63, 292–300, 1976.

Slepian, D., Prolate spheroidal wave functions, Fourier analysis and uncertainty – V. The discrete case. *Bell System Technical J.*, 57, 1371–1430, 1978.

Slepian, D., Some comments on Fourier analysis, uncertainty and modeling. *SIAM Review*, 25, 379–393, 1983.

Slepian, D. and Pollak, H.O., Prolate spheroidal wave functions, Fourier analysis and uncertainty – I. *Bell Syst. Tech. J.*, 40, 43–64, 1961.

Stratton, J.A., Morse, P.M., Chu, L.J., Little, J.D.C. and Corbato, F.J., *Spheroidal Wave Functions*, Wiley, 1956.

Thomson, D.J. Spectrum estimation and harmonic analysis. *Proc. IEEE*, 70, 1055–1096, 1982.

Tomlinson, M., New automatic equalizer employing modulo arithmetic. *Electronics Lett.*, 7, 138–139, 1971.

Tretter, S.A., *Constellation Shaping, Nonlinear Precoding and Trellis Coding for Voiceband Telephone Channel Modems*, Kluwer, Academic publishers, 2002.

Tribus, M. and McIrvine, E.C., Energy and information. *Scientific American*, 179–188, 1971.

Ville, J.A. and Bouzitat, J., Sur un type de signaux pratiquement bornés en temps et en fréquence. *Câbles et Transm.*, 9A, 293–303, 1955.

Ville, J.A. and Bouzitat, J., Note sur un signal de durée finie et d'énergie filtrée maximum. *Câbles et Transm.*, 11A, 102–127, 1957.

Walter, G., Prolate spheroidal wavelets: differentiation, translation, and convolution made easy. *J. Fourier Anal. Appl.*, 11, 73–84, 2005.

Walter, G. and Shen, X., Sampling with prolate spheroidal wave functions. *J. Sampling Theory Sign. Image Proc.*, 2, 25–52, 2003.

Walter, G. and Shen, X., Wavelets based on prolate spheroidal wave functions. *J. Fourier Anal. Appl.*, 10, 73–84, 2004.

Walter, G. and Soleski, T., A new friendly method of computing spheroidal wave functions and wavelets and uncertainty. *J. Appl. Comput. Harmonic Anal.*, 19, 432–443, 2005.

Watson, G.N., *A Treatise on the Theory of Bessel Functions*, 2nd edition, Cambridge University Press, 1958.

Weber, H., HUber die Integration der partiellen Differentialgleichung: $\frac{\partial^2 u}{\partial x^2} + \frac{\partial^2 u}{\partial y^2} + k^2 u = 0$. *Math. Ann.*, 1, 1–36, 1869.

Whittaker, E.T., On the functions which are represented by the expansions of the interpolation theory. *Proc. Roy. Soc. Edinburgh*, 35, 181–194, 1915.

Whittaker, E.T. and Watson, G.N., *A Course of Modern Analysis*, 4th edition, Cambridge University Press, 1927.

Wódkiewicz, K., Quantum Malus' Law. *Phys. Lett.*, 112A, 276–278, 1985.

Xu, Y., Haykin, S. and Racine, R.J., Multiple window time-frequency distribution and coherence of EEG using Slepian sequences. *IEEE Trans. Biomed. Eng.*, 49, 861–866, 1999.

Chapter 5

Polarization and Axis Modulated Ultrawideband Signal Transmission[1,2]

5.1. Introduction to Atmospheric/Ionospheric Transmission

5.1.1. *Overview*

The propagation of beams — whether of RF, Terahertz, light or of any frequency — through turbulent media in the atmosphere or ionosphere can result in energy absorption, scattering and jitter. The range of beam transmission is transmitted energy dependent and is affected by absorption and scatter. Because jitter adversely affects beam pointing, this is a concern to communications and surveillance systems designers. Absorption, scattering and jitter can result in (a) a transmitted beam complete misalignment with a receiving antenna; (b) dropped signals in the transmission; or (c) an unacceptable reduction in range.

Methods are described here — polarization modulation (POLMOD) and azimuth or axis modulation (AXMOD) — that partially defeat atmospheric beam absorption, scattering and jitter. These methods combine two beams — orthogonally polarized in the

[1]This work was supported by Air Force Office of Scientific Research Grant FA9550110274.

[2]Special thanks to Kevin Suter, Christopher Beairsto, David Tellez and Stephen Squires, HELSTF, White Sands Missile Test Range, U.S. Army, New Mexico, for helping to collect the data obtained by these experiments.

case of POLMOD and parallel polarized in the case of AXMOD — of different frequency, into a single beam of rapidly changing polarization (POLMOD) or azimuth (AXMOD) and creates a beat frequency. Viewed in the *time domain,* as in viewing a beam *from the side*, POLMOD results in wave packet polarization modulation (WPPM) or *polarization modulated wave packets* of short-time duration; and AXMOD results in wave packet azimuth modulation (WPAM) or *azimuth modulated wave packets* of short duration.

The term POLMOD is associated with modulated spin angular momentum (SAM). POLMOD is a classical version of the quantum mechanical SAM. One needs to be aware of some confusion in the literature as the term "SAM" is used in reference to both classical and quantum mechanical transmissions. Another component of the total angular momentum of a beam is azimuth or axis-phase modulation (AXMOD) which is associated with modulated orbital angular momentum (OAM). Thus POLMOD and AXMOD, or modulated SAM and modulated OAM, together compose a beam's total angular momentum. OAM is presently under investigation worldwide as a high data-rate modulation scheme and demonstrated to be unaffected by turbulence in kilometer range point-to-point communications.

A POLMOD beam in the *time domain* — viewed *from the side* — can be decomposed into two orthogonal channels. However, an AXMOD (or modulated OAM) beam in the *spatial domain* — as in viewing a beam *head-on or in the beam cross-section* — can potentially be decomposed into an infinite number of orthogonal channels (leaving aside the issue of the creation of noise by the necessity for extremely high sampling rates). For this reason, OAM (conceived classically) is presently the preferred modulation scheme for communications. However, this chapter is not focused on communications, but on defeating a medium's or channel's absorption, scattering and jitter, by challenging the absorption and scattering times of the molecular composition of the medium. This type of modulation was chosen to exploit the duration requirements of the absorption and scattering times of the molecular composition of the medium. Therefore, POLMOD (modulated SAM) was the chosen method to discuss.

5.1.2. *POLMOD theory and applications*

Atmospheric and ionospheric propagation losses can constitute a large fraction of the electromagnetic (EM) wave energy dissipated during EM beam propagation. EM systems from long to short wavelengths such as radio communications transmissions, radars, sensors and laser point-to-point communications can lose a significant fraction of their beamed energy, and lack of jitter control affects beam pointing and targeting. We know how atmospheric losses arise primarily from:

- *Turbulence*, due to vorticity that causes radial and axis air density variations inside beams and results in scattering.
- *Thermal blooming*, due to beam expansion from air heating.
- *Extinction* due to EM energy absorption in dense concentrations of particles (e.g., clouds, fog).

Thus a major objective for communications and surveillance at any wavelength requires minimizing the propagation jitter/absorption effects of turbulent atmospheres. Models of the inner scale of turbulent/distributed volume atmospheres and the performance of adaptive optics designed to mitigate the effects of those atmospheres, are both presently inadequate, although there is ongoing effort to improve them.

This chapter addresses a method to defeat and/or mitigate atmospheric/ionospheric beam jitter, and beam redirection by polarization modulation (POLMOD) at light wavelengths. The method is also applicable at other wavelengths. The method depends on the modulation rate/frequency, $d\varphi/dt$, $d\varphi^n/dt^n$, or rate of phase modulation of interbeam phase, φ, exceeding that of the inverse of molecular relaxation times in the media through which a beam passes. We can produce a polarization modulated beam (POLMOD) by modulating the phase between two *orthogonal* beams that are combined; and an azimuth modulated beam (AXMOD) by modulating the phase between two *similarly polarized* beams that are combined. In both cases, ideally, one of the two beams of the combined beam would be independently phase modulated. The tests discussed here used a substitute method which combined two beams with a wavelength offset $\Delta\lambda$. This substitute method provided phase modulation, but with only (i) a linear phase change, and (ii) a confinement in beat frequency dependence to the availability of lasing lines, λ_1 and λ_2, defining $\Delta\lambda = \lambda_1 - \lambda_2$. In other words, the method discussed here only covers linear phase modulation using the available lasing lines. Despite these limitations, these tests demonstrate substantial mitigation of spatial and temporal beam jitter that results in beam redirection or loss of pointing angle, as well as amplitude loss resulting in range limitations.

It is important to note how the molecular systems which interact with the radiating beams are viewed here as *polarization selective receivers*, in the case of POLMOD, or *axial/azimuth selective receivers* in the case of AXMOD. The dual beam recombination method creates an amplitude modulated beat frequency wave packet, which, from the perspective of a selective receiver, e.g., a molecular system, is sampled as a broadband pulse.

The media substituting for the atmosphere and ionosphere in these studies were: a water vapor chamber, heated air, and static and rotated phase plates. The recording methods were either near-linear power meters or nonlinear CCD cameras. Because two forms of jitter were distinguished — temporal (rapid amplitude changes over time) and spatial (i.e., changes in the pointing direction) — the two recording methods independently addressed temporal and spatial jitter: the power meters addressing temporal jitter and the cameras, spatial. The power meters also addressed amplitude changes. Because no optimal beam combiners were available, a number of arbitrary beam combining methods were used in the tests. No Terahertz (THz) spectrometer was available to investigate the frequency or wavelength dependence (the λ_i dependence) of the results. Therefore, the possible frequency dependence of the results reported here is not known at this time.

Nonetheless, the results, obtained under the nonideal conditions described above, indicate the POLMOD/AXMOD signals reduced

- Temporal jitter up to a maximum of at least 50%.
- Spatial jitter up to a maximum of at least 33%.

- Absorption/scattering up to 85% in *some media* and at *some wavelengths* of combined beams.

The modulation approach used in these tests to improve beam propagation through disturbed atmospheres is applied in transmittance (TX), and may be viewed as an adjunct or an alternative to the adaptive optics method, which is applied on reception (RX). Another difference found in the case of adaptive optics is that for the adaptive optics case *there is an accent on the spatial or beam cross-section.* This accent on the spatial may not be appropriate for robust communications. Whereas in astronomy and imaging, the objective is *to preserve spatial information*, in communications the objective is *to preserve information sequence and point focus.* POLMOD on the TX side offers either an alternative approach to the same problem, or it can be used together with (improved) adaptive optics, on the RX side, each addressing one side of the channel.

Recently, there have been demonstrations of (a) SAM, understood classically and associated with beam polarization, and of (b) OAM, also understood classically, and associated with the beam azimuth phase of the complex electric field, in RF communications.[3] As mentioned above, in the literature, SAM and OAM are somewhat loosely used in both the quantum mechanics and classical mechanics sense, which can result in confusion. POLMOD and AXMOD, both terms used here in the classical mechanics sense, are related to SAM and OAM, also used in the classical sense, in that POLMOD is *modulated* SAM and AXMOD is *modulated* OAM. However, whereas in SAM and OAM communications it is the *spatial modes, e.g.,* Hermite–Gaussian and Laguerre–Gaussian, that are exploited, in the present studies the focus is on *temporal modes.*

The "polarization wave packets" generated by beat methods described here, are composed of frequencies in the Terahertz (THz)

[3]Djordjevic and Arabaci (2010) have proposed exploitation of photon spin angular momentum (SAM) associated with polarization, and orbital angular momentum (OAM) associated with the azimuthal phase of the complex electric field (cf. Djordjevic *et al.* 2007; Djordjevic and Djordjevic, 2009; Djordjevic, 2011; Paterson, 2005; Leach *et al.* 2004; Gibson *et al.* 2004).

range — a neglected range in spectroscopy, and some deem the "last frontier" of spectroscopy. Besides communications and surveillance applications, such pulses or wave packets also have application in the fields of molecular spectroscopy and biomedical imaging.

5.1.3. *POLMOD transmission test objectives and chapter organization*

This chapter is organized according to a model of radiation–molecular interaction/transmission which requires a match between the (a_1) frequency or frequencies and (b_1) the polarization of radiation from a transmitter (TX), with the (a_2) frequency or frequencies and (b_2) the polarization of molecular energy levels conceived as a receiver (RX). In other words, the model is that of radiation–molecule, or of TX→RX, transmission/reception with respect to (a) frequency or frequencies, and (b) polarization. (see Fig. 5.1).

That is the case for TX→RX reception, or radiation–molecular interaction scattering or absorption. However, the ultimate and primary aim of the tests described here is not improving reception at the receiver, but the defeat of scattering/absorption in the channel

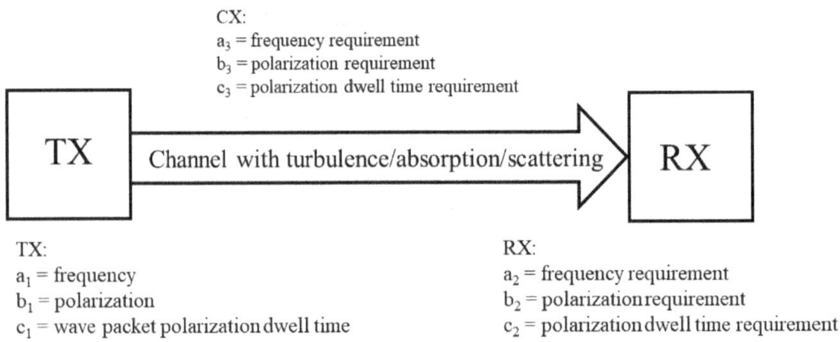

Fig. 5.1. To defeat a turbulent channel's turbulence/absorption/scattering, the transmitter (TX) is of a certain frequency (a_1), polarization (b_1) and wave packet duration or dwell time (c_1), that must be *compatible with* the receiver's absorption requirements of frequency (a_2), polarization (b_2) and wave packet duration or dwell time (c_2), yet *mismatched* to the channel's turbulence, absorption/scattering frequencies (a_3), polarization (b_3) and also less than a wave packet absorption duration or dwell time requirement (c_3).

medium, the requirements for which are frequencies, a_3, polarization, b_3, and absorption or dwell time requirements, c_3. Therefore, the tests address methods to obtain this defeat by challenging the temporal compatibility required of (b_1) radiation — polarization or azimuth — with (c_3) molecular energy levels, i.e., challenging the temporal duration requirement of a c_1:c_3 compatibility.

If, therefore, $a_1 = a_3$, $b_1 = b_3$, and $c_1 \geq c_3$, then scattering/absorption in the channel will most likely occur. But if $c_1 < c_3$, then scattering/absorption may be avoided, in which case it is possible to defeat scattering/absorption and achieve transmission through media conceived as a channel.

In order to obtain $c_1 < c_3$, there is the requirement to not only generate TX wave packets of defined (center) frequency and polarization, i.e., $a_1=a_2=a_3$, $b_1=b_2=b_3$ and $c_1=c_2$, but also $c_1 < c_3$. The method to achieve $c_1 < c_2$, i.e., a TX→CX packet duration mismatch, is described in Sec. 5.1 and the requirement for polarization compatibility for absorption, i.e., $b_1=b_2=b_3$, in Sec. 5.2. It is important to note throughout that *constant polarization* wave packets are *not* addressed, but rather wave packets of *changing polarization*, or *polarization modulated* wave packets.

In Secs. 5.5 and 5.6, the relationship of wave packet polarization modulation (WPPM) to spin angular momentum encoding (SAM) is described and contrasted with orbital angular momentum (OAM) encoding.

Finally, in Secs. 5.2–5.4, experimental results are described that are achieved with laser WPPM propagated through disturbed atmospheres generated in the laboratory and across short desert distances.

5.2. POLMOD Fundamentals

The most familiar signals are covered by classical and *mostly steady-state* electromagnetic theory which covers amplitude, frequency and phase modulated signals. But polarization modulation (POLMOD) raises different issues in signal creation, representation and consideration of the *transient* interactions of an incident wave and molecular media. The classical approach to radio system design considers the *steady-state* action of the signal and media, i.e., it is assumed that the media has a constant attenuation at a given frequency. It is known how *steady-state* assumptions work well for conventional applications and methods, but here the signal transmission/reception effects in the *time domain* are examined and transient effects are considered.

This section presents the theoretical and practical considerations required to generate and receive polarization modulated (POLMOD) and axis modulated (AXMOD) wave packet signals, namely:

- How to construct POLMOD/AXMOD wave packets.
- How to represent POLMOD/AXMOD wave packets on the Poincaré sphere.
- How the Bloch sphere represents the polarization and azimuth of media molecular systems, a representation necessary for describing induced transparency.

5.2.1. *Generating defined frequency packets by constant wavelength wave beating*

Using a simple transmit–receive model, we begin by considering the case of the transmitter (TX) and receiver (RX) being of the same frequency as in Fig. 5.2(a). In this trivial case, all that is required is that the constant frequency of the TX and the center frequency of the RX bandwidth, be the same, i.e., $a_1 = a_3$ as defined above.

The next case, illustrated in Fig. 5.2(b) requires combining 2× TX emitters of identical polarization resulting in a beam in which the polarization is constant. The two wavelengths of the combined transmitted wave, $\lambda_1$and λ_2, are different, defining a $\Delta\lambda = \lambda_1 - \lambda_2$. The combined beam of constant polarization is shown and we notice that wave packets result in the combination of these two emitters — but of constant polarization, i.e., $a_1 = a_3$ and $b_1 = b_3$ as defined above.

The third case, illustrated in Fig. 5.2(c), is identical to the situation shown in Fig. 5.2(b), except that the two constituent

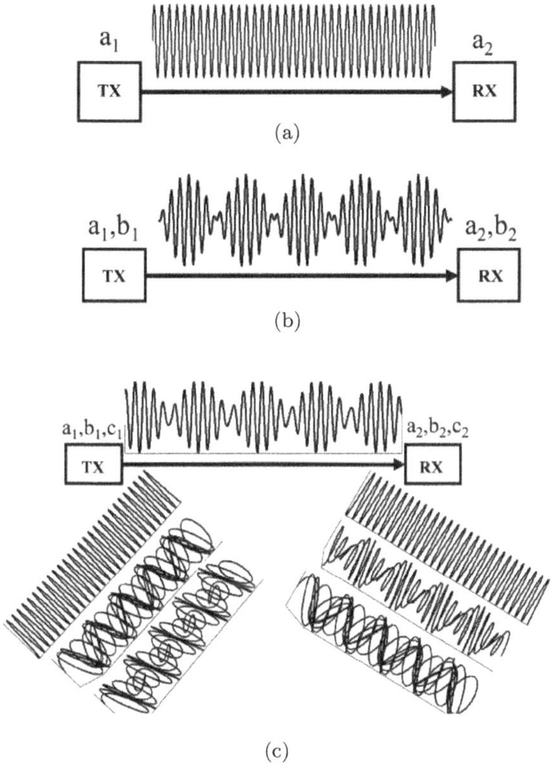

Fig. 5.2. TX-RX Systems: (a) TX–RX constant frequency match, (b) TX–RX packet frequency and polarization match TX–RX, and (c) TX–RX packet frequency match with polarization match time limited. (a) Here, TX and RX are matched in frequency only. (b) Here, TX and RX are matched in frequency and polarization. The polarization is constant. The two wavelengths of the combined transmitted wave, $\lambda_1$and λ_2, are different, defining a $\Delta\lambda = \lambda_1 - \lambda_2$. The combined beam of constant polarization is shown. This is a method for obtaining packets of defined frequency by constant wavelength wave beating. (c) Here, TX and RX are matched in frequency and the polarization is modulated, not constant. The two wavelengths of the combined transmitted wave, λ_1 and λ_2, are different, defining a $\Delta\lambda = \lambda_1 - \lambda_2$, but, unlike (b), in this case the constituent beams are orthogonally polarized, and the result is polarization modulation, not constant polarization. The combined polarization modulated beam is shown at various angles. If a_1=a_3, b_1=b_3, and $c_1 \geq c_3$, then reception/absorption will likely occur. But if a_1=a_3, b_1=b_3, and $c_1 < c_3$, then reception/absorption may not occur, in which case it is possible to defeat reception/absorption and achieve transmission. This is a method of polarization modulation by wave beating for obtaining packets of defined frequency and with transient polarization, or "polarization modulated packets".

beams are orthogonally polarized.[4] In this case, $a_1 = a_2$ and b_3 can be equal to b_3, but only for a short dwell time. This dwell time and the required dwell time for absorption/reception is defined above as c_1 and c_3. The combined polarization modulated beam is shown at various angles in Fig. 5.2(c). At one angle, the pulse nature of the combined beam is shown — as in Fig. 5.2(b), but at other angles the constituent beams can be seen, and at yet other angles, the polarization modulation of the combined beam is seen. If $a_1=a_3$, $b_1=b_3$, and $c_1 \geq c_3$, then reception/absorption will likely occur. But if $a_1=a_2$, $b_1=b_3$, and $c_1 < c_3$, then channel media reception/absorption may not occur, in which case it is possible to defeat reception/absorption in the channel and achieve transmission through an otherwise absorbing and/or jitter affected channel. This method of polarization/axial modulation by wave beating produces packets of defined frequency and with transient polarization or azimuth, or "polarization modulated packets" and "azimuth modulated packets".

5.2.2. *The Poincaré sphere representation of polarization*

To fully comprehend the possibilities offered by POLMOD and AXMOD, it is necessary to appreciate and understand the accepted representation of EM beams. This section shows how the Poincaré sphere can represent polarizations and beam rotations.

But before addressing the Poincaré sphere, we need to indicate and stress the difference between a *statically polarized beam* and a *modulated beam.* "Static polarization" means a beam is, e.g., *always* linearly (or circularly or elliptically) polarized, which means all the waves are either *always* vertically — or horizontally — aligned, or *always* circularly, or *always* elliptically polarized, where in all cases the wave moves in a circular (right or left) or elliptical pattern as it

---

[4] Toptica Photonics AG, Graefelfing/Munich, Germany, has developed Terahertz (100 GHz–10 THz) spectrometers using the method of Fig. 5.2(b). As of writing we know of no spectrometers using the method of Fig. 5.2(c), although adapting a Fig. 5.2(b) spectrometer to Fig. 5.2(c) merely involves the use of orthogonally polarized combined beams.

propagates. Therefore, it can be represented as a *single point* on the surface of a Poincaré sphere. *In contrast*, polarization modulation results in a *movement* from point to point over time and along a trajectory as shown in Fig 5.3(b), below, that movement being at a constant speed, or at varying speeds.

The Poincaré sphere can represent all *static* polarizations of a radiated beam, as well as all *static* rotations about the axis of the beam[5] (Figs. 5.3, and 5.4(a)). Therefore, a vector representation

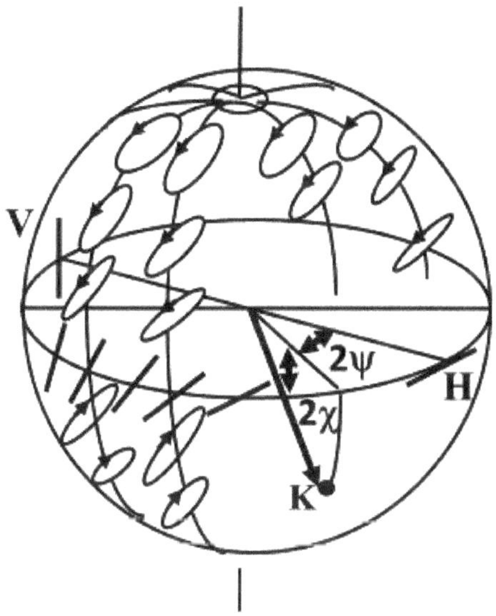

Fig. 5.3. The Poincaré sphere can statically represent all polarizations and azimuth/axial rotations. A circularly polarized corotating beam is represented at the north pole; and a circularly polarized counter-rotating beam is represented at the south pole. Along the longitudinal lines are represented the varieties of elliptical polarizations. Around the equator latitude lines are shown horizontally polarized linear, and vertically polarized linear configurations, as well as the variety of linear polarizations in between these. The angles 2ψ and 2χ are shown for a combined beam. Another angle, δ, is relevant and is the interbeam phase of the combined beam's constituent beams and not represented here. These are static representations. *However, POLMOD is a movement of the K vector on the sphere and is not static.*

[5]Jules Henri Poincaré (1854–1912), French mathematician, theoretical physicist, engineer and philosopher of science.

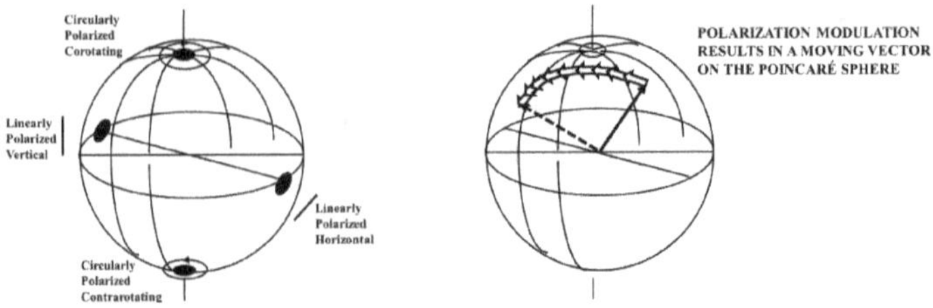

(a) (b)

Fig. 5.4. (a) The Poincaré sphere represents beams of static polarization. The latitude equatorial region (the azimuth or axis) spans linearly polarized beams parallel to perpendicular. The longitudinal lines can depict beams of elliptical or circularly polarization corotating (north pole) to contrarotating (south pole), passing through forms of elliptical polarization depending on latitude. Here, there is a Poincaré sphere representation of 4 ***static*** polarizations. (b) A Poincaré sphere representation for a combined beam which is a ***moving*** representation of a POLMOD beam.

of a beam, centered in the middle of the sphere and pointed to the underside of the surface of the sphere at a location on the surface, will represent a beam's static polarization. That static vector represents the *instantaneous* polarization and rotation angle of the beam. However, Fig. 5.4(b) represents the case of a *polarization modulated* (and/or*axial or azimuth modulated*) beam with the vector tracing a trajectory on the sphere surface over time (i.e., *the vector is moving*) at a *constant or varying velocity* over time. If the beam is sampled by a detector device (which can occur by absorption through the excitation of electronic or vibrational energy levels of a molecular system) at a rate which is *less than* the rate of modulation, then the sampled output from the detector device will be a mixing or mapping of *two components* of the wave, which are continuously changing with respect to each other, into *one snapshot* of the wave, at *one location* on the surface of the sphere and at *one instantaneous polarization and azimuth rotation.* If the modulation $\partial\varphi^n/\partial t^n$ ($\varphi = \psi$ or χ) is according to the first differential of the phase φ (between two combined beams), i.e., $n = 1$, and $\partial\varphi/\partial t$, the trajectory has a constant velocity (Fig 5.3(b)).Thus, from the viewpoint of a device sampling at a rate *less than* the modulation rate, a two-to-one mapping (over time) will have occurred, which is the signature of an $SU(2)/Z_2$ field[6] (Barrett, 1997–2008). With $n > 1$ there are interbeam phase modulations that provide trajectories of nonconstant velocity.

The different trajectories on the Poincaré sphere for (a) two lasers orthogonally polarized of two different wavelengths producing polarization modulation (POLMOD); and (b) two lasers plane polarized, i.e., not orthogonally polarized, of two different wavelengths producing axis (azimuth phase) modulation (AXMOD) is shown in Figs 5.5(a) and 5.5(b). Using three lasers of different frequencies in a combined beam results in a number of different mixed POLMOD and AXMOD trajectories on the Poincaré sphere (Fig. 5.6)

If the polarization-modulated beam, or polarization-modulated-and-rotated beam is viewed head-on, and the beam is sampled by two orthogonally polarized detectors, the **E** field of the combined beam traces out Bowditch[7] or Lissajous[8] patterns (Fig. 5.7) in the

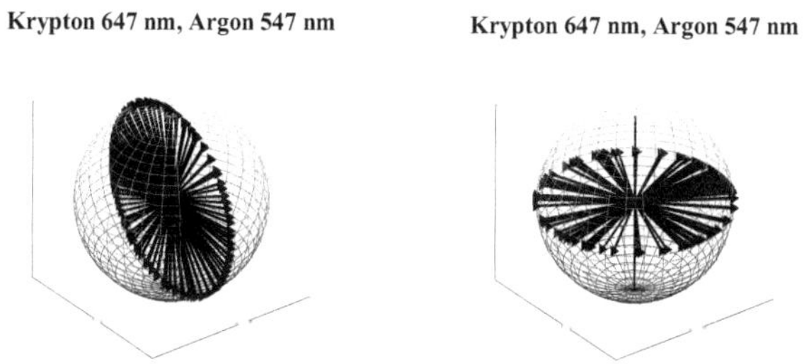

(a) (b)

Fig. 5.5. Poincaré sphere for combined laser signals. (a) An arbitrary POLMOD beam (two lasers ***orthogonally polarized*** of two different wavelengths) Poincaré sphere representation: $\partial\chi/\partial t$. Trajectories on the Poincaré sphere: the Krypton laser at $\lambda = 647$ nm, and the Argon laser at $\lambda = 457$. In comparison with (b), notice that there are only longitudinal changes, while the latitude (axis rotation) is constant. (b) An arbitrary sole axial or azimuth phase modulated (AXMOD) beam (two lasers ***plane polarized*** of two different wavelengths) Poincaré sphere trajectory representation: $\partial\psi/\partial t$. In comparison with (a), notice that there are only latitudinal changes, while the longitude (polarization) is constant.

---

[6]This field is characterized by a generalization of Maxwell's equations — see Barrett (2008).
[7]Nathaniel Bowditch (1773–1838), American mathematician.
[8]Jules Antoine Lissajous (1882–1880), French physicist.

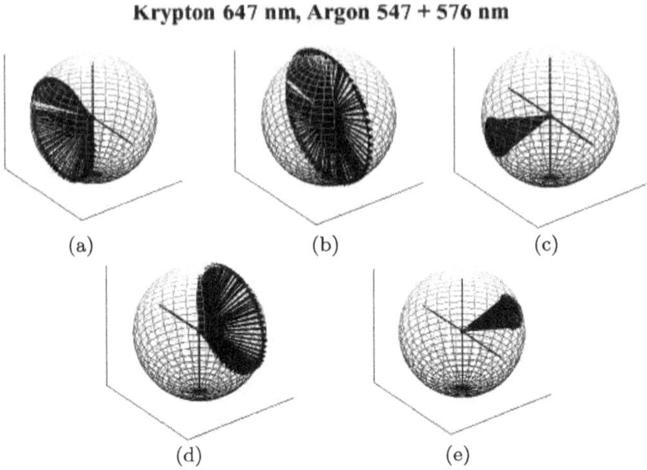

Fig. 5.6. Amplitude control of trajectories on the Poincaré sphere over 1 picosecond, Three laser situation: $\lambda_1 = 647$ nm, $\lambda_2 = 547$ nm and $\lambda_3 = 576$ nm. On one linearly polarized channel: $a_1 \sin(2\pi f_1 t)$. On second, orthogonally polarized channel: $a_2 \sin(2\pi f_2 t) + a_3 \sin(2\pi f_3 t)$. (a) $a_1 = 1; a_2 = 1; a_3 = 1$. (b) $a_1 = 1$; $a_2 = 0.5; a_3 = 0.5$. (c) $a_1 = 1; a_2 = 4; a_3 = 4$. (d) $a_1 = 4; a_2 = 1; a_3 = 1$. (e) $a_1 = 16; a_2 = 1; a_3 = 1$.

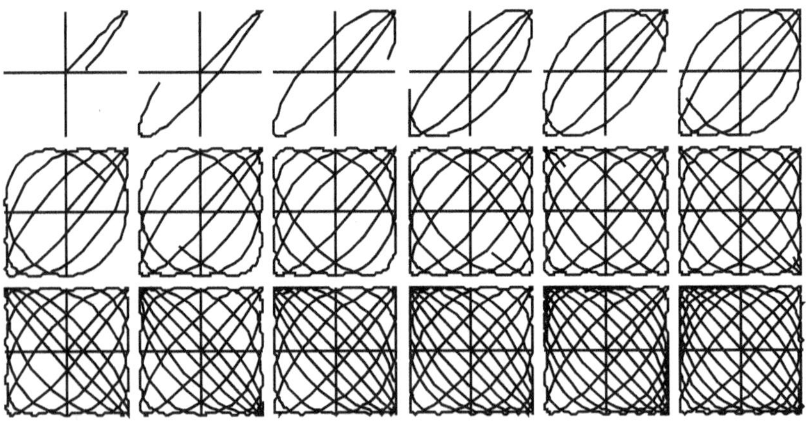

Fig. 5.7. Bowditch or Lissajous patterns representing the polarized electric field over the time of one cycle of the beam, viewed in the plane of incidence, resulting from two orthogonally polarized fields, which are out of phase by the following degrees: 0, 21,42, 64, 85, 106, (top row); 127, 148, 169, 191, 212, 233 (middle row); 254, 275, 296, 319, 339, 360 (bottom row). In these patterns, the plane polarizations (horizontal and vertical) are represented at 45° to the axes, and each pattern is the history of the beam up until one cycle (wavelength) of the beam is traversed. These are *instantaneous* Poynting vector representations.

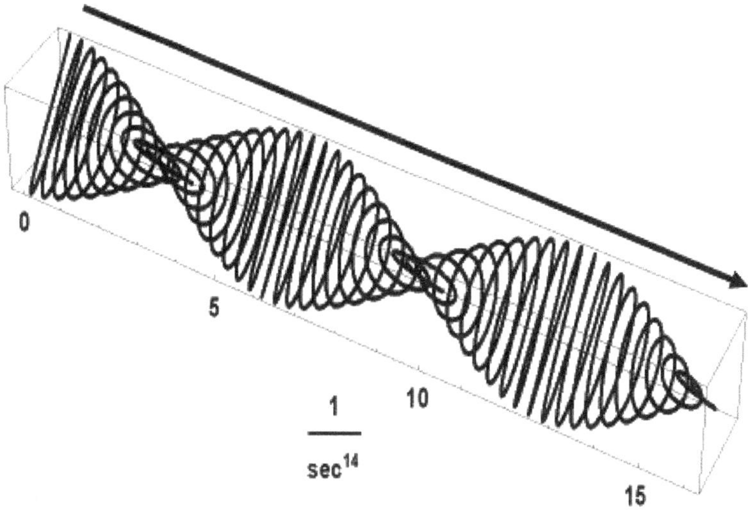

Fig. 5.8. Representation of the **E** field of a polarization-modulated beam ($\lambda_1 =$ 647 nm, $\lambda_2 =$ 547 nm): the z or time direction. *Note*: this does not represent circular polarization. Circular polarization — whether right-handed, or left-handed — is defined as a constant 45° phase difference between the two orthogonal beams. Here, on the other the phase angle is *continuously changing polarization over time as indicated.* Now the beam, *at a special instant*, but only at an instant, will be circularly polarized *if* the two orthogonal beams are 45° out-of-phase. That is a special and unique case and represented at the two poles of the Poincaré sphere. *In the more general case*, the resultant beam, represented by vector **K** in Fig. 5.4(b), can roam, moving anywhere on the Poincaré sphere, not remaining static at the poles of the sphere, which define the static conditions of circularly polarized light, right- and left-handed — see Fig. 5.4(a).

x, y-directions. In the z-direction, the beam would appear as in Fig. 5.8.

Losses in transmission through the atmosphere/ionosphere are due to the rate of polarization modulation being less than the relaxation times of the absorption mechanisms of the atmosphere/ionosphere. Furthermore, the modulations, which result in trajectories of the moving vector K (see Fig. 5.4b) on the sphere, are infinite in number. The moving K (the Stokes parameter, s_0) is not confined to any specific wandering over the sphere. Moreover, those modulations at a rate of *multiples* of 2π result in the return to a pass through, or pass by, a *single* location on the sphere and *are detected* at a frequency of *exactly and only* 2π, *if the detection is by a device or molecular system sampling at a rate of only* 2π. In other words, a relatively slow-rate sampling device or molecular system cannot detect the trajectory of a fast moving beam K, departing from, and arriving back at, the location on the sphere representing the slow-rate detector. *To a relatively slowly sampling device or molecular system, a fast-modulated beam can have "internal energies" quite undetected and unsuspected* (Barrett, 2008).

If the modulated or moving K vector in Fig. 5.4(b) is rotating continuously (i.e., through multiples of 0–2π) at a rate that is greater than, or a multiple of, the sampling rate of a polarization selective detector (statically positioned on the sphere), then observed from the vantage point of that detector, K appears to be static (i.e., statically polarized) — yet it is not. It is as if the detector, positioned at a location on the Poincaré sphere only "opens its aperture to receive" as the K vector passes into and through that location. As a result the static detector samples the K vector as a pulse that is also momentarily and statically polarized at that location — yet it is not, being polarization modulated, moving and passing through that location. Moreover, if the detector requires the K vector to remain stationary at that location on the sphere for a certain period of time (in order that the beam that K represents to be absorbed/sampled), then if K *does not remain for that certain period of time* due to the modulation rate being higher than the detector's sampling rate, the detector may not not detect/sample K (i.e., elastically or inelastically scatter or absorb the beam). "The requirement for K to remain stationary for a certain period" in a real detector or molecular system is due to *the relaxation time* of a detector or molecular system. In other words, it always takes a certain time for energy to be captured, or parametrically up or down converted, or passed to a "heat or thermal sink".

A statically polarized detector is conceived here as *a* $U(1)$ *symmetry unipolarized, constant rotational axis, sampling device.* This kind of detector obeys Maxwell's equations in their basic, i.e., $U(1)$ symmetry form. On the other hand, the fast polarization (and rotation) modulated beam is a *multipolarized, multirotation axis,*

SU(2) *symmetry beam*, that obeys a more complex form of Maxwell's equations (see Barrett, 2008). In principle, the period of modulation can be faster than the electronic or vibrational or dipole relaxation times of any atom or molecule. *In other words, such modulated beams can be modulated at a rate greater than the relaxation or absorption time of electronic and vibrational states, possibly permitting a form of self-induced transparency.* This is the method used in these studies to minimize turbulence and other losses.

More specifically, let a beam be represented as $\sum_n \psi_n = \sum_n \chi_n a_n \exp(i2\pi\nu t_n)$ where χ_n is the polarizability induced by the nth E-vector field $a_n \exp(i2\pi\nu t_n)$. Let an arbitrary polarized field be represented by two such beams but orthogonally polarized:

$$\begin{aligned} I(\tau) &= \Psi\Psi^* = |\psi_1 + \psi_2|^2 \\ &= |\chi_1 a_1 \exp(i2\pi\nu t) + \chi_2 a_2 \exp(i2\pi\nu(t+\tau))|^2 \\ &= \chi_1^2 a_1^2 + \chi_2^2 a_2^2 + (\chi_1\chi_2)\cos(\theta) 2a_1 a_2 \cos(2\pi\nu\tau), \end{aligned} \tag{5.1}$$

where $\theta = 90°$ for orthogonally polarized beams; and $\tau = t_1 - t_2$. This composite beam resulting from the two beams would be*statically polarized* and with a constant τ dependency.

If the same field resulted from beams in which τ is modulated, i.e., there is a $d\tau/dt$, then that resultant would *not be* statically polarized, but *polarization modulated* (POLMOD):

$$\begin{aligned} I\left(\frac{d\tau}{dt} = k\frac{d\varphi}{dt}\right) &= \Psi\Psi^* = |\psi_1 + \psi_2|^2 \\ &= |\chi_1 a_1 \exp(i2\pi\nu t) + \chi_2 a_2 \exp(i2\pi\nu(t + d\tau/dt))|^2 \\ &= \chi_1^2 a_1^2 + \chi_2^2 a_2^2 \\ &\quad + (\chi_1\chi_2)\cos(\theta) 2a_1 a_2 \cos(2\pi\nu\tau + d\tau/dt), \end{aligned} \tag{5.2}$$

where φ is phase. Ideally, this is the way one would obtain POLMOD. Technically, this is quite possible, but requires equipment of appropriate complexity.

An easier, but limited, way to obtain POLMOD is to combine and beat two orthogonally polarized beams of different wavelengths:

$$\begin{aligned} I\left(\frac{d\tau}{dt} = k\frac{d\varphi}{dt}\right) &= \Psi\Psi^* = |\psi_1 + \psi_2|^2 \\ &= |\chi_1 a_1 \exp(i2\pi\nu_1 t) + \chi_2 a_2 \exp(i2\pi\nu_2 t)|^2 \\ &= ((\chi_1\chi_2)\cos(\theta)\, 2a_1 a_2 \cos(2\pi\nu_3 t)\sin(2\pi\nu_4 t))^2, \end{aligned} \tag{5.3}$$

where $\nu_3 = \frac{\nu_1 - \nu_2}{2}$ and $\nu_4 = \frac{\nu_1 + \nu_2}{2}$ and the beat frequency is $\nu_{\text{beat}} = \nu_1 - \nu_2$.

If the time constant required for both (i) polarization compatibility between incident radiation and molecule and (ii) for energy transfer to the detecting molecule to occur, is greater than $1/\nu_{\text{beat}}$, there will be a situation similar to that of pulse/transient self-induced transparency (i.e., SIT, see Sec. 6.5.1), with the incident radiation captured by the detecting molecule, but reradiated back into the radiating beam. The difference is that pulse self-induced transparency achieves transparency by challenging the relaxation time of the detecting molecule by using shorter time interval pulse than that relaxation time. In the present constant wavelength self-induced transparency, the relaxation time of the detecting molecule is challenged by the shorter time interval of polarization compatibility of radiation and receiving molecular system.

As an example, Fig. 5.9 shows the spectra of two constituent beams, λ = 532 and 457 nm and their resultant combination: a beat wave carrier and its envelope. If the two constituent beams are orthogonally polarized and combined, the resulting combined beam is a POLMOD beam. Alternatively, if the two constituent beams are similarly polarized and combined, the combined beam is an AXMOD beam. In which case, a molecular system detects the combined beam by the conjoint simultaneous dipole stimulation/sampling, and this molecular sampling is set by the excited state relaxation time.

In Sec. 5.1.2, we mentioned that in the literature, SAM and OAM are used in both the quantum mechanics (QM) and classical mechanics (CM) sense. POLMOD and AXMOD, both terms used

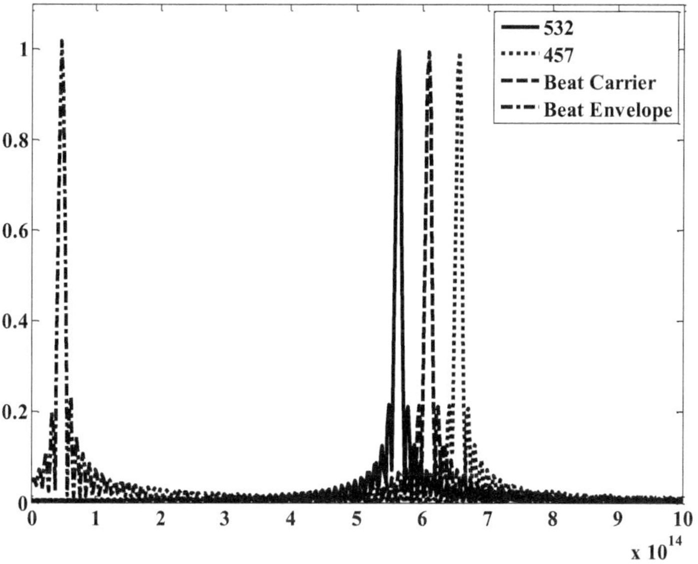

Fig. 5.9. Spectra of two constituent beams at $\lambda = 532$ and 457 nm, the resulting beat wave carrier and the envelope. This representation does not address different polarizations. If the two beams are orthogonally polarized and combined, the combined beam is a POLMOD beam. If the two beams are similarly polarized and combined, the combined beam is an AXMOD beam. A molecular system detects the combined beam by the conjoint simultaneous dipole stimulation/sampling, and the molecular sampling is set by the excited state relaxation time.

here only in the CM sense, are related to SAM and OAM, used in both the CM sense, in that POLMOD is *modulated* SAM (CM) and AXMOD is *modulated* OAM (CM). These somewhat confusing distinctions are laid out in Table 5.1.

The major differences between classical mechanical spin and quantum mechanical spin are as follows.

(1) In the case of classical spin, which is related to angular momentum, a complete rotation restores a field back to where it commenced, while in the case of quantum mechanical spin, a *double rotation* is necessary to restore a quantum particle back to where it commenced.
(2) Whereas classical spin can take on many different values, quantum mechanical spin may take on only one of two possible values.

Table 5.1. Classical and quantum mechanical associations.

Classical-independent variable	Classical mechanics entity	Associated quantum mechanics entity
Poincaré sphere: changing 2χ	Polarization Modulation (POLMOD) Modulated spin angular momentum (SAM)	Spin quantum, a property that results in a quantum particle respond, like a magnet, to an external magnetic field
Poincaré sphere: changing 2ψ	Axis or Azimuth Modulation (AXMOD) Modulated orbital angular momentum (OAM)	Orbital quantum: the region within an atom that encloses where an electron is likely to be 90% of the time.

(Furthermore, we only passingly mention that the quantum entities, bosons, only take on spin quantum numbers that are integers, and the quantum entities, fermions, only spin quantum numbers that are half integers.)

5.2.3. *Method for obtaining "polarization packets" by polarization modulation*

A POLMOD beam is formed from two orthogonally polarized beams. Therefore, each constituent beam acts in the manner of an additional magnetic field to the other beam; or, stated differently, each magnetic field of one beam acts as an electric field to the other. The total energy density is the sum of the densities of the field of each beam:

$$\omega = \omega_1 + \omega_2 = \frac{1}{2}\epsilon E_1^2 + \frac{1}{2}\varepsilon E_2^2 \text{ joules}/m^2 \tag{5.4}$$

For a small volume, Δv, the decrease in energy as a function of time is:

$$-\frac{\partial}{\partial t}\left(\frac{1}{2}\epsilon E_{21}^2 + \frac{1}{2}\varepsilon E_2^2\right)\Delta v = \int_S S \cdot ds \text{ watts} \tag{5.5}$$

where S = energy per unit area passing per unit time through the surface of the volume Δv in units of watts/m^2. Dividing by Δv and taking the limit obtains:

$$\nabla \cdot S = - - \frac{\partial}{\partial t}\left(\frac{1}{2}\epsilon E_1^2 + \frac{1}{2}\varepsilon E_2^2\right) \tag{5.6}$$

Using Maxwell equation substitutions, the result is

$$S = E_1 \times E_2 \tag{5.7}$$

Using the fact that E_2 is the effective magnetic field to E_1 and substituting gives

$$S = E \times H \text{watts/m}^2 \tag{5.8}$$

Thus, the Bowditch or Lissajous pattern representations of Fig. 5.7 are the differential, or instantaneous Poynting vector representations; and the combined E field representation Fig. 5.8, *when integrated over time*, is the Poynting vector.

In Fig. 5.10, (a) shows a combined radiated POLMOD beam traveling in the z (time) direction; (b) the same combined beam as a function of the two orthogonally polarized constituent beams; and (c) the changing frequencies of the first and second constituent beams in the x and y directions. Figure 5.10(c) indicates that a statically polarized molecular detector will experience radiation changing in the x and y direction at the rates shown.

If a horizontally polarized beam of 647 nm (463 THz) is combined with a perpendicularly polarized beam of 514 nm (583 THz), and if a molecular detector's wavelength and polarization preference is 523 THz ((583+463)/2), horizontally polarized, then the detector will be polarization-compatible with the radiation, only for durations that are a function of the beat frequency indicated in dotted line (Fig. 5.11). The polarization-compatible incident combined beam is indicated as a carrier. An objective of POLMOD beam design was to establish the beat frequency such that the relaxation time of atmospheric molecules is defeated.

We refer now back to the Poincaré sphere in Fig. 5.3, which represents *statically*, all polarizations and rotations. Reviewing the previous discussion: circularly polarized co-rotating is represented at the north pole; circularly polarized counter-rotating is represented

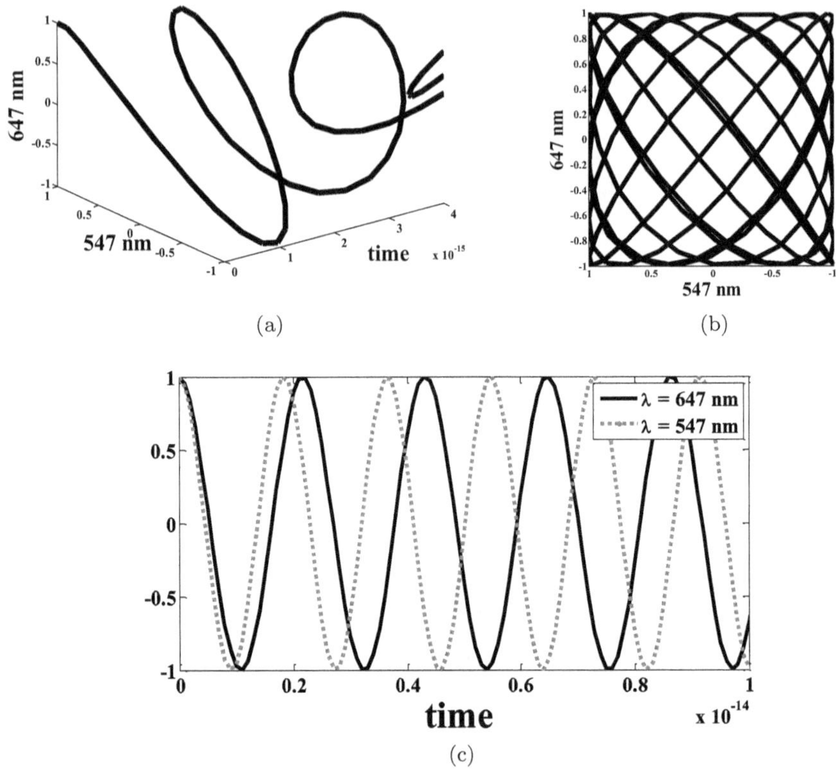

Fig. 5.10. Combined beam representations. (a) A combined E field radiated POLMOD beam traveling in the z (time) direction. (b) The same E field combined beam as a function of the two orthogonally polarized constituent beams — a Lissajous pattern — an instantaneous Poynting vector representation. (c) The constituent E field beams in the x- and y-directions superposed.

at the south pole. The longitudinal lines represent the varieties of elliptical polarizations. Around the equator latitude line are shown horizontally polarized linear, and vertically polarized linear configurations, as well as the variety of linear polarizations in between these. The angles 2ψ and 2χ are shown for a combined beam. Another angle, δ, is relevant and is the interbeam phase of the combined beam's constituent beams and not represented here. Therefore, POLMOD is a movement of the K vector on the sphere and is not static. Djordjevic

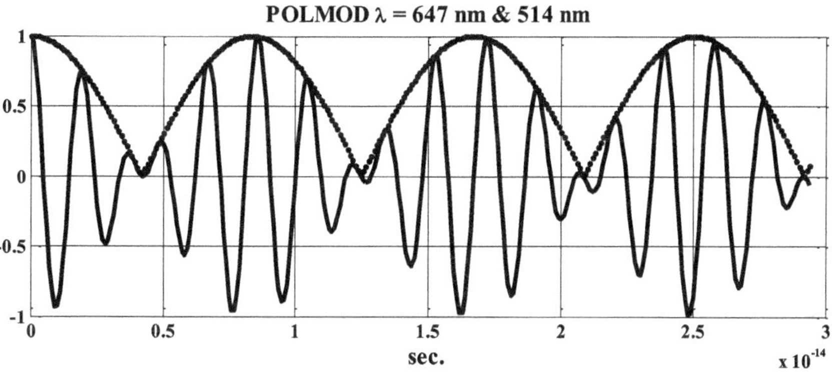

Fig. 5.11. POLMOD wave packets. If a horizontally polarized beam of 647 nm (463 THz) is combined with a perpendicularly polarized beam of 514 nm (583 THz), and if a molecular detector's wavelength and polarization preference is 523 THz ((583 + 463)/2), horizontally polarized, then the detector will be polarization-compatible with the radiation, only for durations that are a function of the beat frequency indicated in dotted line. The polarization-compatible incident combined beam is indicated as a carrier. An objective of the tests conducted was to establish the beat frequency repetition interval (here 3^{-15} s) such that the relaxation time of atmospheric molecules is defeated.

and Arabaci (2010) have proposed exploitation of photon spin angular momentum (SAM) associated with polarization, and orbital angular momentum (OAM) associated with the azimuth phase of the complex electric field.[9] POLMOD is related to modulated SAM, and AXMOD to modulated OAM, but used in the present instance for purposes of mitigating channel/atmospheric/ionospheric effects, rather than communications. Both can be represented on the Poincaré sphere.

POLMOD requires a changing angle 2χ. AXMOD requires a changing angle 2ψ. A *moving* representation of a POLMOD beam is shown in Fig. 5.4(b). Both are related to the angular momentum of radiation.

[9]cf. Djordjevic *et al.*, (2007); Djordjevic and Djordjevic (2009); Djordjevic (2011); Leach *et al.*, (2004); Gibson *et al.*, (2004); Paterson (2005).

The angular momentum of a beam can be given by (Jackson, 1999, p. 350)

$$\mathbf{L} = \frac{1}{\mu_0 c^2} \int d^3 \left[\mathbf{E} \times \mathbf{A} + \sum_{j=1}^{3} E_j \left(\mathbf{x} \times \nabla \right) A_j \right]. \tag{5.9}$$

The first term is identified with the "spin" of the photon; and the second with "orbital" angular momentum because of the presence of the operator $\mathbf{x} \times \nabla$ term. It follows, therefore, that POLMOD is defined as

$$\mathbf{dL}_{\mathrm{POLMOD}}/\mathbf{dt} = \frac{1}{\mu_0 c^2} \left[\mathbf{E} \times \mathbf{A} \right], \tag{5.10}$$

and AXMOD as:

$$\mathbf{dL}_{\mathrm{AXLMOD}}/\mathbf{dt} = \frac{1}{\mu_0 c^2} \left[\sum_{j=1}^{3} E_j \left(\mathbf{x} \times \nabla \right) A_j \right]. \tag{5.11}$$

We turn now to the method for obtaining polarization modulation and continue to the main results of the present series of tests reported which are:

- jitter mitigation;
- amplitude loss mitigation;
- beam pointing movement mitigation.

We can indicate that the frequency and amplitude dependence of these reported effects have yet to be investigated due to the unavailability of the frequency control of both the modulating and modulated beams.

5.2.4. *Method for achieving a transient match of polarization packets to molecular polarization*

As stated in Sec. 5.1.3, the aim is to obtain the temporal duration of the polarization dwell time of incident radiation (c_1) to be less than the duration required of a molecular system for channel jitter/absorption/scattering to be defeated and a TX→RX reception/absorption to occur (c_3), i.e., $c_1 < c_3$. It is important to note, again, that we are not addressing *constant polarization* wave packets, or*constant azimuth wave packets*, but wave packets of *changing polarization* (POLMOD) or *changing azimuth* (AXMOD).

This raises an important question: what evidence is there that molecular systems actually do, as stated in Sec. 5.1.2, function as "polarization selective receivers" or "axial selective receivers"; or Why do molecular systems receive or interaction optimally with radiation of a certain polarization? The answer comes from available evidence in the theoretical background to the field of Raman spectroscopy, which exhibits preferential scattering by molecular systems depending on the polarization of the incident radiation.

The relevant Raman spectroscopy background tells us: if (a) exciting radiation is near resonance with an electronic transition, or (b) there are electronic Raman transitions (Mortensen and Konigstein, 1968), or (c) there is a Jahn–Teller active state (Child and Longuet-Higgins, 1961), then antisymmetric as well as symmetric components of the Raman tensor are predicted (Plazcek, 1934; McClain, 1971). If the scattered radiation is due only to an induced electric dipole, a depolarization ratio has three components when antisymmetric occurs:

(1) the isotropy α^2;
(2) the symmetric isotropy γ_s^2;
(3) the anti-symmetric isotropy γ_{as}^2.

The depolarization ratio in terms of these components is

$$\rho = \frac{I_{\|}}{I_{\perp}} = \frac{3\gamma_s^2 + 5\gamma_{as}^2}{45\alpha^2 + 4\gamma_s} \tag{5.12}$$

and in terms of the elements of the Raman tensor $\alpha_{ij}, (ij = x, y, z)$:

$$\begin{aligned}
\alpha^2 &= (\alpha_{xx} + \alpha_{yy} + \alpha_{zz})^2,\\
\gamma_s &= -\frac{1}{2}\left[(\alpha_{xx} - \alpha_{yy})^2 + (\alpha_{yy} - \alpha_{zz})^2 + (\alpha_{zz} - \alpha_{xx})^2\right]\\
&\quad -\frac{3}{4}\left[(\alpha_{xy} + \alpha_{yx})^2 + (\alpha_{xz} + \alpha_{zx})^2 + (\alpha_{yz} + \alpha_{zy})^2\right],\\
\gamma_{as} &= -\frac{3}{4}\left[(\alpha_{xy} - \alpha_{yx})^2 + (\alpha_{xz} - \alpha_{zx})^2 + (\alpha_{yz} - \alpha_{zy})^2\right].
\end{aligned} \tag{5.13}$$

For the following equations, we define

I_{PARA} = incident radiation linearly polarized parallel to scattered radiation;
I_{PERP} = incident radiation linearly polarized perpendicular to scattered radiation;
I_{CO} = incident radiation circularly polarized corotating to scattered radiation;
I_{CONTRA} = incident radiation circularly polarized contrarotating to scattered radiation.

Then a reversal coefficient for scattered radiation at 90° to incident radiation is

$$r = \frac{6\gamma_s^2}{45\alpha^2 + \gamma_s^2 + 5\gamma_{as}^2} = \frac{I_{\text{CO}}}{I_{\text{CONTRA}}}, \tag{5.14}$$

and a general reversal coefficient dependent on the scattering angle, ν, is (Mortensen and Hassing, 1980):

$$R(\nu) = \frac{1 - \frac{1-\rho}{2(1+\rho)}\sin^2(\nu) - \frac{1-r}{1+r}\cos^2(\nu)}{1 - \frac{1-\rho}{2(1+\rho)}\sin^2(\nu) + \frac{1-r}{1+r}\cos^2(\nu)}. \tag{5.15}$$

This general reversal coefficient is a function of the three invariants: the isotropy α^2, the symmetric isotropy γ_s^2 and the antisymmetric isotropy γ_{as}^2, and the intensities in four polarization spectra are related to the three invariants by

$$\begin{aligned} I_{\text{PERP}} &= 3\gamma_s^2 + 5\gamma_{as}^2, \\ I_{\text{PARA}} &= 45\alpha^2 + 4\gamma_s^2, \\ I_{\text{CO}} &= 6\gamma_s^2, \\ I_{\text{CONTRA}} &= 45\alpha^2 + \gamma_s^2 + 5\gamma_{as}^2. \end{aligned} \tag{5.16}$$

Therefore, in the case of scattered radiation due only to an induced electric dipole:

$$I_{\text{PERP}} + I_{\text{PARA}} = I_{\text{CO}} + I_{\text{CONTRA}} \tag{5.17}$$

and only three measurements are needed, the fourth providing a check.

The invariants are related to the intensity data by

$$
\begin{aligned}
6\gamma_s^2 &= I_{\mathrm{CO}}, \\
5\gamma_{as}^2 &= I_{\mathrm{PERP}} - \frac{1}{2}I_{\mathrm{CO}}, \\
45\alpha^2 &= I_{\mathrm{PARA}} - \frac{2}{3}I_{\mathrm{CO}}.
\end{aligned}
\tag{5.18}
$$

This is the simple picture if there are moments higher than an induced electric dipole that are not involved. This is not the case with optically active molecules (Barron, 1978, 1982). In these cases, besides the polarizability tensor, scattering is defined in terms of optically active tensors defining electric and magnetic dipoles. The Hamiltonian of a system of charged particles interacting with an electromagnetic field is a multipole expansion (Fiutak, 1963), in which the semiclassical Kramers–Heisenberg dispersion equation is demonstrated to be identical with the corresponding quantum mechanical description (Göppert-Mayer, 1931). The Hamiltonian describing a molecule excited by a field is[10]

$$
\mathcal{H} = \mathcal{H}_0 - d_\alpha E_\alpha - \frac{1}{3}\Theta_{\alpha\beta}E_{\alpha\beta} - m_\alpha - 1/2\chi_{\alpha\beta}H_\alpha H_\beta \tag{5.19}
$$

where

- $\mathcal{H}_0$ is the Hamiltonian for the free molecule;
- $d_\alpha = \sum\limits_i e_i r_i$ or the electric dipole moment;
- $\Theta = \frac{1}{2}\sum\limits_i (3r_{i_\alpha} r_{i_\beta} - r_i^2\delta_{\alpha\beta})$ or the quadrupole moment;
- $m_\alpha = \sum\limits_i (e_i/2m)\varepsilon_{\alpha\beta\gamma} r_{i_\beta} p_{i_\gamma}$ or the magnetic moment operator for momentum p;
- E_α and $E_{\alpha\beta}$ are the electric field and the field gradient at the origin due to external charges;
- $\chi_{\alpha\beta}$ is the nonlinear susceptibility;
- H_α and H_β denote the magnetic field;

[10] Hertzfeldt and Göppert-Mayer (1936); Fiutak (1963); Buckingham (1967).

- e_i is the ith element of charge at the point r_i relative to an origin fixed at some point on the molecule;
- $\alpha\beta$ denote vector or tensor components and can be equal to x, y, z.

This steady-state Hamiltonian multipole expansion is only tangentially relevant to the work reported here, as the expansion does not address the relaxation times (transient state) challenged by polarization modulated beams. The duration of the transient polarization "window" for a variety of gases is as follows.

The Maxwell–Boltzmann distribution law for molecular velocities provides the mean square and mean velocities as:

$$\begin{aligned}\bar{\nu}^2 &= \frac{3fT}{m},\\ \bar{\nu} &= \left(8\bar{\nu}^2/3\pi\right)^{1/2}.\end{aligned} \tag{5.20}$$

The kinetic theory of gases, which assumes that molecules interact like hard spheres, gives

$$\eta = (5/16\sigma^2)(mkT/\pi)^{1/2}, \tau = l/\nu = 4\eta/5p, \tag{5.21}$$

where

$l = m/(\pi\rho\sigma^2\sqrt{2})$,

$k =$ Boltzmann′s constant,	$\sigma =$ molecular diameter,
$T =$ absolute temperature,	$m =$ mass of molecule,
$p =$ pressure,	$l =$ mean free path,
$\rho =$ density,	$\tau =$ mean time between collisions.
$\eta =$ viscosity,	

The 9.4-μm CO_2 absorption band has a band interval from 850(cm^{-1}) = 12 μm (25.48 Terahertz) to 1250(cm^{-1}) = 8 μm (37.47 Terahertz), i.e., 11.99 Terahertz. The relaxation time is thus on the order of 1/11.99e12 = 83 femtoseconds. Another limiting factor on the window is the molecular inter-collision time. The mean time between collisions is 108 picoseconds — three orders of magnitude greater than the relaxation time.

Thus, inelastic light scattering (Raman effect) (e.g., Barrett, 1979a,b, 1980, 1981a,b, 1982, 1984) demonstrates and exhibits a molecular polarization dependence or preference (or incident radiation-molecular system polarization correspondence). This polarization preference is exhibited also in electronic conductivity states — (e.g., Barrett, 1983a,b). We thus refer to molecular systems as "polarization selective receivers".

5.2.5. *The Bloch sphere*

The Poincaré sphere provides a representation of the polarization and rotation of radiation and we are treating molecular systems as polarization selective receivers. Similarly, the Bloch sphere represents the polarization and rotation of molecular systems. The Bloch sphere (Bloch, 1946, Fig. 5.12) is a geometrical representation of the state-space of a two-level quantum system and the Poincaré sphere, which we have discussed, is an example of this sphere. Futhermore, the Bloch sphere is a unit 2-sphere, with each pair of antipodal points corresponding to mutually orthogonal state vectors. The north and south poles of the Bloch sphere are typically chosen to correspond to the standard basis vectors $|0\rangle$ and $|1\rangle$. Substituting polarization states for quantum states gives the Poincaré sphere. Each point on the Bloch sphere represents a linear combination of the states with spin up and spin down. In the same way, an arbitrary polarization vector can always be expressed as a linear combination of left- and right handed circular polarization, which is represented by a point on the Poincaré sphere.

If the incident radiation on a molecular system is a (POLMOD) "pulse" or "polarization packet", then at the end of the "pulse" the Bloch vector has been rotated through an angle:

$$A(z,t) = \theta(z,t) = \int_{-\infty}^{t} \Omega\left(z,t'\right) dt', \tag{5.22}$$

where θ is the tipping angle or the angle through which the Bloch vector is rotated in y' and z' plane about the x'-axis. The angle A is also referred to as the area of the pulse, as it is proportional to the area under the pulse envelope in the time domain.

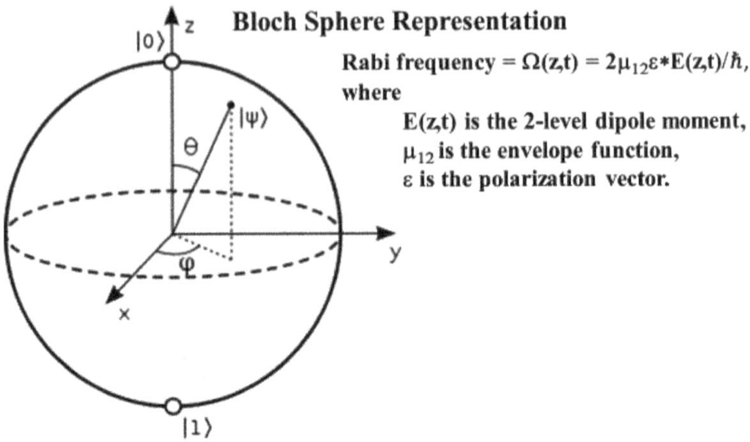

Fig. 5.12. The Bloch sphere and the Rabi frequency. The Rabi frequency is the frequency of population oscillation for a given atomic transition in a given light field. It is associated with the strength of the coupling between the light and the transition. Rabi flopping between the levels of a 2-level system illuminated with resonant light, will occur at the Rabi frequency. The Rabi frequency is a semi-classical concept as it is based on a quantum atomic transition and a classical light field. If the incident radiation is a pulse, then at the end of the pulse the Bloch vector has been rotated through an angle: $A(z,t) = \theta(z,t) = \int_{-\infty}^{t} \Omega\left(z,t'\right) dt'$, where θ is the tipping angle or the angle through which the Bloch vector is rotated in the y' and z' plane about the x'-axis. The angle A is also referred to as the area of the pulse, as it is proportional to the area under the pulse envelope in the time domain. For the special case of an intense light pulse producing atomic excitation, and if the pulse area is: $A = 2n\pi, n = 1, 2, 3, \ldots$. The pulse propagates through the medium without attenuation of its area, i.e., the medium appears transparent to the light pulse, *despite that its midfrequency coincides with an atomic resonance.*

For the special case of an intense light pulse producing atomic excitation, and if the pulse area is

$$A = 2n\pi, \quad n = 1, 2, 3, \ldots, \tag{5.23}$$

the pulse propagates through the medium without attenuation of its area, i.e., the medium appears transparent to the light pulse, *despite the fact that its midfrequency coincides with an atomic resonance.*

A simple representation of an ion is as an electron cloud attached to a fixed ion. If the attachments are represented in the x, y and z directions as "springs", and those springs have different spring constants in the x, y and z directions, then polarization is more apt to occur in the direction or axis of the weakest spring constant. Therefore:

- Induced polarization may *not be parallel* to the direction of the inducing field.
- Susceptibility is not a scalar.

But differences in refractive index (birefringence) and absorption (dichroism) depend on the polarization:

$$\mathbf{P} = \varepsilon_0 \chi \mathbf{E} = \varepsilon_0 \left(\bar{\varepsilon} - 1\right) \mathbf{E}, \tag{5.24}$$

where:

$$1 + x = \overline{\varepsilon} = \overline{\eta}^2 = \left(\eta + i\frac{c}{2\omega}\alpha\right)^2. \tag{5.25}$$

Therefore, a difference in the refractive index, e.g., in the x- and y-directions (birefringence) or the absorption coefficient (dichroism) implies a difference in the susceptibility, e.g., in the x and y directions, which is due to the induced polarization. As previously noted, the appropriate representation of the susceptibility is thus not as a vector, but as a tensor:

SUSCEPTIBILITY TENSOR

$$\mathbf{P} = \varepsilon_0 \begin{bmatrix} \chi_{11} & \chi_{12} & \chi_{13} \\ \chi_{21} & \chi_{22} & \chi_{23} \\ \chi_{31} & \chi_{32} & \chi_{33} \end{bmatrix} \mathbf{E}. \tag{5.26}$$

Atoms are polarized by applied fields:

$$\begin{aligned} \mathbf{D} &= \varepsilon_0 \mathbf{E} + \mathbf{P}, \\ \mathbf{P} &= \varepsilon_0 \chi \mathbf{E}, \\ \mathbf{D} &= \varepsilon_0 \left(1 + \chi\right) \mathbf{E}; \end{aligned} \tag{5.27}$$

and the conventionally defined Poynting vector is orthogonal to the incident field:

$$\mathbf{S} \cdot \mathbf{E} = 0, \tag{5.28}$$

but the wave vector is orthogonal to the electric displacement:

$$\mathbf{k} \cdot \mathbf{D} = 0. \tag{5.29}$$

Therefore, because in anisotropic media $\mathbf{E}$ and $\mathbf{D}$ are not necessarily parallel initially at time t_0, the Poynting vector and the wavevector may diverge.

In the Lorentz harmonically bound classical particle picture:

$$m\ddot{x} = -\frac{dV}{dx} = -m\omega_0^2 x, \tag{5.30}$$

the potential V is anharmonic for large displacements:

$$V(x) = \frac{m\omega_0^2}{2} x^2 + bx^3 + cx^4 + \cdots. \tag{5.31}$$

Therefore, the polarization varies nonlinearly with the field:

$$\mathbf{P} = \varepsilon_0 \left\{ \chi^{(1)}\mathbf{E} + \chi^{(2)}\mathbf{E}^2 + \chi^{(3)}\mathbf{E}^3 + \cdots \right\}. \tag{5.32}$$

A one-dimensional electromagnetic wave propagating in the z direction through a medium in which there is a macroscopic polarization is represented by

$$\left(\frac{\partial^2}{\partial z^2} - \frac{n^2 \partial^2}{c^2 \partial t^2} \right) \mathbf{E}(z,t) = \frac{1}{\varepsilon_0 c^2} \frac{\partial^2 \mathbf{P}(z,t)}{\partial t^2}, \tag{5.33}$$

where $\mathbf{P}(z,t)$ is a macroscopic polarization. However, the picture offered by this equation neglects the fact that $\mathbf{E}$ *is a vector and* $\mathbf{P}$ *is a tensor*, and that $\mathbf{P}$ *does not spring immediately, and instantaneously, into full amplitude.*

There is also the claim that the *dielectric constant is a tensor* (Beth, 1936). Therefore, the field intensity $\mathbf{E}$ is, in general, not parallel to the polarization, $\mathbf{P}$, nor to the electric displacement, $\mathbf{D}$. The torque is then given as

$$\mathbf{L} = \mathbf{P} \times \mathbf{E}. \tag{5.34}$$

As the so-called Rabi frequency is associated with the Bloch sphere picture, we turn now to explain why the polarization selective picture is not concerned with this frequency. The Rabi frequency is the frequency of population oscillation for a given atomic transition in a given light/radiation field. It is associated with the strength of the coupling between the radiation and the transition. Rabi flopping between the levels of a 2-level system illuminated with resonant light will occur at the Rabi frequency. But the Rabi frequency is a semi-classical concept as it is based on a quantum atomic transition and a classical radiation field. The Rabi frequency represented in the Bloch picture also *does not represent the transient polarization state of a rapidly changing polarization modulated wave, nor does it account for the tensor nature of polarization* **P**. This is an important distinction as we are addressing transient effects.

5.3. Kolmogorov Turbulence Theory and Modulation Methods

In this TX →RX picture, atmospheric turbulence may be considered as a form of channel noise. This characterizes the interference target for defeat by spin angular momentum (SAM) and orbital angular momentum (OAM). To introduce a discussion of these modulation methods: the applications, or in some cases, the misapplications, of Kolmogorov turbulence theory.

5.3.1. *Kolmogorov turbulence theory*

To appreciate this section, you must know that turbulence creates a medium where the physical parameters such as velocity and viscosity defy adequate descriptive theory. The problem is that one cannot adequately model or predict the media conditions at a given point and time instant. This means models make assumptions and rely on statistical descriptions to get any usable or understandable model at all. The trick is to make the model close enough to physical reality to accomplish the objective. Richardson[11] described turbulent fluid behavior with this couplet:

> *"Big whirls have little whirls that heed on their velocity,*
> *And little whirls have littler whirls and so on to viscosity."*

---

[11]English scientist and philosopher Lewis Fry Richardson (1881–1953).

For historical background we turn to Kolmogorov.[12]

The Kolmogorov turbulence theory (KT) assumes statistical homogeneity, isotropy and no preferential direction (Kolmogorov, 1941a,b, 1962; Obukhov, 1962). These assumptions imply that the mean value of the field is constant and the correlations between random fluctuations in the field from point to point are independent of the chosen observation. KT assumes that small scale structures are statistically homogeneous, isotropic and independent of large scale structures. KT is also limited to the inertial subrange $l \ll L_o$, (where l is the length of an eddy and L_0 is the flow length scale of the large eddies). It is assumed that the rate at which energy is supplied to the largest possible scale is equal to that dissipated at the shortest scale. In some sense, KT requires that eddies "know" how big they are, at which rate energy is supplied to them, and at which rate they must supply it to the next smaller eddies in the cascade. External energy is assumed to be input only on the largest scales and only dissipated on the smallest scales.

The optical field associated with the propagation through atmospheric turbulence of radiation from a monochromatic point source has a randomly modulated amplitude and a randomly modulated phase. Where the amplitude goes to zero the phase manifests a spatial dependence that indicates the presence of a branch point (Fried, 1998). A standard adaptive optics system, in particular one utilizing a least mean square error wave-front reconstructor, will not be able to properly determine the phase of the optical field in the vicinity of such a branch point.

Oesch and Sanchez *et al.*[13] showed that fields with branch points are indicators of photons with orbital angular momentum (OAM) that can be created in optical waves propagating through disturbed

---

[12]Soviet scientist Andrey Nikolaevich Kolmogorov (1903–1987) who made significant contributions to the mathematics of probability theory, topology, intuitionistic logic, turbulence, classical mechanics, algorithmic information theory and computational complexity.

[13]Sanchez and Oesch (2011a,b); Oesch and Sanchez (2012); Oesch *et al.*, (2012, 2013).

turbulence and the creation appears to be governed by the inner scale of turbulence.

While conventional phase-conjugate adaptive-optics provide good correction for weak-scintillation conditions, under strong scintillation conditions there is a significant degradation in correction as the scintillation increased. The presence of branch points in the phase appears to be the primary reason for the degradation in correction as the scintillation increases (Primmerman *et al.* 1995). Furthermore, Levine *et al.* (1998) is the first study of statistics over near-ground paths *that also provide information on temporal fluctuations instead of probability density functions.* That study showed that near ground turbulence effects *are not consistent with the assumption of a Kolmogorov power spectrum.* For this and the other reasons described above, Kolmogorov theory has limited applicability.

5.3.2. *Spin angular momentum (SAM) modulation*

Any electromagnetic radiation — radio, visible, etc. — can have spin angular momentum (SAM) and orbital angular momentum (OAM). One recalls that there is a fundamental difference between the term "spin" as applied in classical mechanics and as applied — in quantum mechanics. The classical explanations offered here, although often referring to light, also apply to the radio frequencies.

Light beams that are not plane waves will possess orbital angular momentum (OAM).[14] Fields with branch points are indicators of photons with OAM that can be created in optical waves propagating through disturbed turbulence and the creation appears to be governed by the inner scale of turbulence.[15] Jackson (1999, p. 350) has long provided a classical formulation of angular momentum with respect to SAM and OAM contributions.

As previously noted, POLMOD is essentially *modulated* SAM and axis or azimuthal modulation (AXMOD) is *modulated* OAM, but used in the present instance for mitigating

[14]Beth (1950); Allen *et al.*, (1992), Allen and Padgett (2000).

[15]Sanchez and Oesch, (2011a,b); Oesch and Sanchez (2012); Oesch *et al.* (2012, 2013).

channel/atmospheric/ionospheric effects, rather than communications.

The total mechanical angular momentum, **J**, of the electromagnetic field can be separated into an external and an internal part:

$$\mathbf{J} = \mathbf{L} + \mathbf{S}, \tag{5.35}$$

where the external part, **L**, is the angular momentum, and the internal part, **S**, is the "spin" part. The vector **J** can be separated into two gauge-invariant parts **L** and **S**, which are often termed the "orbital" and the "spin" part. The former quantity is similar to external angular momentum, as it is defined relative to a reference point. The "spin" part is known as the intrinsic part, because it is independent of the choice of this point. For the classical free electromagnetic field both quantities **L** and **S** are conserved (Van Enk and Nienhuis, 1994).

Allen *et al.* (1992) recognized that the phase cross-section of a helically phased light beam carries an orbital angular momentum that is described by $\exp(il\phi)$, with a value of $l\hbar$ per photon. In contrast, a spin angular momentum for a left and right circularly polarized beam is $\sigma\hbar = \pm 1$ per photon (Beth, 1936). The OAM signature is the amplitude cross-section of the beam and is independent of SAM. Furthermore, whereas OAM is associated with an unbounded state space (the l in $\exp(il\phi)$ can take any integer value (Franke-Arnold *et al.* 2008)), SAM has only two orthogonal states. For this reason, OAM is favored for RF communications. It is important to note that in OAM communications demonstrations it is the spatial, not the temporal modes that are exploited.

In the paraxial approximation, the Poynting vector representing radiation current or momentum density can be decomposed into orbital and spin contributions (Berry, 2009). This vector, to which SAM is associated, has an azimuth component to which OAM is associated (Allen and Padgett, 1996). Furthermore, with respect to spatial, i.e., cross-sectional, modes, whereas the natural basis set for SAM beams is a Hermite–Gaussian (HG) mode expansion, it was determined that the natural basis set for describing beams with OAM is a Laguerre–Gaussian (LG) mode expansion. In other words, Hermite–Gaussian modes are SAM eigenstates and Laguerre–Gaussian modes are OAM eigenstates. A Jones matrix formulation can be given for both SAM and OAM (Allen *et al.* 1999). Whereas both HG and LG beams can carry SAM, only LG beams can carry OAM (Padgett *et al.* 2004).

5.3.3. *Orbital angular momentum (OAM) modulation*

Beams with planar wavefronts are characterized in terms of Hermite–Gaussian modes that have rectangular symmetry described in terms of two mode indices, m and n, which give the number of nodes in the x and y directions. In contrast, beams with helical wavefronts are characterized in terms of Laguerre–Gaussian modes described by the indices, l, the number of intertwined helices, and p, the number of radial modes (Padgett and Allen, 2000).

The amplitude form of expansion formulas for a Hermite–Gaussian (HG) and a Laguerre–Gaussian (LG) modes propagating along the z-axis are (Beijersbergen *et al.* 1993):

$$\begin{aligned} u_{\mathrm{nm}}^{\mathrm{HG}} &= C_{\mathrm{nm}}^{\mathrm{HG}} \left(\frac{1}{w}\right) \exp\left[-ik(x^2+y^2)/2R\right] \\ &\quad \times \exp\left[-(x^2+y^2)/w^2\right] \exp\left[-i(n+m+1)\psi\right] \\ &\quad \times H_n\left(x\sqrt{2}/w\right) H_m\left(y\sqrt{2}/w\right), \end{aligned} \tag{5.36}$$

$$\begin{aligned} u_{\mathrm{nm}}^{\mathrm{LG}}(r,\phi,z) &= C_{\mathrm{nm}}^{\mathrm{LG}} \left(\frac{1}{w}\right) \exp\left(-\frac{ikr^2}{2R}\right) \exp\left(-\frac{r^2}{w^2}\right) \\ &\quad \times \exp\left[-i(n+m+1)\psi\right] \exp\left[-i(n-m)\phi\right] \\ &\quad \times (-1)^{\min(n,m)} \left(\frac{r\sqrt{2}}{w}\right)^{|n-m|} \\ &\quad \times L_{\min(n,m)}^{|n-m|}\left(\frac{2r^2}{w^2}\right), \end{aligned} \tag{5.37}$$

where

- $R(z) = \frac{(z_R^2+z^2)}{z}$,
- $\frac{1}{2}kw^2(z) = \frac{(z_R^2+z^2)}{z_R}$,

- $\psi(z) = \arctan\left(\frac{z}{z_R}\right)$, or the Gouy phase that a beam undergoes when passing through a waist,
- $H_n(x)$ is the Hermite polynomial of order n,
- L_p^l is the generalized Laguerre polynomial,
- k is the wavenumber,
- $z_R = \frac{\pi w_0^2}{\lambda}$ is the Rayleigh range of the mode (half the confocal parameter),
- w_0 is the beam width;
- $N = n + m$ is the order of the mode,
- $p = \min(n, m)$;
- $l = n - m$.

Normalization achieves $\int dxdy\, |u|^2 = 1$ and gives:

$$C_{\mathrm{nm}}^{\mathrm{HG}} = \left(\frac{2}{\pi n!m!}\right)^{1/2} 2^{-N/2}, \tag{5.38}$$

$$C_{\mathrm{nm}}^{\mathrm{LG}} = \left(\frac{2}{\pi n!m!}\right)^{1/2} \min(n, m) \tag{5.39}$$

HG and LG modes can be decomposed into each other by a mode converter.[16] However, as before stated, whereas a LG beam carries OAM, an HG beam does not.[17] The optical conversion of HG modes to LG modes is described by Beijersbergen *et al.* (1993) and a further analysis of this relationship and conversion has been provided by Van Enk and Nienhuis (1994).

A dual-symmetric formulation of classical electromagnetism (Bliokh *et al.* 2013) provides a separation of spin and orbital degrees of freedom. With the separation of angular momentum into orbital and spin parts, as above:

$$\mathbf{J} = \mathbf{L} + \mathbf{S}, \tag{5.40}$$

[16] Beijersbergen *et al.* (1993); Kimel and Elias (1993).
[17] Allen *et al.*, (1992).

this formulation permits:

$$\mathbf{L} = \frac{1}{2}\left[\mathbf{E}\cdot(\mathbf{r} \times \nabla)\,\mathbf{A} + \mathbf{B}\cdot(\mathbf{r} \times \nabla)\,\mathbf{C}\right] = \mathbf{r} \times \mathbf{P}_0, \tag{5.41}$$

$$\mathbf{S} = \frac{1}{2}\left[\mathbf{E} \times \mathbf{A} + \mathbf{B} \times \mathbf{C}\right], \tag{5.42}$$

where $\mathbf{P} = \mathbf{E} \times \mathbf{B} = \mathbf{P_0} + \mathbf{P_S}$ is an energy–momentum current (Poynting vector), with $\mathbf{P_0}$ the orbital part and $\mathbf{P_S}$ the spin part.

A and **C** are the magnetic and electric vector potentials:

$$\nabla \times \mathbf{A} = \mathbf{B},$$
$$\nabla \times \mathbf{C} = -\mathbf{E}$$

$r^\alpha = (\mathrm{t}, r)$ represents Minkowski spacetime with signature $(-,+,+,+)$.

In the case of OAM, transverse modes can be divided into sets of the same mode order, N, and p is the radial mode index describing the number of radial modes in a field distribution. If $\mathbf{L} = \pm 1$ & $\mathbf{P} = 0$, there is a two-dimensional subspace which can be represented on a type of Poincaré sphere. Another situation that permits a Poincaré sphere form representation is: $\mathbf{L} = 0$ & various $\mathbf{P}$ values and $\mathbf{L} = 2$ & $\mathbf{P} = 0$. All other OAM modes cannot be represented on a single sphere with: (a) a Poincaré sphere representation of SAM; (b) a sphere representation of the OAM condition. Such representations are shown in Fig. 5.13 (Padgett and Courtial, 1999; Romero *et al.* 2013).

5.4. Transport of OAM Through a Turbulent Atmosphere: Applications in Communications and Biomedical Imaging

The angular momentum flux, or flow, together with angular momentum density expresses the conservation of angular momentum (Barnett, 2002). The angular momentum flux about the direction of propagation has spin and orbital parts. As stated previously, the

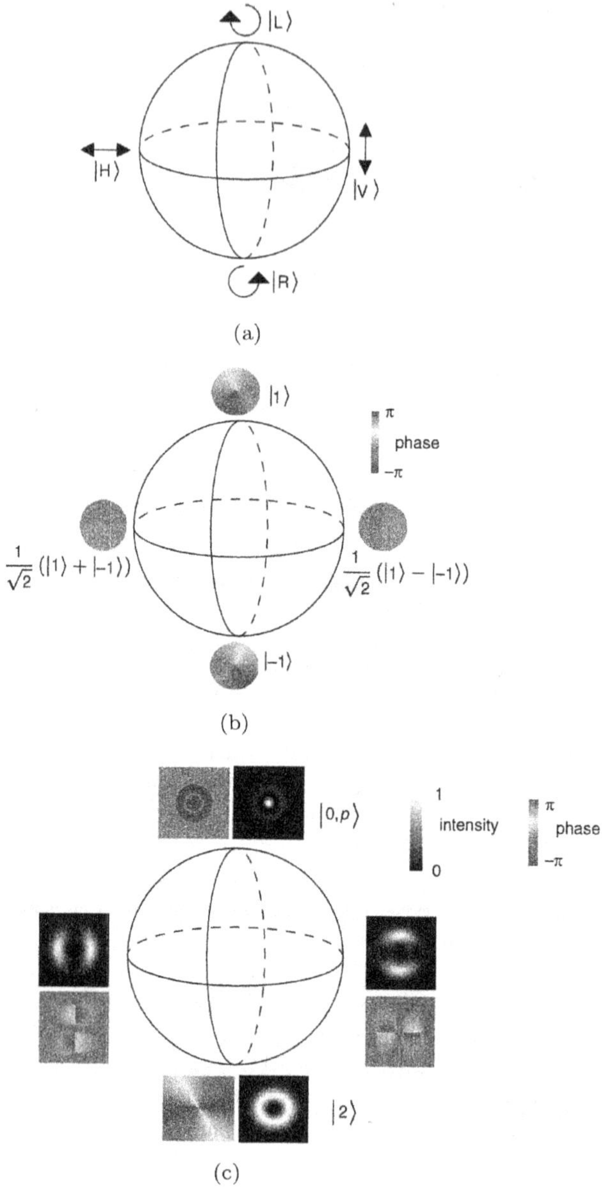

Fig. 5.13. POLMOD (SAM) and OAM signal representations. (a) Poincaré sphere representation of polarization states as in SAM. (b) Analogous sphere representation for OAM with the poles $| \pm l >$. (c) Another OAM sphere representation with the north pole $l = 0$ & *various* p and the south pole $l = 2$ and $p = 0$. From Romero *et al.* (2013) by permission, M.J. Padgett, and Cambridge University Press.

spin angular momentum (SAM) is associated with polarization, and orbital angular momentum (OAM) is associated with the azimuth phase.

There have been numerous reports of OAM RF communications[18] and communications through turbulent atmospheres.[19] OAM beams of light have been demonstrated at data rates of 2.56 Terabits/s., RF beams at 32 Gigabits/sec. across 2.5 m, and 1.6 Terabits/s. over optical fiber.[20] However, critics of the possibility of applications in applications at RF frequencies believe that impractical large antennas would be required; and even proponents have indicated the problems involved with signal recovery (Willner, 2016).

An OAM radiation beam can be decomposed into an unlimited number of Laguerre–Gaussian spatial, or $x-y$ cross-sectional, orthogonal modes, but a SAM radiation can be decomposed into only two spatial $x-y$ cross-sectional, orthogonal modes.[21] Therefore, as noted previously, OAM is presently preferred over SAM in communication demonstrations, because, it is claimed, each orthogonal mode, offers an unlimited number of data streams. Each mode becomes doubled in frequency and transformed to a higher mode (Dholakia *et al.* 1996). However, this claim is made without reference to sampling rate requirements and production of noise. Because spatial pattern recognition generally requires mode identification, there are possible tradeoffs between the number of orthogonal modes/data streams on the one hand and sampling rate and noise levels on the other, which precludes an unlimited number of data streams.

Interestingly, perhaps for astrophysicists, there is an angular Doppler effect associated with OAM.[22]

---

[18]Thidé *et al.* (2007, 2015); Wang *et al.* (2012); Tamburini *et al.* (2012); Bozinovic *et al.* (2013); Krenn *et al.* (2014, 2016); Hui *et al.* (2015); Willner *et al.* (2015); Willner (2016, 2017); Xie *et al.* (2017a,b).

[19]Sanchez and Oesch (2011); Pors *et al.* (2011a).

[20]Chant (2014); Willner (2016).

[21]Molina-Terriza *et al.* (2007); Fernandez-Corbaton *et al.* (2012).

[22]Garetz (1981); Simon *et al.* (1988); Nienhuis (1996); Bialynicki-Birula and Bialynicki-Birula (1997); Courtial *et al.* (1998).

5.4.1. *Applications in communications, molecular spectroscopy and biomedical imaging*

OAM has been used successfully in transmissions across Vienna, Austria in a 3 km data link and through turbulent air using 16 OAM spatial modes identified by pattern recognition.[23] These authors also found that the relative phase of the superposition modes was not affected by the atmosphere. These successful demonstrations of turbulence defeat stand in contrast to experiments showing significant degradation in mode quality after transmission in the laboratory through simulated turbulence.[24] The success of the Krenn *et al.* demonstrations is likely due to their incoherent detection scheme which employed adaptive pattern recognition. This method avoids coherent phase-dependent measurements.

The dimensionality of OAM spatial modes has been studied[25]and as well as their relations to Shannon dimensionality and quantum entanglement.[26]

The POLMOD and AXMOD beams generated by beat methods described here have frequencies in the Terahertz (THz) range. Besides communications and surveillance, such pulses or wave packets have application in the fields of molecular spectroscopy and biomedical imaging (Jacoby, 2015). These methods can also be applied in high-speed OAM acoustic communications.[27]

5.4.2. *Relationship to adaptive optics*

Adaptive optics is a system for controlling the spatial phase of a light beam to compensate optical aberrations, or to change the light beam spatial phase to create a desired effect. Some alternative technologies are: deconvolution, spatial filtering, null corrector plates, nonlinear

---

[23]Krenn *et al.* (2014, 2016).

[24]Pors *et al.* (2011a); Malik *et al.* (2012); Ibrahim *et al.* (2013); Rodenburg *et al.* (2014); Ren *et al.* (2013).

[25]Pors *et al.* (2008a, 2011b).

[26]Mair *et al.* (2001); Barreiro *et al.* (2008); Pors *et al.* (2008b).

[27]Renn *et al.* (2016); Shi *et al.* (2017).

phase conjugation. *There is an accent with these technologies on the spatial or beam cross-section.* This is also the case with OAM communications, as we have seen above. SAM can also be approached in its spatial mode aspect, but with modulated SAM, i.e., POLMOD, *the focus is on the temporal domain or modes.*

Preserving spatial information is the objective of astronomy and imaging. However, SAM and modulated SAM (POLMOD) can be a preferred technique if the main objective is to preserve information sequence and point focus. Furthermore, whereas adaptive optics addresses *the receive side* of a channel, POLMOD addresses *the transmit side.* Therefore, POLMOD offers either an alternative approach to the same problem, or can be used together with (improved) adaptive optics, each addressing one side of a channel.

5.5. POLMOD Testing

5.5.1. *Test objectives*

The basic test objectives described here were to determine the mitigation of (a) spatial and (b) temporal jitter, and (c) absorption/scattering, by comparing these effects on beams sent through disturbing media. To record results, special CCD video cameras recorded changes in (a) spatial jitter, and power meters recorded changes in (b) temporal jitter and (c) absorption/scattering.

5.5.2. *Transmission test media*

The laboratory generated transmission media were of three types: (1) water vapor in a chamber (Fig. 5.14); (2) heated air (Fig. 5.15); and (3) a medium of changing refractive index: either a compact disk cover, or a pseudo-random phase plate (Fig. 5.16). The random phase plate was attached to a rotary stage and operated at 5, 20 and 40 rpm. Tests were also conducted in desert air over a short ranges — 80 m and 160 m (Fig. 5.16). Table 5.2 summarizes the test media types.

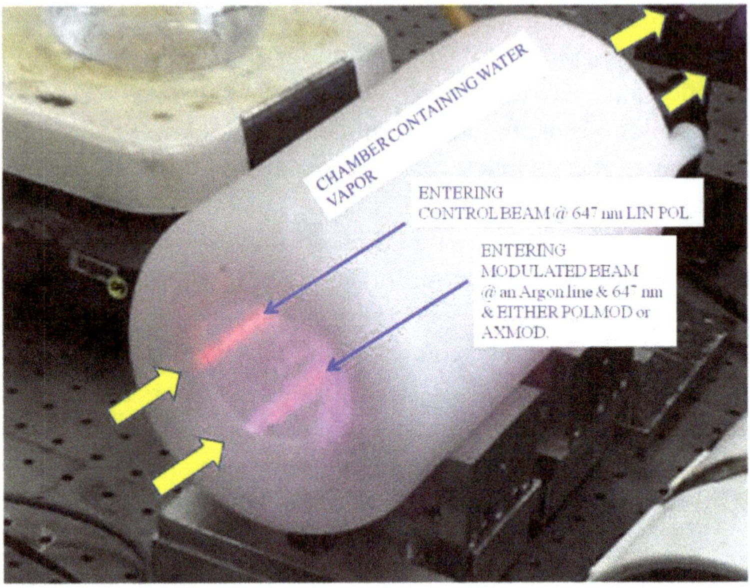

Fig. 5.14. Test medium: water vapor arrangement. Water vapor (steam) was passed into the chamber through which the modulated and control beams were passed simultaneously.

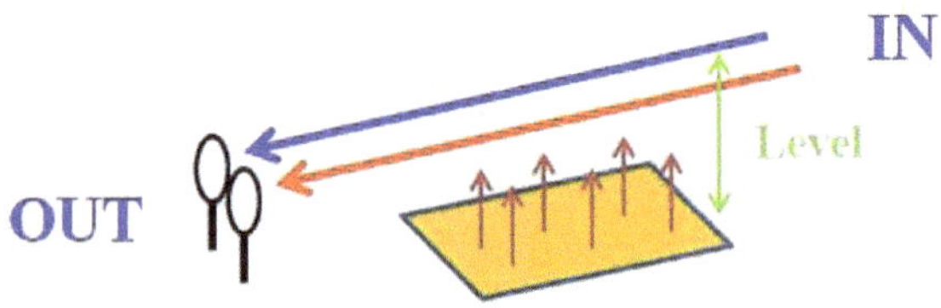

Fig. 5.15. Test arrangement: Hotplate providing heated air. A hotplate heated air, creating a turbulent condition. Simultaneously, both the test beam (POLMOD) and the control beam were sent over the hotplate at known heights of (1) 2.86 cm, (2) 3.49 cm and (3) 4.76. Two photodiodes recorded the beams entering the turbulent medium (IN) and exiting the turbulent medium (OUT).

5.5.3. *Laboratory test configurations*

Table 5.3 lists the media test arrangements and Table 5.4 lists the test combination designs. The designs indicate the possible ways that these studies of jitter and absorption/scattering defeat could proceed in the future.

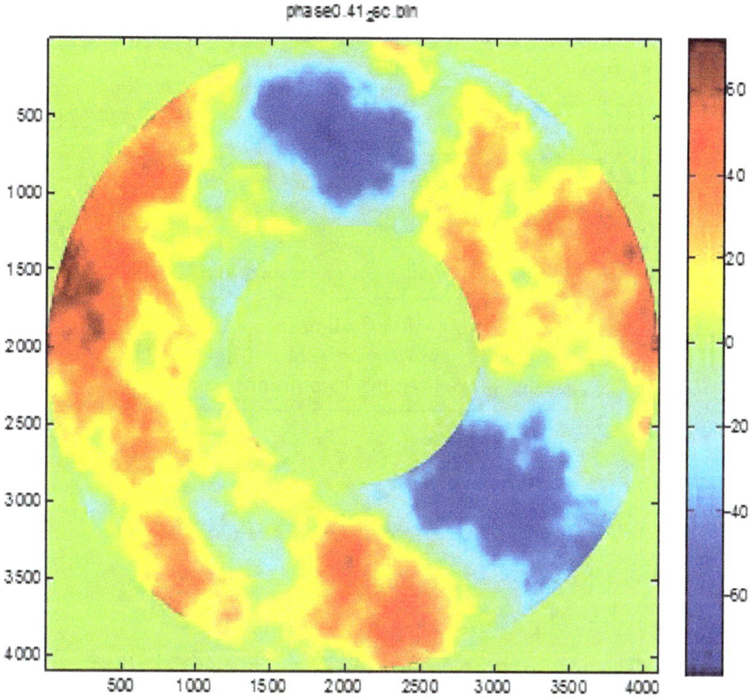

Fig. 5.16. Lexitek pseudorandom phase plate, 4-inch square and a 4096 × 4096 array. There is −78.7204 to +72.1974 radians of phase at the design wavelength $\lambda = 632.8$ nm as indicated on the right. 0.0008-inch grid spacing yielding a 3.277-inch (83.25 mm) diameter. Plate was attached to a rotary stage and was operated at 5, 20 and 40 rpm. Beams were propagated at left (225°) at radius 3 inches (0.0762 m). Therefore 5 rpm = 2.39 m/s; 20 rpm = 9.58 m/s; and 40 rpm = 19.15 m/s.

Table 5.2. Transmission test media summary.

No.	Medium	Figure
1	Water vapor in chamber	5.14
2	Heated air	5.15
3	Media of changing refractive index	
3.1	Pseudo-random phase plate rotated at 5, 20 and 40 rpm	5.16
3.2	Phase plate: compact disk (CD) cover	
4	Atmospheric tests — Desert air	
4.1	80 m	
4.2	160 m	

Table 5.3. Media test arrangements used.

Heated air and water vapor — optical configurgation
Atmospheric tests — Desert air
Heated air
Rotated pseudo — random phase plate

Table 5.4. Test combination designs.

Laser combination designs
Designs for two laser combinations
Designs for three laser combinations

5.6. POLMOD Turbulent Media Test Results

5.6.1. *Data analysis methods*

Table 5.5 lists the data analysis methods used.

One particular analysis method is described here for later reference. The two CCD cameras provided x and y spatial centroid measures for the laser beams entering the medium and the exiting the medium. Therefore, a difference measure, d, can be calculated as a measure of spatial jitter (Fig. 5.17), and improvements over expected values determined (Fig. 5.18).

5.6.2. *Data analysis results*

This section provides the evidence of jitter and absorption/scattering mitigation and the quantitative results under the conditions and test configurations described in the previous section. The publisher's website provides a data analysis of a library of data sets.

A representative set of results is shown in Figs. 5.19–5.29. The analysis of the complete library of data sets shows that POLMOD/AXMOD reduced temporal jitter up to a maximum of at least 50%, spatial jitter up to a maximum of at least 33%, and absorption/scattering up to 85% in some media and at some combination of wavelengths of combined beams.

Table 5.5. Data analysis methods.

1	Centroid amplitude difference: POLMOD vs. unmodulated beam
2	Mean and standard deviation: increase/decrease
3	x and y centroid histograms and probability densities
4	Centroid d distances: input vs. output of media
5	Video camera output: standard deviation extraction method
6	Power meter output: standard deviation extraction method
7	Centroid d distance expectancy
8	Entropy of randomness

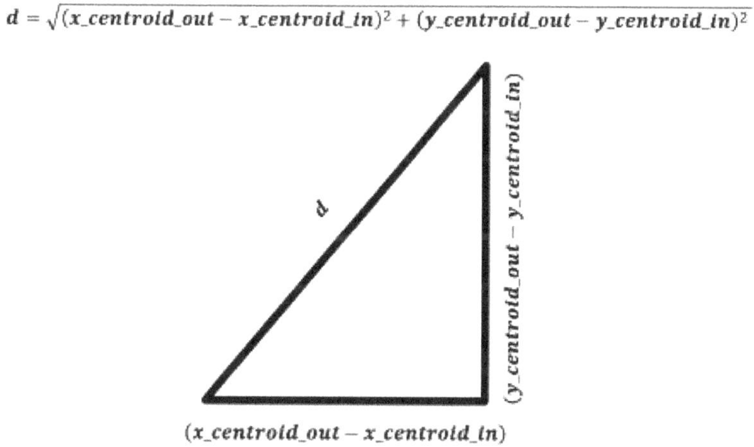

Fig. 5.17. The means of the x and y centroids for the input and output to the CD medium provided by the two CCD cameras permitted calculating the change, d, as effected by the medium.

It should also be noted that these positive results were obtained under the nonideal conditions for beam combining and with limited availability in laser wavelengths.

5.6.3. *Test summary and conclusions*

Evidence was described above indicating that spatial and temporal jitter and absorption/scattering are mitigated under the nonoptimum conditions and test configurations described. This mitigation is clearly evident in the analyzed data. However, due to the nonoptimal data collection conditions, an exact measure of how much such

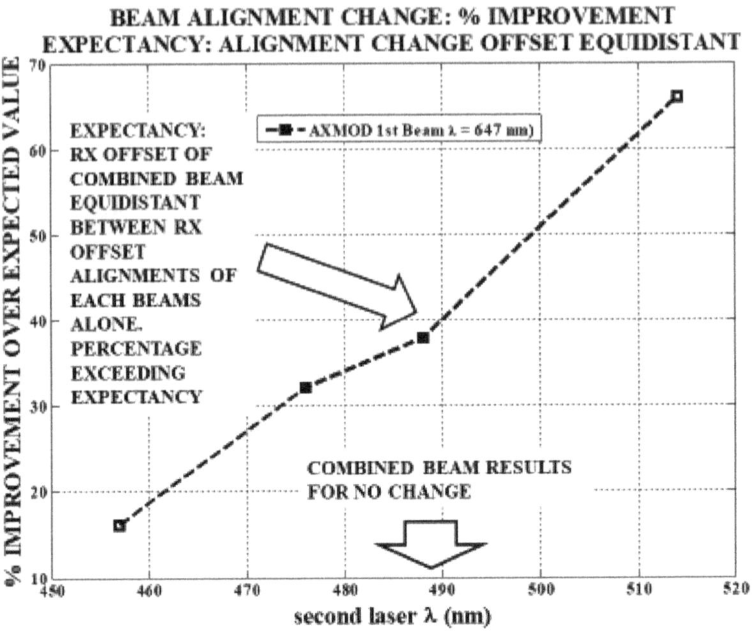

Fig. 5.18. Data analysis method. Using the d measures shown in Fig. 5.17, a no-effect expectancy was calculated based on the return signal (RX) offset d alignment of the combined beam (POLMOD or AXMOD) being equidistant between the RX offset d alignment of the two control beams. This expectancy is the null result condition. Then a percentage improvement, i.e., percentage deviation from the expectancy, could then be calculated.

jitter and absorption/scattering was mitigated and how much such mitigation is frequency dependent was not provided. Yet such mitigation has been shown here and from a visual inspection of all the analyzed data, the conclusion is that POLMOD/AXMOD reduced temporal jitter up to a maximum of at least 50%, spatial jitter up to a maximum of at least 33%, and absorption/scattering up to 85% in some media and some combination of wavelengths of combined beams. Further tests with optimum wavelength combinations are expected to provide improvements in these extremely promising results.

These design combinations are relevant to any procedures to mitigate jitter, absorption and scattering of transmissions through the atmosphere/ionosphere.

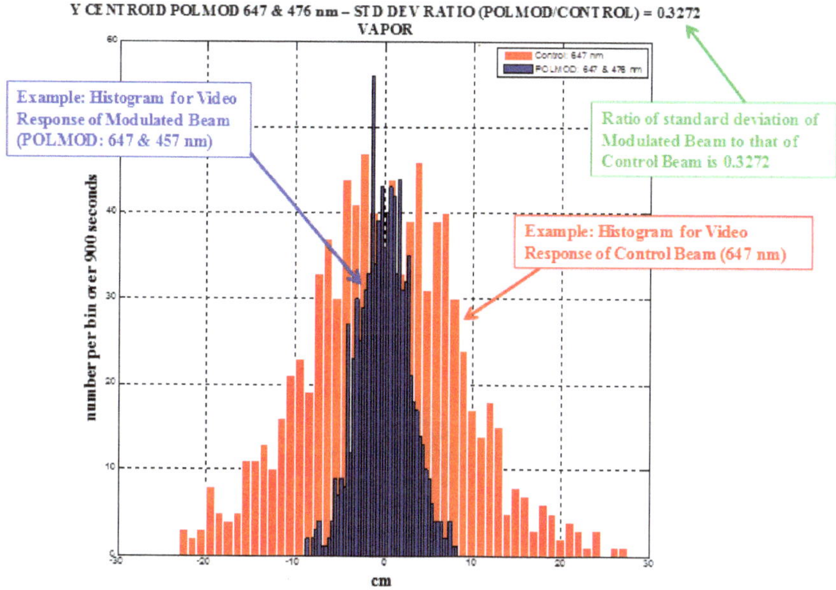

Fig. 5.19. Test of spatial jitter. Medium: water vapor. Video capture. An example of the analysis procedure for video camera capture. The statistical measure in this example is the x centroid and histogram bin counts were calculated for both control (the 647 nm beam — see label) and the modulated beam (the 647 and 457 nm beam — see label) — POLMOD in this case. The result for the modulated beam is overlaid on that for the control beam. The standard deviations were taken from the original data, and the ratio — modulated beam to control beam calculated. In this case it is 0.3272, indicating a substantial decrease (67%) in spatial jitter due to the modulated beam. Both beams, modulated and unmodulated, were measured to be of equal power before entering the medium.

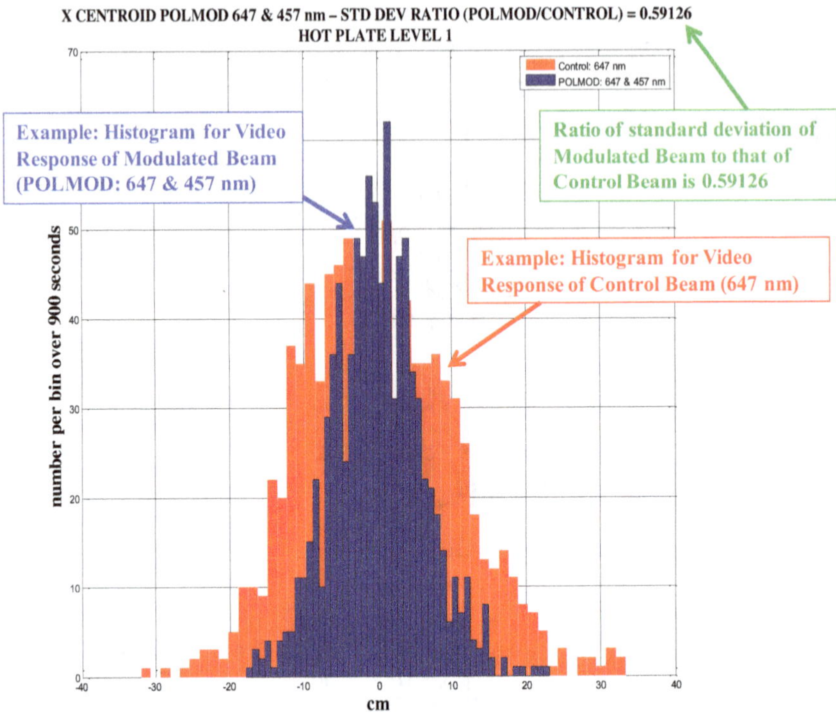

Fig. 5.20. Test of spatial jitter with medium, heated air, and with video capture. The statistical measure is the x centroid and histogram bin counts were calculated for both control (the 647 nm beam — see label) and the modulated (the 647 and 457 nm beam — see label) — POLMOD in this case. The result for the modulated beam is overlaid on that for the control beam. The standard deviations were taken from the original data, and the ratio — modulated beam to control beam-calculated. In this case it is 0.59126, indicating a decrease in spatial (41%) jitter for the modulated beam. Both beams, modulated and unmodulated, were measured to be of equal power before entering the medium.

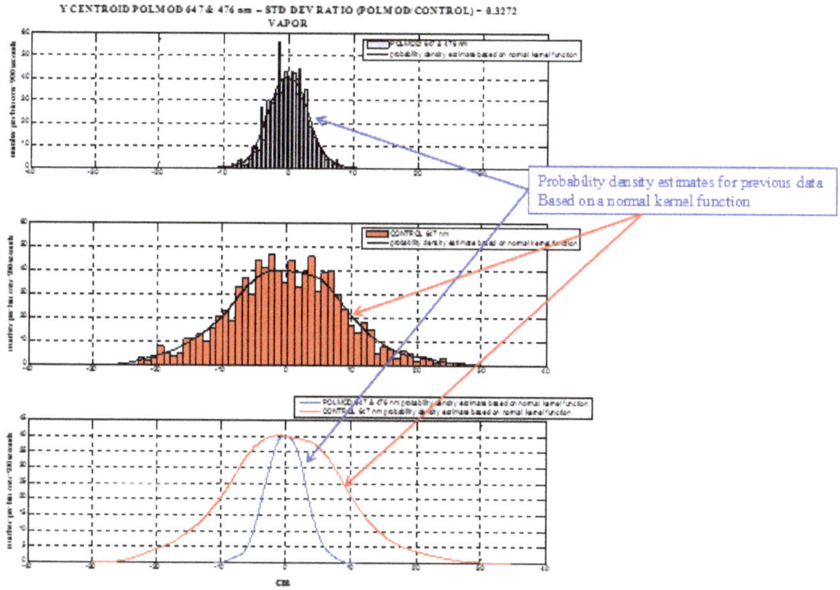

Fig. 5.21. Test of spatial jitter continued. Medium: water vapor. Probability density measures were calculated based on a normal kernel function for the data of Fig 5.19, above. An easily visualized comparison of the control and modulated beam is obtained — lower figure. Abscissa: as in Fig 5.19, the x centroid and histogram bin counts in cm.

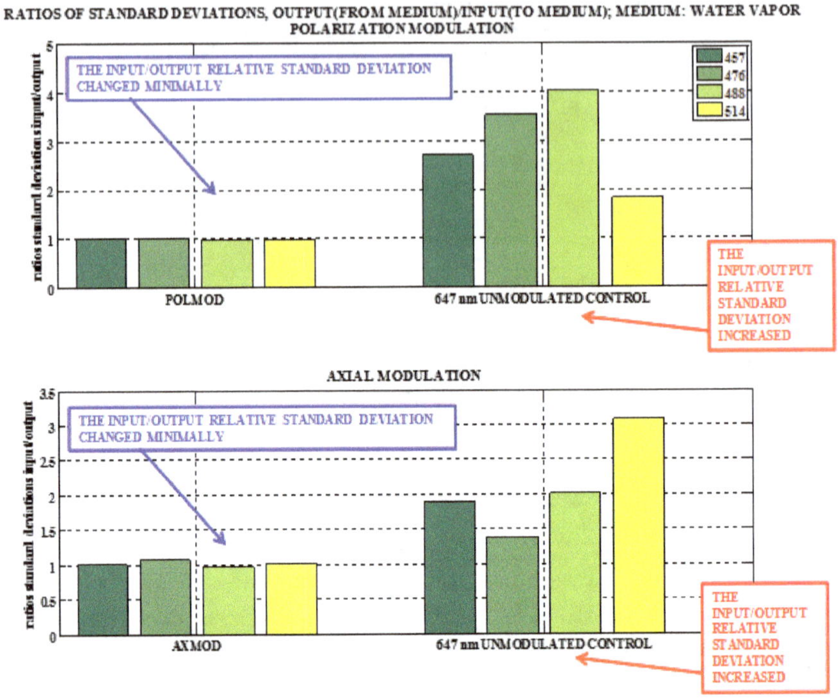

Fig. 5.22. Test of spatial jitter — Standard Deviations. Medium: water vapor. Above: POLMOD. Below: AXMOD. Ordinate: Ratio of standard deviations (output/input), of modulated beams to control beam — a value of 1.0 indicates no change in the standard deviations, input to the medium, vs. output from the medium. The result is clear. The ratio increases in the case of the control (unmodulated) beams shown on the right, but remains remarkably constant near 1.0 in the case of the modulated (POLMOD and AXMOD) beams shown on the left. Both beams, modulated and unmodulated, were measured to be of equal power before entering the medium.

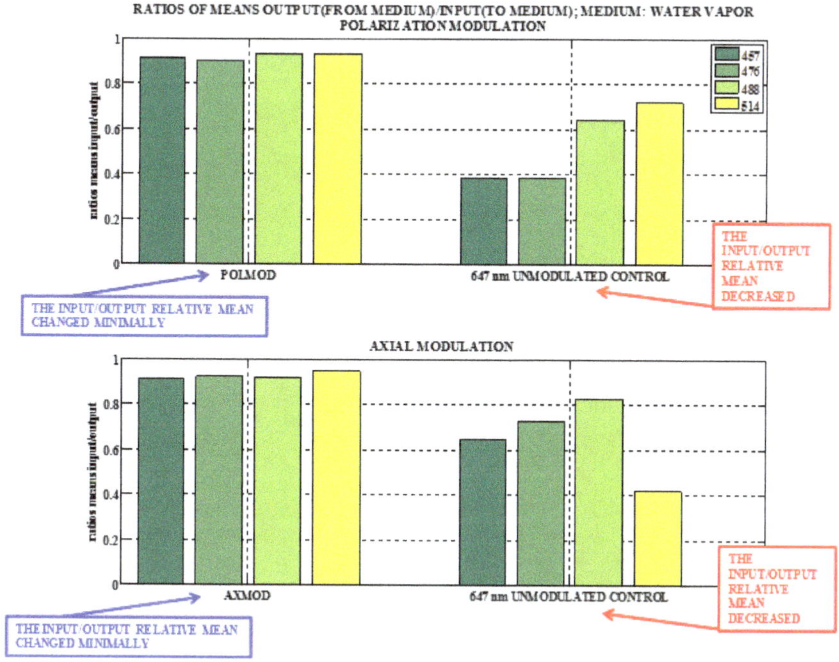

Fig. 5.23. Test of absorption/scattering — Means. Medium: water vapor. Ordinate: Ratio of means (output/input), modulated beams vs. a control beam. A value of 1.0 indicates no change in the means of power input to the medium, vs. power output from the medium. The result is again clear. The ratio decreases in the case of the control (unmodulated) beam, but remains remarkably constant near 0.9 in the case of the modulated (POLMOD and AXMOD) beams. Both beams, modulated and unmodulated, were measured to be of equal power before entering the medium.

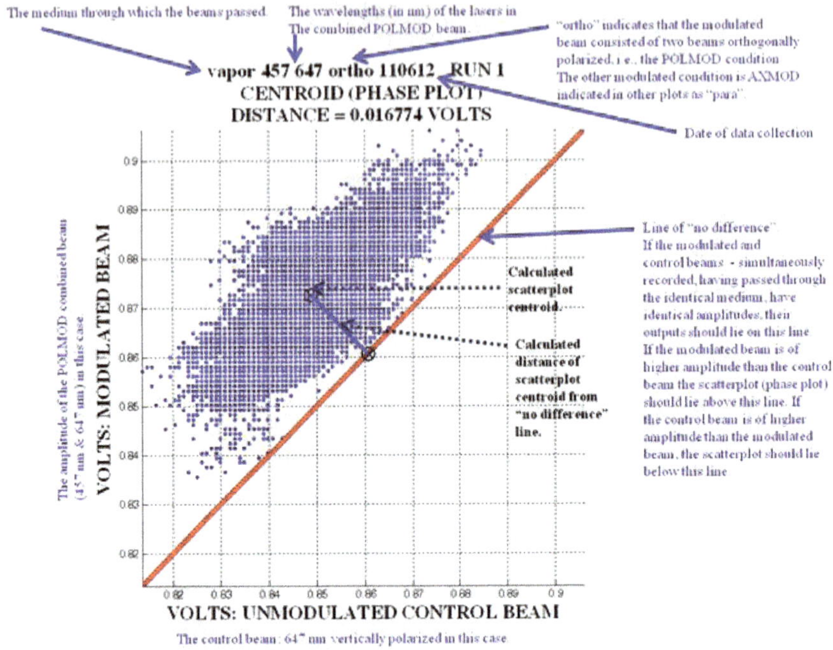

Fig. 5.24. Test of absorption/scattering. Medium: water vapor. Besides recordings from the modulated (here: 457 nm and 647 orthogonally polarized) and the control (here: a single beam at 647 nm, linearly polarized) beams exiting the medium, the "background" — or no input — condition, and the "input" — the beam amplitudes prior to entering the medium — condition, were also recorded. The two beams, modulated and unmodulated or control — were aligned at the same amplitude. The data for the two conditions — modulated and control — were recorded at 30 Hz exiting the medium. The data were further treated by (1) subtracting the mean of the background from the medium-exiting or output amplitude, and (2) dividing by the mean of the medium-entering or input amplitude. This treated data were displayed in a phase, or scatterplot, shown here. The scatterplot thus removes the time variable. The centroid for the data, and the distance from the center line of "no difference" between the modulated and control conditions were then calculated. If the modulated and control beams have identical amplitudes, having passed through water vapor, then the data points would fall along the red line. If the control beam is of higher amplitude than the modulated beam, having passed through water vapor, the data points would fall below the red line. Finally, if the modulated beam is of higher amplitude than the control beam having passed through water vapor — indicating a positive result of POLMOD modulation — the data points would fall, and as shown above do fall, above the red line. There is a clear mitigation of absorption/scattering in the modulated (POLMOD) beam case.

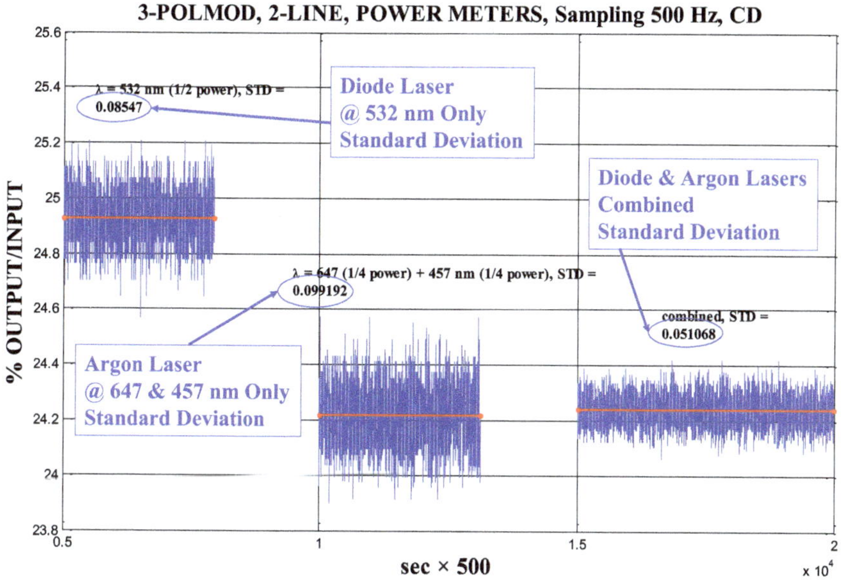

Fig. 5.25. Test of temporal jitter. Medium: Phase Plate. Power Meters: 3 Laser design, standard deviation ratios of entering medium vs. exiting. Equal power levels entering medium for all conditions. Abscissa: 100–400 nanoseconds. (1) Diode $\lambda = 532$nm; (2) Argon laser a $\lambda = 647+457$ combination; (3) Combined Diode orthogonal to Argon laser: a $\lambda = 647+457+$ (orthog) 532 nm combination. The temporal jitter is reduced as shown in (3) compared with (1) and (2). All beams, modulated and unmodulated, were measured to be of equal power before entering the medium.

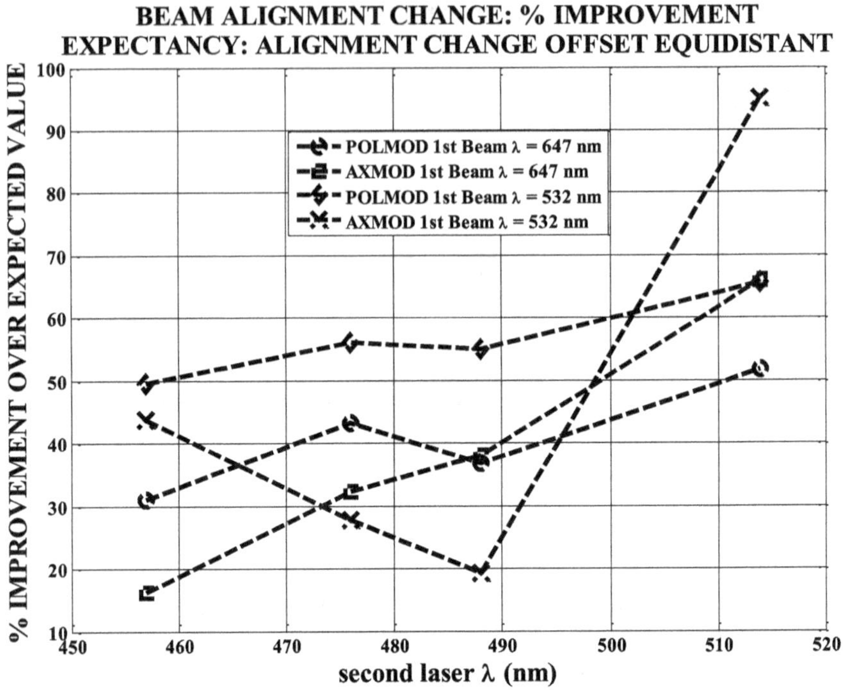

Fig. 5.26. Test of spatial jitter. Medium: Phase Plate. The resulting percentage d deviations from expectancy measures for conditions (I) $\lambda_1 = 647$ nm and (II) $\lambda_1 = 532$ nm, under the POLMOD and AXMOD conditions, and for modulating frequencies, $\lambda_2 = 457$, 476, 488 or 514 nm. The legend shows the wavelength λ_1 of a first laser of a combined beam and whether POLMOD or AXMOD. The abscissa is the wavelength λ_2 of the second laser of a combined beam. A trend can be seen that indicates the highest improvement in refractive index-induced deviation in beam alignment occurs when $(\lambda_1 - \lambda_2)$ is least.

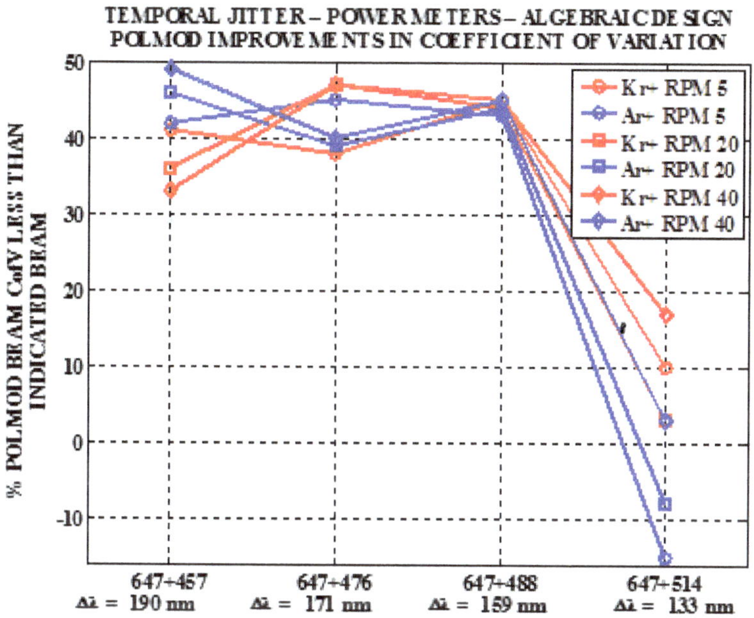

Fig. 5.27. Test of spatial jitter. Medium: Rotating Phase Plate (see Fig. 5.15 above). Temporal Jitter results, 2 Line Algebraic Design. POLMOD improvements in the coefficient of variation, at phase plate rotations 5, 20 and 40 rpm, and with orthogonal beam wavelength separations of $\Delta\lambda = 190$ nm (647 + 45 nm), 171 nm (647 + 476 nm), 159 nm (647 + 488 nm) and 133 nm (647 + 514 nm). Note the decline in coefficient of variation improvement at 133 nm (647 + 514 nm).

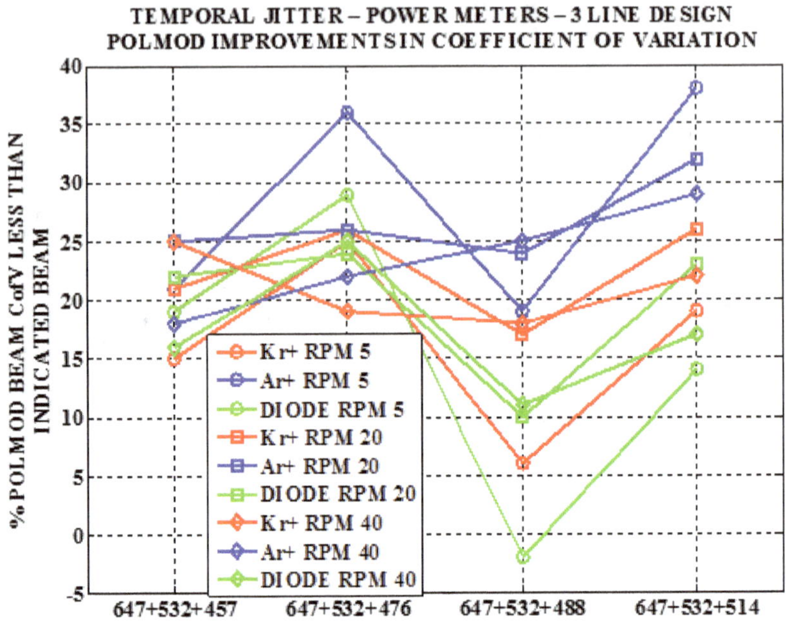

Fig. 5.28. Test of temporal jitter. Medium: Rotating Phase Plate (see Fig. 5.16 above). 3 Beam Design. The abscissa shows the wavelengths λ's of the three lasers in the combined beam. Percentages that POLMOD improves the coefficient of variation, at phase plate rotations 5, 20 and 40 rpm (see legend). Note the percentage improvements (ordinate), but also the decline in the coefficient of variation improvement at all rpm's with combination 647 + 532 + 488 nm. This result indicates that there are optimum wavelength combinations in a combined beam for jitter mitigation in specific media.

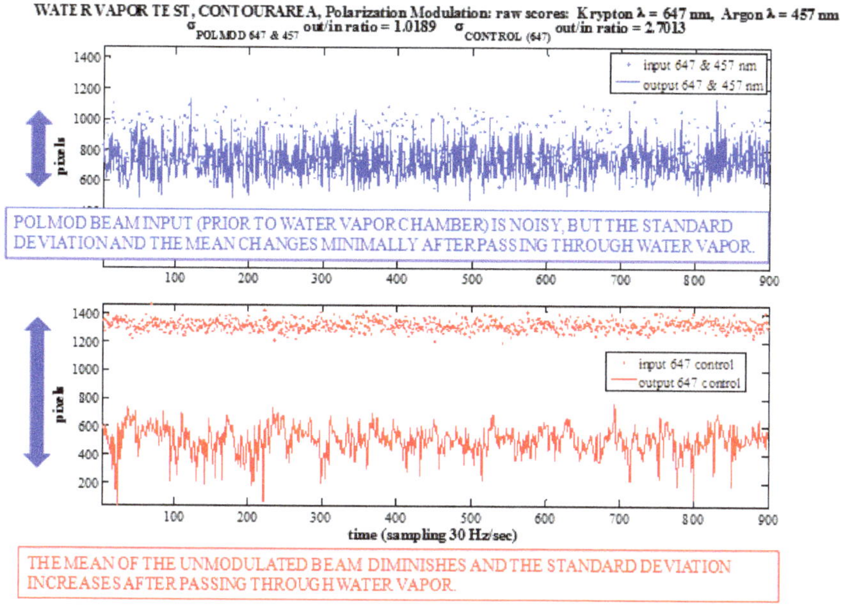

Fig. 5.29. Test of absorption scattering and temporal jitter. Medium: water vapor. As shown here, (i) the relative reduction in the mean of the entering (input) to the chamber vs. exiting (output) beam from the chamber, was greater for the control beam, and (ii) the relative increase in the standard deviation of the entering (input) to the chamber vs. exiting (output) beam from the chamber, was also greater for the control beam. Thus, while the modulated beam was noisier regardless of the medium, the beam was remarkably stable in mean (absorption/scattering mitigation) and standard deviation (temporal jittermitigation) after passing through the medium.

Bibliography

Allen, L., Beijersbergen, M.W., Spreeuw, R.J.C. and Woerdman, J.P., Orbital angular momentum of light and the transformation of Laguerre-Gaussian laser modes. *Phys. Rev. A*, 45, 8185–8189, 1992.

Allen, L., Courtial, J. and Padgett, M.J., Matrix formulation for the propagation of light beams with orbital and spin angular momentum. *Phys. Rev. E*, 60, 7497–7503, 1999.

Allen, L. and Padgett, M.J., Optical tweezers and optical spanners with Laguerre-Gaussian modes. *J. Modern Optics*, 43, 2485–2491, 1996.

Allen, L. and Padgett, M.J., The Poynting vector in Laguerre-Gaussian beams and the interpretation of their angular momentum density. *Optics Commun.*, 184, 67–71, 2000.

Barnett, S.M., Optical angular-momentum flux. *J. Opt. B: Quantum Semiclass.*, 4, 67-S16, 2002.

Barrett, T.W., A difference in polarizability between normal and sickle cell oxyhemoglobin as revealed by resonance Raman spectroscopy. *J. Raman Spectroscopy*, 8, 122–124, 1979a.

Barrett, T.W. and Peticolas, W.L., Laser Raman inelastic light scattering investigations of hyaluronic acid primary and secondary structure. *J. Raman Spectroscopy*, 8, 35–38, 1979b.

Barrett, T.W., pH-induced modification of NAD and NADH solutions detected by Raman spectroscopy. *J. Raman Spectroscopy*, 9, 130–133, 1980.

Barrett, T.W., Resonance Raman complete polarization studies of oxyhemoglobin in the long- and short-wavelength regions: Evidence of anomalous results. *Chem. Phys. Lett.*, 78, 125–128, 1981a.

Barrett, T.W., Laser Raman spectra of mono-, oligo-, and polysaccharides in solution. *Spectrochimica Acta*, 36A, 233–239, 1981b.

Barrett, T.W., Resonance Raman complete polarization studies of ferrocytochrome C in the long and short wavelength regions: Evidence of magnetic resonance. *Chem. Comm. J. Chem. Soc.*, 1, 9–10, 789, 1982.

Barrett, T.W., Wohltjen, H. and Snow, A., A new method of modulating electrical conductivity in phthalocyanines using light polarization. *Nature*, 301, 694–695, 1983a.

Barrett, T.W., Modulation of electrical conductance of phthalocyanine films by circularly polarized light. *Thin Solid Films*, 102, 231–244, 1983b.

Barrett, T.W., Freitas, J.A. and Snow, A., Resonance Raman excitonic profile of polydiacetylene described by linear susceptibility theory. *Chem. Phys. Lett.*, 109, 291–295, 1984.

Barrett, T.W., Wohltjen, H. and Snow, A., A new method of modulating electrical conductivityin phthalocyanines using light polarization. Chapter 24, in *Molecular Electronic Devices II*, F.L. Carter (ed), Marcel Dekker, New York, NY, pp. 475–505, 1987a.

Barrett, T.W., Physical properties of ordered ultrathin organic films. *NSF Workshop "The Molecular Engineering of Ultrathin Polymeric Films"*, University of California at Davis. *Thin Solid Films*, 152, 67–98, 1987b.

Barrett, T.W., Physical properties of ordered ultrathin organic films, Chapter 5, in P. Stroeve & E. Ranses (eds) *Molecular Engineering of Ultrathin Polymeric Films*, Elsevier, New York, pp. 67–98, 1987c.

Barrett, T.W., A Polarization-rotation modulated, spread polarization-rotation, wide-bandwidth communications system, U.S. Patent 5,592,177 dated January 7th, 1997.

Barrett, T.W., Topology and the physical properties of the electromagnetic field. *Apeiron*, 7, 3–11, 2000.

Barrett, T.W., Topological foundations of electromagnetism. *Ann. Fondation Louis de Broglie, special edition: En hommage á George Lochak pour son 70ème anniversaire*, 26, 55–79, 2001.

Barrett, T.W., Topological approaches to electromagnetism, in M. Evans (ed.) *Modern Nonlinear Optics*, 2nd edition, Wiley-Interscience, NY, pp. 699–734, 2001.

Barrett, T.W., *Topological Foundations of Electromagnetism*, World Scientific, 2008.

Barreiro, J.T., Wei, T.-C. and Kwiat, P.G., Beating the channel capacity limit for linear photonic superdense coding. *Nature*, 4, 282–286, 2008.

Barron, L.D., Raman spectroscopy, Chapter 6 in R.J.H. Clark & R.E. Hester (eds.) *Advances in Infrared and Raman Spectroscopy*, Vol. 4, Heyden, London, pp. 271–331, 1978.

Barron, L.D., *Molecular Light Scattering and Optical Activity*, Cambridge University Press, 1982.

Beijersbergen, M.W., Allen, L., van der Veen, H.E.L.O. and Woerdman, J.P., Astigmatic laser mode converters and transfer of orbital angular momentum. *Optics Comm.*, 96, 123–132, 1993.

Berry, M.V., Optical currents. *J. Optics A*, 11, 094001, 2009.

Beth, R.A., Mechanical detection and measurement of the angular momentum of light. *Phys. Rev.*, 50, 115–125, 1950.

Bialynicki-Birula, I. and Bialynicki-Birula, Z., Rotational frequency shift. *Phys. Rev. Lett.*, 78, 2539–2542, 1997.

Bliokh, K.Y., Bekshaev, A.Y. and Nori, F., Dual electromagnetism: Helicity, Spin, Momentum and angular momentum. *New J. Phys.*, 15, 033026, 2013.

Bloch, F. Nuclear induction. *Phys. Rev.*, 70, 7–8, 1946.

Bozinovic, N., Yue, Y., Ren, Y., Tur, M., Kristensen, P., Huang, H., Willner, A.E. and Ramachandran, S., Terabit-scale orbital angular momentum mode division multiplexing in fibers. *Science*, 340, 1545–1548, 2013.

Buckingham, A.D., Permanent and induced molecular moments and long-range intermolecular forces. *Adv. Chem. Phys.*, 12, 107–142, 1967.

Chant, I., "Twisted" radio beams data at 32 Gigbits per second. *IEEE Spectrum*, 2014.

Child, M.S. and Longuet-Higgins, H.C., *Phil. Trans. Roy. Soc.*, A254, 259, 1961.

Courtial, J., Dholakia, K., Robertson, D.A., Allen, L. and Padgett, M.J., Measurement of the rotational frequency shift imparted to a rotating light beam possessing orbital angular momentum. *Phys. Rev. Lett.*, 80, 3217–3219, 1998.

Dholakia, K., Simpson, N.B., Padgett, M.J. and Allen, L., Second-harmonic generation and the orbital momentum of light. *Phys. Rev.*, 54, R3742–R3745, 1996.

Djordjevic, I.B., Deep-space and near-earth optical communications by coded orbital angular momentum (OAM) modulation. *Optics Express*, 19, 14277–14289, 2011.

Djordjevic, I.B. and Arabaci, M., LDPC-coded orbital angular momentum (OAM) modulation for free-space optical communication. *Optics Express*, 18, 24722–24728, 2010.

Djordjevic, I.B. and Djordjevic, G.T., On the communication over strong atmospheric turbulence channels by adaptive modulation and coding. *Optics Express*, 17, 18250–18262, 2009.

Djordjevic, I.B., Vasic, B. and Neifeld, M.A., LDPC coded OFDM over the atmospheric turbulence channel. *Optics Express*, 15, 6336–6350, 2007.

Fernandez-Corbaton, I., Zambrana-Puyalto, X. and Molina-Terriza, Helicity and angular-momentum: a symmetry-based framework for the study of light–matter interactions. *Phys. Rev.*, A 86, 042103, 2012.

Fiutak, J., The multipole expansion in quantum theory. *Can. J. Phys.*, 41, 12–20, 1963.

Franke-Arnold, S., Allen, L. and Padgett, M., Advances in optical angular momentum. *Laser & Photonics Reviews*, 2, 299–313, 2008.

Fried, D.L., Branch point problem in adaptive optics. *J. Opt. Soc. Am. A.*, 15, 2759–2768, 1998.

Garetz, B.A., Angular Doppler effect. *J. Opt. Soc. Am.*, 71, 609–611, 1981.

Gibson, G., Courtial, J., Padgett, M., Vasnetsov, M., Pas'ko, V., Barnett, S. and Franke-Arnold, S., Free-space information transfer using light beams carrying orbital angular momentum. *Opt. Express*, 12, 5448–5456, 2004.

Göppert-Mayer, M., Über Elementarakte mit zwei Quantensprüngen. *Ann. Physik*, 9, 273–294, 1931.

Herzfeldt, K.F. and Göppert-Mayer, M., On the theory of dispersion. *Phys. Rev.*, 49, 332–339, 1936.

Hui, X., Zheng, S., Chen, Y., Hu, Y., Jin, X., Chi, H. and Zhang, X., Multiplexed millimeter wave communication with dual orbital angular momentum (OAM) mode antennas. *Sci. Rep.*, 5, 10148, 2015.

Ibrahim, A.H., Roux, F.S., McLaren, M., Konrad, T. and Forbes, A., Orbital angular momentum entanglement in turbulence. *Phys. Rev. A*, 88, 012312, 2013.

Jackson, J.D., *Classical Electrodynamics*, 3rd edition, Wiley, New York, 1999.

Jacoby, M., Medical imaging turns to oft-neglected part of light spectrum. *Chem. & Eng. News*, 93, 44, 10–14, 2015.

Kimel, I. and Elias, L.R., Relation between Hermite and Laguerre Gaussian Modes. *IEEE J. of Quantum Electronics*, 29, 2562–2567, 1993.

Kolmogorov, A.N. The local structure of turbulence in incompressible viscous fluid for very large Reynolds numbers. *Proc. USSR Acad. Sci.*, 30, 299–303, 1941a (in Russian), translated into English by V. Levin: Kolmogorov, Andrey Nikolaevich (July 8, 1991). "The local structure of turbulence in incompressible viscous fluid for very large Reynolds numbers". *Proc. Royal Soc. A*, 434, 9–13, 1991. Bibcode:1991RSPSA.434....9K. doi:10.1098/rspa.1991.0075.

Kolmogorov, A.N. Dissipation of energy in the locally isotropic turbulence. *Proc. USSR Acad. Sci.* 32, 16–18, 1941b (in Russian), translated into English by Kolmogorov, Andrey Nikolaevich (July 8, 1991). "The local structure of turbulence in incompressible viscous fluid for very large Reynolds numbers". *Proc. Royal Soc. A*, 434, 15–17, 1980. Bibcode:1991RSPSA.434...15K. doi:10.1098/rspa.1991.0076.

Kolmogorov, A.N., A refinement of previous hypotheses concerning the local structure of turbulence in a viscous incompressible fluid at high Reynolds number. *J. Fluid Mech.*, 13, 82–85, 1962.

Krenn, M., Fickler, R., Fink, M., Handsteiner, J., Malik, M., Scheidl, T., Ursin, R. and A. Zeilinger, A., Communication with spatially modulated light through turbulent air across Vienna. *New J. Phys.*, 1–11, 2014.

Krenn, M., Handsteiner, J., Fink, M., Fickler, R., Ursin, R., Malik, M. and Zeilinger, A., Twisted light transmission over 143 km. *Proc. Natl. Acad. Sci.*, doi/10.1073/pnas.1612023113, 2016.

Leach, J., Courtial, J., Skeldon, K., Barnett, S.M., Franke-Arnold, S. and Padgett. M.J., Interferometric methods to measure orbital and spin, or the total angular momentum of a single photon. *Phys. Rev. Lett*, 92, 013601, 2004.

Levine, B.M., Martinsen, E.A., Wirth, A., Jankevics, A., Toledo-Quinones, M., Landers, F. and Bruno, T.L., Horizontal line-of-sight turbulence over near-ground paths and implications for adaptive optics corrections in laser communications. *Applied Optics*, 37, 4553–4560, 1998.

Mair, A., Vaziri, A., Weihs, G. and Zeilinger, A., Entanglement of the orbital angular momentum states of photons. *Nature*, 412, 313–316, 2001.

Malik, M., *et al.*, Influence of atmospheric turbulence on optical communication using orbital angular momentum. *Optics Express*, 20, 13195–13200, 2012.

McClain, W.M., Polarization dependence of three-photon phenomena for randomly oriented molecules. *J. Chem. Phys.*, 57, 2264–2272, 1972.

Molina-Terriza, G., Torres, J.P. and Torner, L., Twisted photons. *Nature*, 3, 305–310, 2007.

Mortensen, S. and Hassing, S. Kramers-Kronig relations and resonance Raman scattering, in *Advances in Infrared and Raman spectroscopy*, Vol. 6, R.T.H. Clark & R.E. Hester (eds), Heyden, London, pp. 1–60, 1980.

Mortensen, S. and Konigstein, J.S. Electronic Raman spectra II': Symmetry of the scattering tensor for electronic Raman transitions. *J. Chem. Phys.*, 48, 3971, 1968.

Nienhuis, G., Doppler effect induced by rotating lenses. *Optics Comm.*, 132, 8–14, 1996.

Obukhov, A.M., Some specific features of atmospheric turbulence. *J. Fluid Mech.*, 13, 77–81, 1962.

Oesch, D.W. and Sanchez, D.J., Creating well-defined orbital angular momentum states with a random turbulent medium. *Optics Express*, 20, 12292–12302, 2012.

Oesch, D.W. and Sanchez, D.J., Tewksbury-Christle, C.M., Aggregate behavior of branch points — persistent pairs. *Optics Express*, 2, 1046–1059, 2012.

Oesch, D.W., Sanchez, D.J., Gallegos, A.L., Holzman, J.M., Brennan, T.J., Smith, J.C., Gibson, W.J., Farrell, T.C. and Kelly, P.R., Creation of photonic orbital angular momentum by distributed volume turbulence. *Optics Express*, 21, 5440–5445, 2013.

Padgett, M. and Allen, L., Light with a twist in its tail. *Contemp. Phys.*, 41, 275–285, 2000.

Padgett, M. and Courtial, J., Poincaré-sphere equivalent for light beams containing orbital angular momentum. *Optics Lett.*, 24, 430–432, 1999.

Padgett, M., Courtial, J. and Allen, L., Light's orbital angular momentum. *Phys. Today*, May, 35–40, 2004.

Paterson, C. Atmospheric turbulence and orbital angular momentum of single photons for optical communication. *Phys. Rev. Lett.*, 94, 153901, 2005.

Plazcek, G., in *Rayleigh and Raman Scattering*, UCRL Trans. No. 526L, *Handbuch der Radiologie* (E. Marx, (ed.)) Leipzig, Akademische Verlagsgesellschaft VI, 2, 209, 1934.

Pors, B-J., Aiello, A., Oemrawsingh, S.S.R., van Exeter, M.P., Eliel, E.R. and Woerdman, J.P., Angular phase-plate analysis for measuring the dimensionality of multimode fields. *Phys. Rev. A*, 77, 033845, 2008a.

Pors, B-J., Oemrawsingh, S.S.R., Aiello, A., van Exeter, M.P., Eliel, E.R., 't Hooft, G.W. and Woerdman, J.P., Shannon dimensionality of quantum channels and its application to quantum entanglement. *Phys. Rev. Lett.*, 101, 120502, 2008b.

Pors, B-J., Monken, C.H., Eliel, E.R. and Woerdman, J.P., Transport of orbital–angular-momentum entanglement through a turbulent atmosphere. *Optics Express*, 19, 6671–6683, 2011a.

Pors, B-J., *Entangling Light in High Dimensions*. Ph.D. Thesis, Casimir Research School, University of Delft-Leiden, The Netherlands, 2011b.

Primmerman, C.A., Price, T.R., Humphreys, R.A., Zollars, B.G., Barclay, H.T. and Herrmann, J. Atmospheric-compensation experiments in strong-scintillation conditions. *Appl. Optics*, 34, 2081–2088, 1995.

Ren, Y., *et al.*, Atmospheric turbulence effects on the performance of a free space optical link employing orbital angular momentum multiplexing. *Optics Letters*, 38, 4062–4065, 2013.

Ren, Y., Li,L., Wang, Z., Kamali, S.M., Arbabi, E., Arbabi, A., Zhao, Z., Xie, G., Cao, Y., Ahmed, N., Yan, Y., Liu, C., Willner, A., Ashrafi, S., Tur, M., Faraon, A. and Willner, A.E., Orbital angular momentum-based space division multiplexing for high-capacity underwater optical communications. *Sci. Rep.*, 6(33306), 1–10, 2016.

Rodenburg, B., *et al.*, Simulating thick atmospheric turbulence in the lab with application to orbital angular momentum communication. *New J. Phys.*, 16, 033020, 2014.

Romero, J., Giovannini, D., Franke-Arnold, S. and Padgett, M.J., An experimentalist's introduction to orbital angular momentum for quantum optics. Chapter 12, in Andrews, D.L. & Babiker, M., *The Angular Momentum of Light*, Cambridge University Press, 2013.

Sanchez, D.J. and Oesch, D.W., Localization of angular momentum in optical waves propagating through turbulence. *Optics Express*, 19, 25388–25396, 2011a.

Sanchez, D.J. and Oesch, D.W., Orbital angular momentum in optical waves propagating through distributed turbulence. *Optics Express*, 19, 24596–24608, 2011b.

Shi, C., Dubois, M., Wang, Y. and Zhang, X., High-speed acoustic communication by multiplexing orbital angular momentum. *Proc. Natl. Acad. Sci.*, doi/10.1073/pnas.1704450114, 1–4, 2017.

Simon, R., Kimble, H.J. and Sudarshan, E.C.G., Evolving geometric phase and its dynamical manifestation as a frequency shift: An optical experiment. *Phys. Rev. Lett.*, 61, 19–22, 1988.

Tamburini, F., Mari, E., Sponselli, A., Thidé, B., Bianchini, A. and Romanato, F., Encoding many channels on the same frequency through radio vorticity: First experimental test. *New J. Physics*, 14, 033001, 2012.

Thidé, B., Then, H., Sjöholm, Palmer, K., Bergman, J., Carozzi, T.D., Istomin, Y.N., Ibragimov, N.H. and Khamitova, R., Utilization of photon orbital angular momentum in the low-frequency radio domain. *Phys. Rev. Lett.*, 99, 087701, 2007.

Thidé, B., Tamburini, F., Then, H., Someda, C.G. and Ravanelli, R.A., The physics of angular momentum radio, arXiv:1410.4268v3, 29 May, 2015.

Van Enk, S.J. and Nienhuis, G., Commutation rules and eigenvalues of spin and orbital angular momentum of radiation fields. *J. Modern Optics*, 41, 963–977, 1994.

Wang, J., Yang, J-Y., Fazal, I.M., Ahmed, N., Yan, Y., Huang, H., Ren, Y., Yue, Y., Dolinar, S., Tur, M. and Willner, A.E., Terabit free-space data transmission employing orbital angular momentum multiplexing. *Nature Photonics*, 6, 488–496, 2012.

Willner, A.E., Huang, H., Yan, Y., Ren, Y., Ahmed, N., Xie, G., Bao, C., Li, L., Cao, Y., Zhao, Z., Wang, J., Lavery, M.P.J., Tur, M., Ramachandran, S., Molisch, A.F., Ashrafi, N. and Ashrafi, S. Optical communications using orbital angular momentum beams. *Adv. Optics Photonics*, 7, 66–106, 2015.

Willner, A.E., Twisted light could dramatically boost data rates. *IEEE Spectrum*, August, 2016.

Willner, A.E., Orbital angular momentum beams generated by passive dielectric phase masks and their performance in a communication link. *Optics Lett.*, 42, 2746–2749, 2017.

Wong, K.C., *Liquid Crystal Overview*, ECE-E443, 2005.

Xie, G., Liu, C., Li, L., Ren, Y., Zhao, Z., Yan, Y., Ahmed, N., Wang, Z., Willner, A.J., Bao, C., Cao, Y., Liao, P., Ziyadi, M., Almaiman, A., Ashrafi, S., Tur, M. and Willner, A.E., Spatial light structuring using a combination of multiple orthogonal orbital angular momentum beams with complex coefficients. *Optics Lett.*, 42, 991–994, 2017a.

Xie, G., Li, L., Ren, Y., Yan, Y., Ahmed, N., Zhao, Z., Bao, C., Wang, Z., Liu, C., Song, H., Zhang, R., Pang, K., Ashrafi, S., Tur, M. and Willner, A.E, Localization from the unique intensity gradient of an orbital-angular-momentum beam. *Optics Lett.*, 42, 395–398, 2017b.

Chapter 6

Geometric (Clifford) Algebra: Transmission Through Disturbed Media and Transient Wave States[1]

6.1. Introduction

In the previous chapter, evidence was provided that beams modulated in polarization (POLMOD) and azimuthal phase (AXMOD) can mitigate the effects of disturbing media. Time varying atmospheres, ionospheres and anisotropic media can result in temporal and spatial jitter, absorption and scattering of a beam. Simply stated, under such circumstances a light or radar beam does not follow a straight path to where it is pointed. Furthermore, a radar return signal may not arrive back at a receiving antenna. These conditions may limit effectiveness in communications and remote sensing. This chapter describes the mathematics and physics of these modulated beams in disturbing media.

Specifically, combined orthogonally polarized beams of different frequency, as in the POLMOD case; and combined beams of the same polarization but of different frequency as in the AXMOD case, require a new approach to analysis.

The previous chapter described experimental results in which beams to a great extent penetrated media normally considered

[1]This work was supported by Air Force Office of Scientific Research Grant FA9550110274.

impenetrable to conventional unmodulated beams. These results are examples of an *induced transparency.* This chapter will relate these results to two different well-known examples of induced transparency: (i) self-induced transparency (SIT), and (ii) electromagnetically-induced transparency (EIT).

As in the previous chapter, the molecular receiving medium can be modeled as a polarization and frequency selective receiver which filters the radiating beam with respect to both polarization, frequency *and time — distinguishing the steady-state* (*time*) *response from the transient.* Furthermore, the dual orthogonal beam recombination method creates amplitude modulated beat frequency wave packets. From the selective receiver perspective, i.e., as a polarization/frequency/temporal filter, the combined steady-state beam is sampled as broadband pulses. As these wave packets or pulses have transient characteristics that challenge the temporal characteristics of the molecular system receivers, describing such selective receiver molecular interactions requires mathematical and physical formulations which address the *transient nature* of the transmitted wave and media dynamics, rather than the *steady-state* nature used in most analyses. This requires some departures from "conventional" steady-state thinking and a knowledge of geometric (Clifford) algebra.

6.2. Geometric (Clifford[2]) Algebra

Polarization modulation finds an appropriate description in geometric (Clifford) algebra, rather than the conventionally used algebra of Gibbs[3] and Heaviside.[4] Here, we provide a short overview of this algebra.[5]

---

[2]William Kingdon Clifford (1845–1879).

[3]Josiah Willard Gibbs (1839–1903).

[4]Oliver Heaviside (1850–1925).

[5]Hestenes and Sobczyk (1984); Chisholm and Common (1986); Jancewicz (1988); Baylis (1999); Doran and Lasenby (2003); Arthur (2011); Chappell *et al.* (2014); Joot (2021).

We commence by considering examples of two beams, each generated by combining two beams:

$$\begin{aligned}
&\text{(A)} \quad \text{beam } 1 = a_1 \cos\left(2\pi \times 29 \times 10^{12} \times t\right);\\
&\qquad\ \text{beam } 2 = a_2 \cos(2\pi \times 29 \times 10^{12} \times t + \pi/2);\ a_1 = a_2 = 1;\\
&\text{(B)} \quad \text{beam } 1 = a_1 \cos\left(2\pi \times 32 \times 10^{12} \times t\right);\\
&\qquad\ \text{beam } 2 = a_2 \cos\left(2\pi \times 29 \times 10^{12} \times t\right);\\
&\qquad\ a_1 = a_2 = 1.
\end{aligned} \tag{6.1}$$

(*Note*: the amplitudes a_1 and a_2 of two constituent beams forming a combined beam can also be modulated, but for purposes of simplifying this exposition, they are not here.) The combined beam (A) of Eq. (6.1) is formed by the combining of two beams of *the same* frequency with an interbeam phase difference of $\pi/2$. Such a combined beam will be of static, and unchanging, circular polarization. In contrast, the combined beam (B) is formed by the combining of two beams of *different frequency*. If the two beams both have the same polarization, the result will be an axially modulated combined beam (AXMOD). If the two beams are orthogonally polarized, the result will be a polarization modulated combined beam (POLMOD).

Now, the basis of geometric algebra is the *geometric product*, and the inner and outer products are derived constructions. For example,

$$\text{the geometric product of two vectors is: } xy = x \wedge y + x \cdot y$$

where "$\wedge$" signifies the outer product and the antisymmetric part of the geometric product, and "$\cdot$" signifies the inner product and symmetric part, e.g.,

$$\text{the outer product of two vectors is : } x \wedge y = \frac{1}{2\,(xy - yx)};$$

and "$\cdot$" signifies the inner (dot) product and the symmetric part of the geometric product, e.g.:

$$\text{the inner product of two vectors is: } x \cdot y = 1/2\,(xy + yx);$$

It is noteworthy that whereas the inner and outer products are not invertible, the geometric product is invertible.

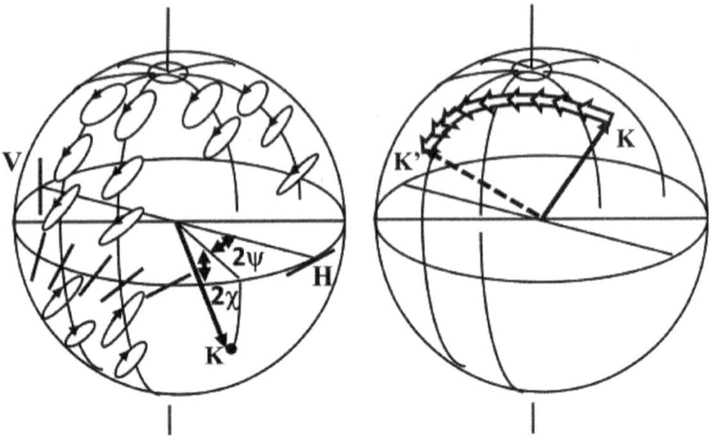

Fig. 6.1. The Poincaré sphere, representing: (a) All polarizations and rotations shown statically. Circularly polarized corotating is represented at the north pole; circularly polarized counter-rotating is represented at the south pole. Along the longitudinal lines are represented the varieties of elliptical polarizations. Around the equator latitude line are shown horizontally polarized linear (H), and vertically polarized linear configurations (V), as well as the variety of linear polarizations in between these. The angles 2ψ and 2χ are shown for a combined beam. Another angle, δ, is relevant and is the interbeam phase of the combined beam's constituent beams and not represented here. The K vector points to a location on the sphere, here shown as a dot, that indicates one particular polarization–rotation. These are *static representations.* (b) In contrast, POLMOD is a movement of the K vector on the sphere and *is not static.*

For ease of reference, we again show the Poincaré sphere and the angle χ which is changing over time in the case of POLMOD, and ψ which is changing over time in the case of AXMOD (Fig. 6.1).

We will address those waveforms that are not static in phase but modulated. In other words, we seek an algebra that describes not just positions on the Poincaré sphere (static polarizations), but movements on the sphere (polarization modulation). Geometric algebra provides this description, because the geometric product is an operator for both a rotation and a dilation (an increase) of one vector to another, with the rotation defined as

$$R = y/x = \frac{yx}{x \cdot y}. \tag{6.2}$$

For ease of exposition, this equation is defined only over the x- and y- axes — but can be developed to include the z- or t- axis. Thus, given (i) a combined wave of *instantaneous* (i.e., unmodulated) *polarization* defined as at some position x at some instant t, and (ii) a combined wave *instantaneous* (i.e., unmodulated) *polarization* defined as some position y at some instant $t+1$, we make the substitution in this instant, i.e., $x = y$, repeating the process for $t+2$, etc. Thus R defines the polarization modulation *trajectory* of the combined beam.

We further define the geometric product as the *graded sum* and introduce the concepts of a *bivector* and a *trivector*. As stated above, the geometric product is

$$xy = x \wedge y + x \cdot y, \tag{6.3}$$

where

(A) $x \wedge y$ is the *outer exterior product*, which is *antisymmetric*, *associative*, a *bivector* and a *directed plane (area) segment*[6];
(B) $x \cdot y$ is the *dot or inner product* and is *symmetric*;
(C) the outer product of a vector and a *bivector* produces a *trivector*, **V**, which is *an oriented volume element* (see Figs. 6.1 and 6.2).

We next turn to describe polarization fields and magnetization fields. For the *steady state*, the polarization field, **P**, and the electric field, **E**, are related by the electric susceptibility tensor, χ, and the vacuum permittivity, ε_0:

$$\begin{pmatrix} P_x \\ P_y \\ P_z \end{pmatrix} = \varepsilon_0 \begin{pmatrix} \chi_{xx} & \chi_{xy} & \chi_{xz} \\ \chi_{yx} & \chi_{yy} & \chi_{yz} \\ \chi_{zx} & \chi_{zy} & \chi_{zz} \end{pmatrix} \begin{pmatrix} E_x \\ E_y \\ E_z \end{pmatrix}. \tag{6.4}$$

However, Eq. (6.4) describes a steady-state condition, and as the tests we address involve plane waves or beams, we can dispense with the continuous z spatial plane, substituting a sampling time, t,

[6]In contrast, the Gibbs cross product is $\mathbf{E} \times \mathbf{D}$, which is a vector, not associative, and is a directed line, not plane, segment.

and for the molecular receiver interacting with a beam, a *molecular orientation time*, τ. Revising Eq. (6.4), we have

$$\begin{pmatrix} P_x \\ P_y \\ P_t \end{pmatrix} = \varepsilon_0 \begin{pmatrix} \chi_{xx} & \chi_{xy} & \chi_{xT} \\ \chi_{yx} & \chi_{yy} & \chi_{yT} \\ \chi_{Tx} & \chi_{Ty} & \chi_{TT} \end{pmatrix} \begin{pmatrix} E_x \\ E_y \\ E_t \end{pmatrix}. \tag{6.5}$$

A similar relation exists between the magnetization, or magnetic dipole moment, **M**, the *volume* magnetic susceptibility, χ^m, and μ_0 the vacuum permeability and the magnetic induction field, **B**:

$$\begin{pmatrix} M_x \\ M_y \\ M_t \end{pmatrix} = \mu_0^{-1} \begin{pmatrix} \chi_{xx}^m & \chi_{xy}^m & \chi_{xT}^m \\ \chi_{yx}^m & \chi_{yy}^m & \chi_{yT}^m \\ \chi_{Tx}^m & \chi_{Tx}^m & \chi_{TT}^m \end{pmatrix} \begin{pmatrix} B_x \\ B_y \\ B_t \end{pmatrix}. \tag{6.6}$$

We pause, here, to make two observations, the first commonly neglected:

(1) Whereas the **E** field is a conventional polar vector, the **B** field is an axial vector, or *bivector*, or a *directed area*. The algebra of conventional electromagnetic theory is inappropriate for describing this difference.
(2) In Figs. 6.2–6.10 the *polar* **E** field and the accompanying *axial* (bivector) **B** field *which arises when the* **E** *field departs from linear polarization* are shown. A rapid rate of change in the **E** and **B** fields challenges the orientation and relaxation times of the electric susceptibility tensor, χ, and the *volume* magnetic susceptibility, χ^m.

In Figs. 6.2–6.10, a trivector directed volume, representation of the propagating combined beam is shown. Table 6.1 indicates the significance of each figure. Each combined beam is composed of two constituent beams that are either:

(A) *orthogonally* polarized, of the *same* frequency but with an interbeam phase of $\pi/2$, resulting in static and unchanging circular polarization, or
(B) *orthogonally* polarized, of *different* frequencies, resulting in polarization modulation (POLMOD).

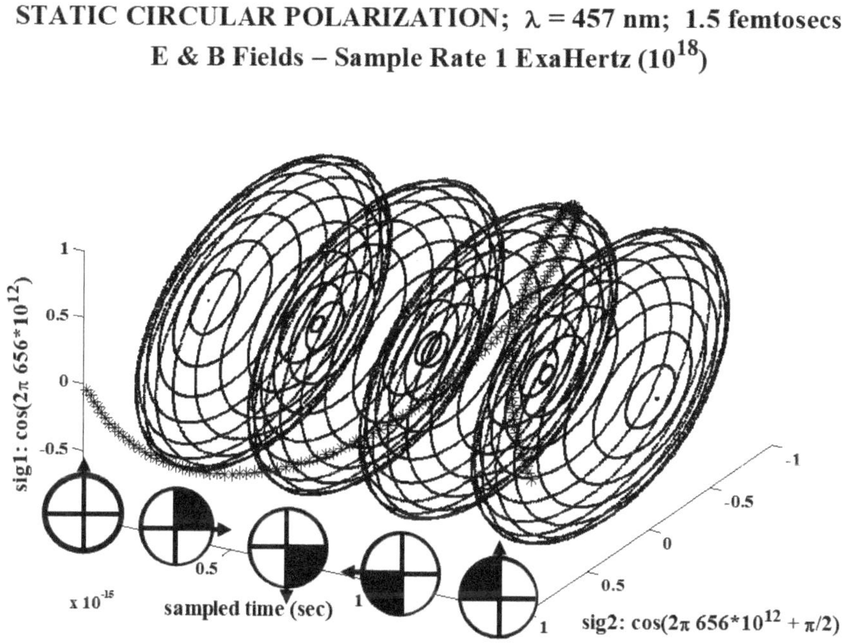

Fig. 6.2. Trivector representation of one cycle of static circular polarization. It represents a trivector formed by the outer product (or wedge product) of three independent vectors: e.g., e1 $\wedge$ e2 $\wedge$ e3, and is a directed volume element or oriented volume. In the present instance, a first linearly polarized **E** beam (signal 1) is represented along e1; a second linearly polarized **E** beam (signal 2), but orthogonally polarized to the first, is represented along e2; and time, or the number of sampling intervals is represented along e3. Both constituent beams are of the same frequency but the interbeam phase is $\pi/2$ resulting in circular polarization — but static. The termination of the **E** beam is shown with a hashed line. The **B** field is shown as a series of bivector areas bounded by solid lines. As the constituent beams that form the combined beam are of the same frequency but orthogonally polarized with $\pi/2$ interbeam phase, the **B** bivector areas form four groupings for every Lissajous pattern cycle. A signed scalar denotes the volume of a trivector. This figure represents one cycle of a circularly polarized **E** field corresponding to four expansions and contractions of the **B** field and to one Lissajous pattern cycle of the **E** field.

There is a third condition, not represented in these figures:

(C) both constituent beams are of the *same* polarization, of *different* frequencies, resulting in azimuthal or axial modulation (AXMOD).

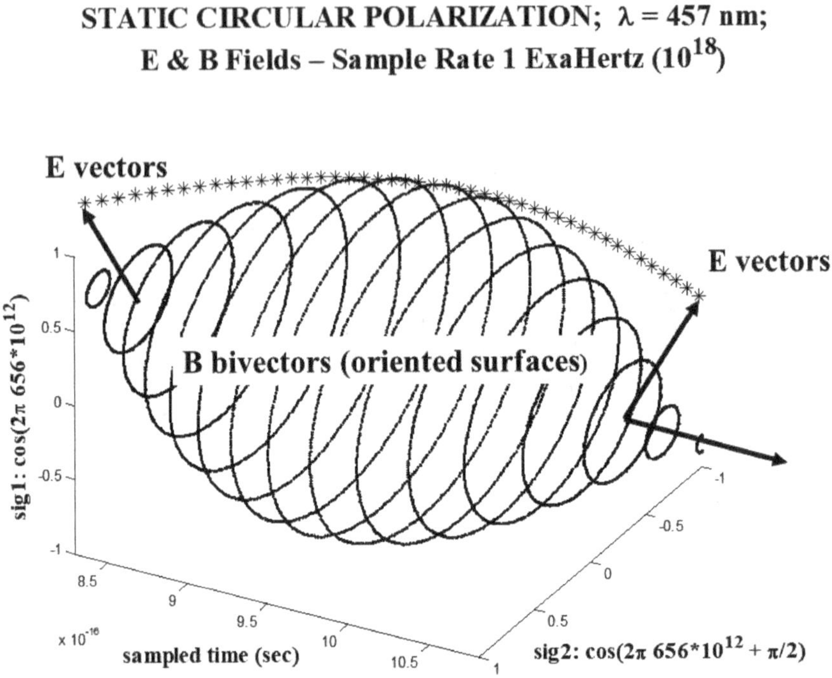

Fig. 6.3. Trivector representation. Focusing on part of the previous figure, this shows a partial representation of a trivector showing one expansion and contraction of the **B** field (oriented surface areas) and corresponding to 1/4 of a Lissajous pattern cycle of the **E** field vectors.

We have thus shown that the trivector or directed volume form of representation adequately describes the polar vector, **E** field, and the axial vector **B** field, under conditions of polarization modulation.

6.3. Knotted Beams

In previous sections, the generation of wave packets by beam combining for media penetration and mitigation of jitter and absorption/scattering was addressed. To achieve the same in complex and composite media requires more complex beam combinations. One such complex combination is knotted vortex beams, which is a

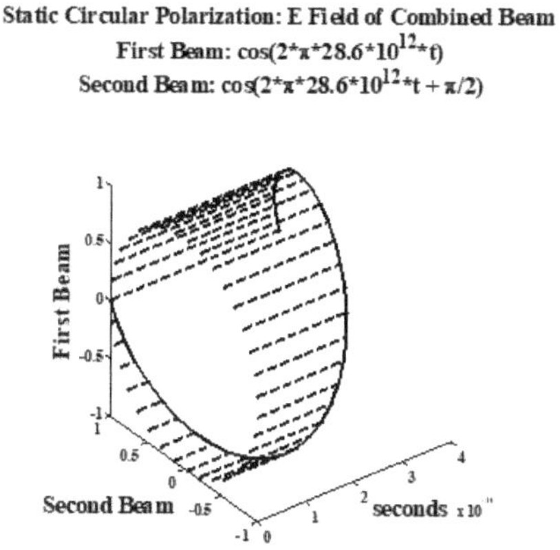

(A1)

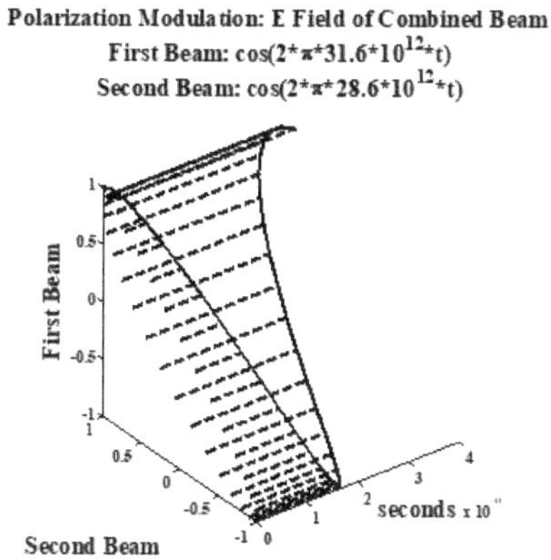

(B1)

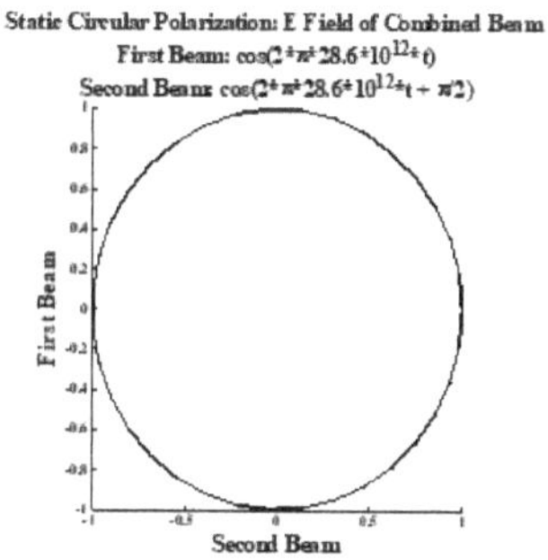

(A2)

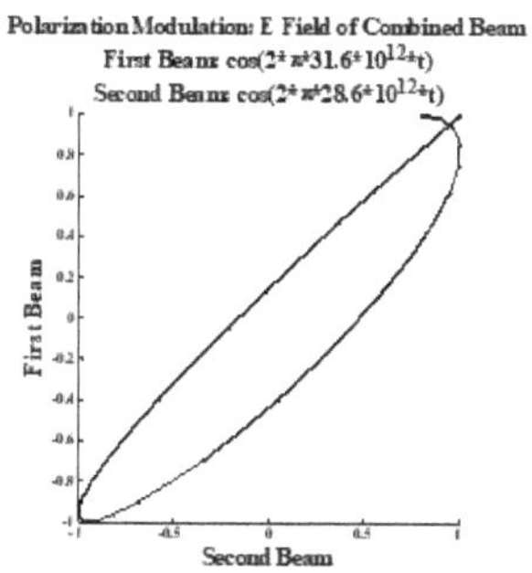

(B2)

Fig. 6.4. Combined beam E field trajectory trivector representation. This figure shows the x, y, t trivector (A1) and (B1); and also viewed head-on (A2) and (B2). In both (A) and (B), 36 sampling points (= 36 femtoseconds) of the resultant summed trajectory of two orthogonally polarized beams (solid line) with the amplitude of the first beam represented on the x-axis, and the amplitude of the second beam on the y-axis are shown. The dashed lines indicate the build over time of so-called polarization ellipses (or x–y plots, or Lissajous figures). In (A1) and (A2), the orthogonal beams are of the same frequency but offset by 90° — therefore, there is *static (unchanging) circular polarization*. In B1 and B2, the orthogonal beams are of different frequencies resulting in *polarization modulation*.

In (A) beam 1 $= a_1 \cos\left(2\pi \times 29 \times 10^{12} \times t\right)$; beam 2 $= a_2 \cos(2\pi \times 29 \times 10^{12} \times t + \pi/2)$; $a_1 = a_2 = 1$.

In (B) beam 1 $= a_1 \cos\left(2\pi \times 32 \times 10^{12} \times t\right)$; beam 2 $= a_2 \cos(2\pi \times 29 \times 10^{12} \times t)$; $a_1 = a_2 = 1$.

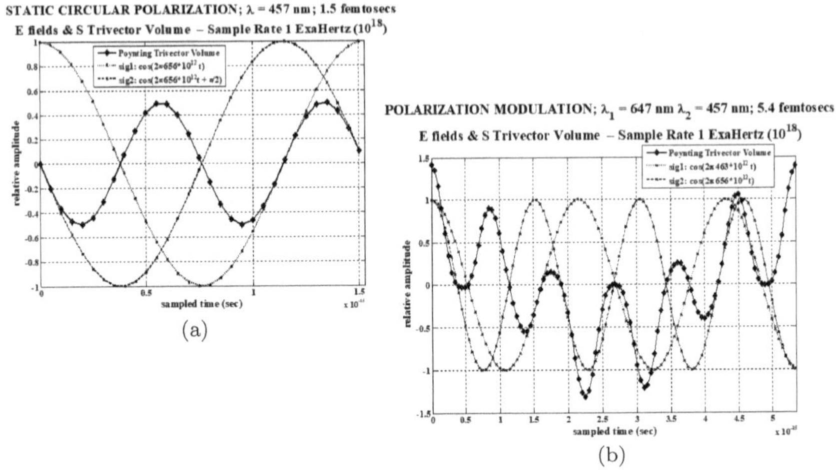

Fig. 6.5. Trivector representation of the two **E** fields and the **S** trivector volume. (a) static (unchanging) circular polarization; and (b) polarization modulation.

comparatively new field:

> "In a light beam, the flow of light through space is similar to water flowing in a river. Although it often flows in a straight line — out of a torch, laser pointer, etc. — light can also flow in whirls and eddies, forming lines in space called 'optical vortices'." ... "Along these lines, or optical vortices, the intensity of the light is zero (black). The light all around us is filled with these dark lines, even though we can't see them "...." The study of knotted vortices was initiated by Lord Kelvin back in 1867 in his quest for an explanation of atoms. This work opens a new chapter in that history." Dennis, M., Tying light in knots, *Phys. Org*, Jan. 17, 2010.

Acting on a theoretical prediction of Helmholz,[7] in 1867, Tait[8] showed that interacting rings of an ideal fluid with no friction would retain their shape even with a change in size and velocity His demonstrations inspired Lord Kelvin[9] to attempt to explain atoms in terms of knot theory (Thomson, 1867). In the case of

[7]Hermann von Helmholtz (1821–1894).
[8]Peter Guthrie Tait (1831–1901).
[9]William Thomson, Lord Kelvin (1824–1907).

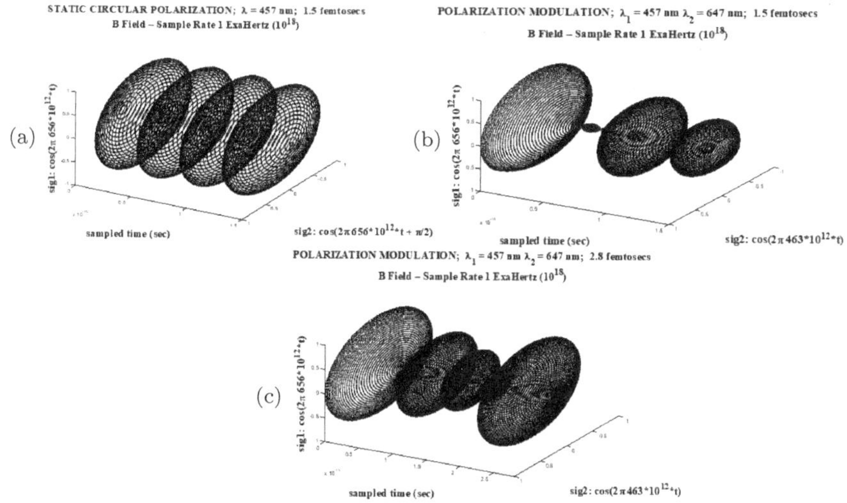

Fig. 6.6. Trivector representation of the **B** field with the x, y, t trivectors sampled at 1 ExaHertz (10^{18} Hz). The magnetic field, **B**, is a bivector and represented as a surface, $\mathbf{B} = e1 \wedge e2$, where, in the present instance, the first beam is represented along the e1-axis, and the second beam along the e2-axis and orthogonal to first. The **B** field surface geometry is arbitrary and is shown here with a circular boundary. In (a) (static circular polarization) the representation is over 1.5 femtosecs; in (b) and (c) (polarization modulation) the representation is over 1.5 femtosecs and 2.8 femtosecs, respectively.

In (a) beam 1 $= a_1 \cos(2\pi \times 29 \times 10^{12} \times t)$; beam 2 $= a_2 \cos(2\pi \times 29 \times 10^{12} \times t + \pi/2)$; $a_1 = a_2 = 1$.

In (b) &(c) beam 1 $= a_1 \cos\,(2\pi \times 32 \times 10^{12} \times t)$; beam 2 $= a_2 \cos\,(2\pi \times 29 \times 10^{12} \times t)$; $a_1 = a_2 = 1$.

light and electromagnetic radiation, the preservation of shape, and as was later shown, the topological structure of field lines (Berry and Dennis, 2001), is due to Maxwell's equations permitting solutions of beams which have linked and knotted field lines. The conservation of topological structure, or the degree of linking and knotting between field lines, has been shown to be due to the conservation of helicity (Rañada, 1989) and that electrical and magnetic fields always are orthogonal (Irvine, 2010). Due to helicity conservation, knotted

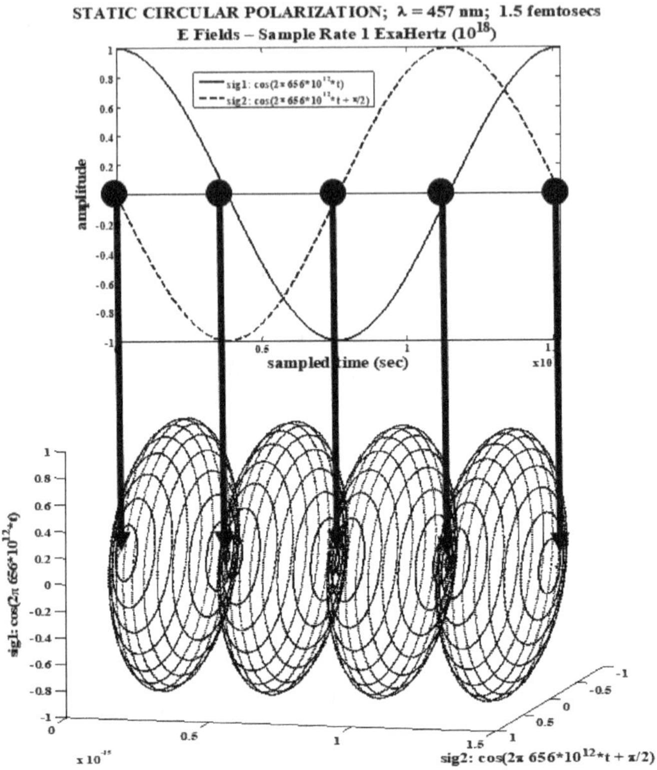

Fig. 6.7. Static circular polarization trivector representation. The progression through the four quarter sections of a complete Lissajous circular polarization cycle is made explicit from an examination of the two orthogonally polarized constituent beams of the combined beam. Both constituent beams are of the same frequency, but orthogonally polarized with $\pi/2$ interbeam phase, as in Fig. 6.2 above.

beams are the most general form of polarization modulated wave packets. This generality is the reason for our interest.

Knotted beam descriptions are based on tensor algebra.[10] However, tensor algebra is coordinate specific and a coordinate free algebra is required with such provided by the Clifford/Geometric algebra. We

[10]Rañada (1989,1992a,b); Rañada and Trueba, (1995–1998, 2001); Irvine (2010); Irvine and Bouwmeister (2008); Dennis *et al.* (2010); Kleckner and Irvine (2013); Kedia *et al.* (2013).

view(24,16)

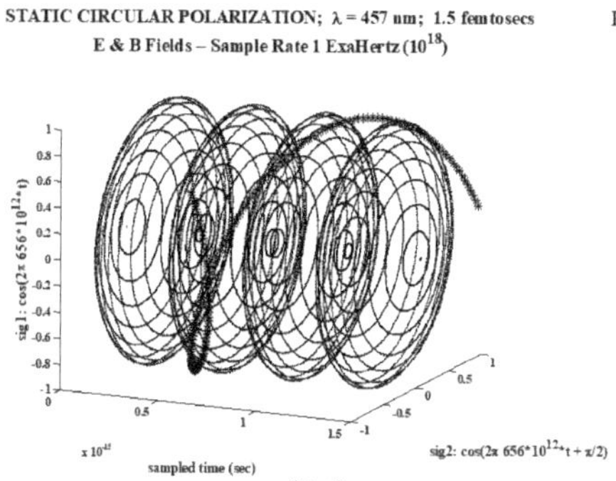

(A1)

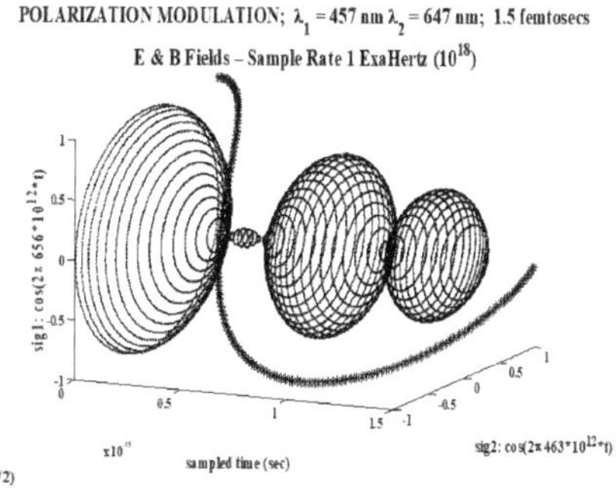

(A2)

view(-29,-36)

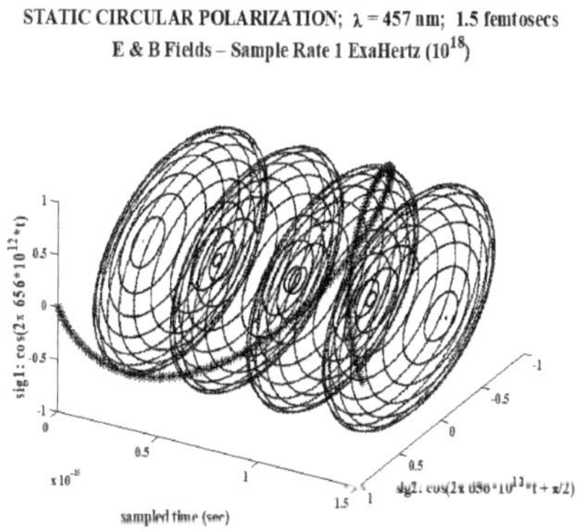

(B1)

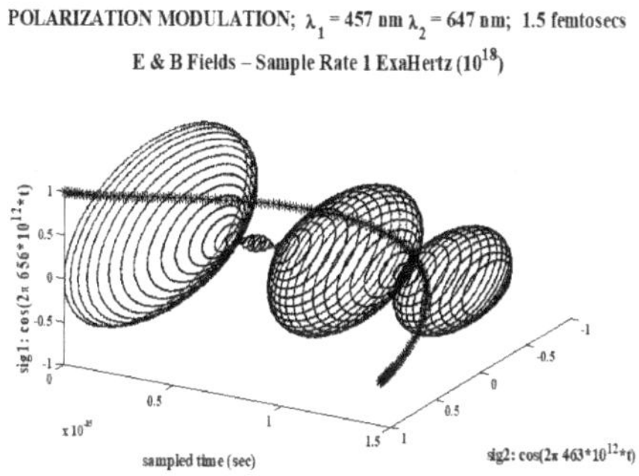

(B2)

view(90,0)

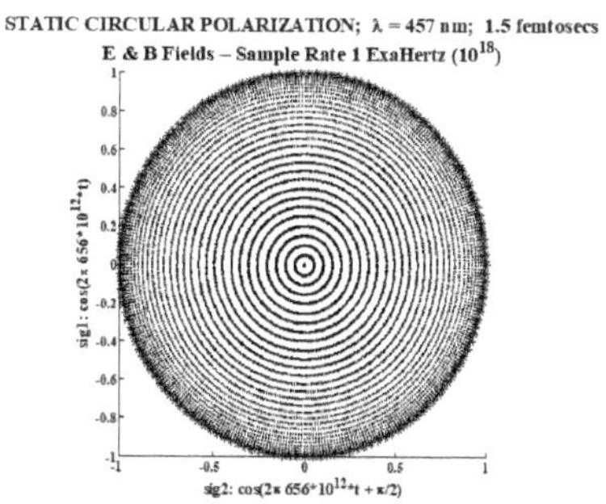

(C1)

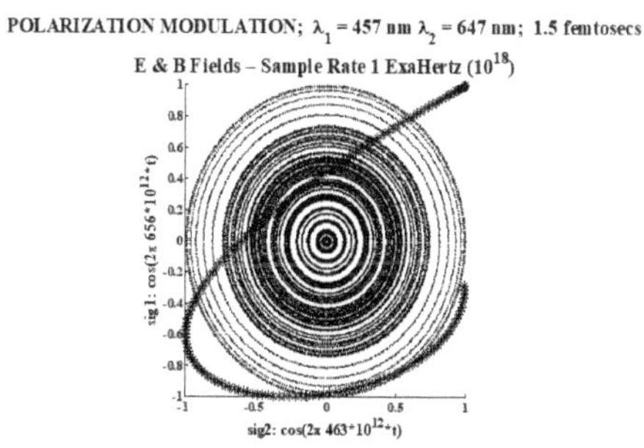

(C2)

Fig. 6.8. (*Continued*)

Fig. 6.8. (A) **E** polar vector and **B** bivector with trivector representations, as in Figs 6.8(B) 6.8(C), but at viewing angle (24°, 16°). The termination of the **E** vector is shown as a hatched line. The **B** field is shown as a series of bivector areas bounded by solid lines. Sample rate is 1 ExaHertz (10^{18} Hz). The representation is over 1.5 femtoseconds.
A static circular polarization is shown in Fig. 6.8(A1) at three different viewing angles and in comparison, a polarization modulation is shown in Fig. 6.8(A2) at the same three different viewing angles.

In (A1): $\text{beam}\,1 = a_1 \cos\left(2\pi \times 29 \times 10^{12} \times t\right)$; $\text{beam}\,2 = a_2 \cos(2\pi \times 29 \times 10^{12} \times t + \pi/2)$; $a_1 = a_2 = 1$.

In (A2): $\text{beam}\,1 = a_1 \cos\left(2\pi \times 32 \times 10^{12} \times t\right)$; $\text{beam}\,2 = a_2 \cos(2\pi \times 29 \times 10^{12} \times t)$; $a_1 = a_2 = 1$.

(B) **E** polar vector and **B** bivector with trivector representations as in Fig. 6.8(A), above and Fig. 6.8(C), but at viewing angle (−29°, −36°). The termination of the **E** vector is shown as a hatched line. The **B** field is shown as a series of bivector areas bounded by solid lines. Sample rate is 1 ExaHertz (10^{18} Hz). The representation is over 1.5 femtoseconds. As in Fig 6.8(A), a static circular polarization is shown in Fig. 6.8(B1) at three different viewing angles and in comparison, a polarization modulation is shown in Fig. 6.8(B2) at the same three different viewing angles.

In (A1): $\text{beam}\,1 = a_1 \cos\left(2\pi \times 29 \times 10^{12} \times t\right)$; $\text{beam}\,2 = a_2 \cos(2\pi \times 29 \times 10^{12} \times t + \pi/2); a_1 = a_2 = 1$.

In (A2): $\text{beam}\,1 = a_1 \cos\left(2\pi \times 32 \times 10^{12} \times t\right)$; $\text{beam}\,2 = a_2 \cos(2\pi \times 29 \times 10^{12} \times t)$; $a_1 = a_2 = 1$.

(C) **E** polar vector and **B** bivector with trivector representations as in Figs 6.8(A) and 6.8(B), but at viewing angle (90°, 0°). The termination of the **E** vector is shown as a hatched line. The **B** field is shown as a series of bivector areas bounded by solid lines. Sample rate is 1 ExaHertz (10^{18} Hz). The representation is over 1.5 femtoseconds. As in Figs. 6.8(A) and 6.8(B), a static circular polarization is shown in Fig. 6.8(C1) at three different viewing angles and in comparison a polarization modulation is shown in Fig. 6.8(C2) at the same three different viewing angles.

In (A1), (B1) and (C1): $\text{beam}\,1 = a_1 \cos(2\pi \times 29 \times 10^{12} \times t)$; $\text{beam}\,2 = a_2 \cos (2\pi \times 29 \times 10^{12} \times t + \pi/2)$; $a_1 = a_2 = 1$.

In (A2), (B2) and (C2): $\text{beam}\,1 = a_1 \cos\left(2\pi \times 32 \times 10^{12} \times t\right)$; $\text{beam}\,2 = a_2 \cos \left(2\pi \times 29 \times 10^{12} \times t\right); a_1 = a_2 = 1$.

The E field in C1 and C2 provides a Lissajous pattern. As to be expected, the Lissajous pattern shown in C1 is circular and associated with a magnetic field.

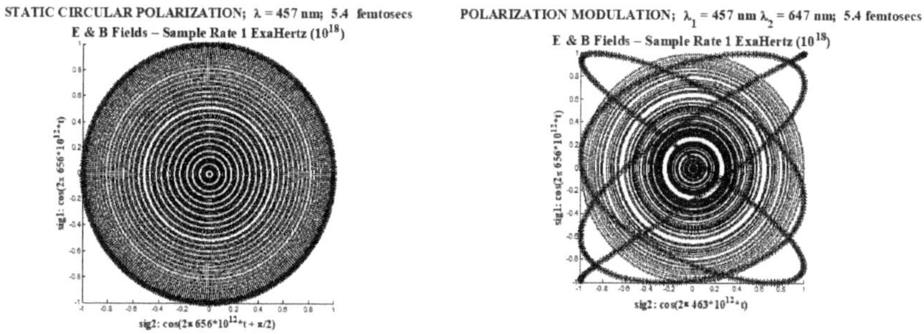

(a) (b)

Fig. 6.9. Trivector representation. **E** and **B** fields of Figs. 6.8(A)–(C), shown head-on as Lissajous patterns. The termination of the **E** vector is shown as a hatched line. The **B** field is shown as a series of bivector areas bounded by solid lines. Sample rate is 1 ExaHertz (10^{18} Hz).
A Static circular polarization; B: Polarization modulation. Both representations are over 5.4 femtosecs.

In (a): beam 1 $= a_1 \cos\left(2\pi \times 29 \times 10^{12} \times t\right)$; beam 2 $= a_2 \cos(2\pi \times 29 \times 10^{12} \times t + \pi/2)$; $a_1 = a_2 = 1$.

In (b): beam 1 $= a_1 \cos\left(2\pi \times 32 \times 10^{12} \times t\right)$; beam 2 $= a_2 \cos(2\pi \times 29 \times 10^{12} \times t)$; $a_1 = a_2 = 1$.

Compare Fig. 6.9(b) with Fig. 6.8(C2). The difference is that the previous figure is calculation over 1.5 femtoseconds and the present figure over 5.4 femtoseconds.

thus seek to establish a correspondence between one algebra — tensor algebra which is coordinate specific — and another algebra — the Clifford/Geometric algebra[11] which is coordinate free.

It is critical to note that a pair of complex scalar fields of a knotted beam, φ, θ, are considered to be dual, as there exists in three dimensions a duality between a vector basis which is contravariant, and a bivector basis which is covariant. (However, in four dimensions the duality is between vectors and trivectors.) With that relationship noted, then it is possible to compare and contrast the two algebras,

[11] More complete descriptions of geometric algebra can be found in Hestenes and Sobczyk (1984), Doran and Lasenby (2003), Arthur (2011) and Chappell *et al.* (2014).

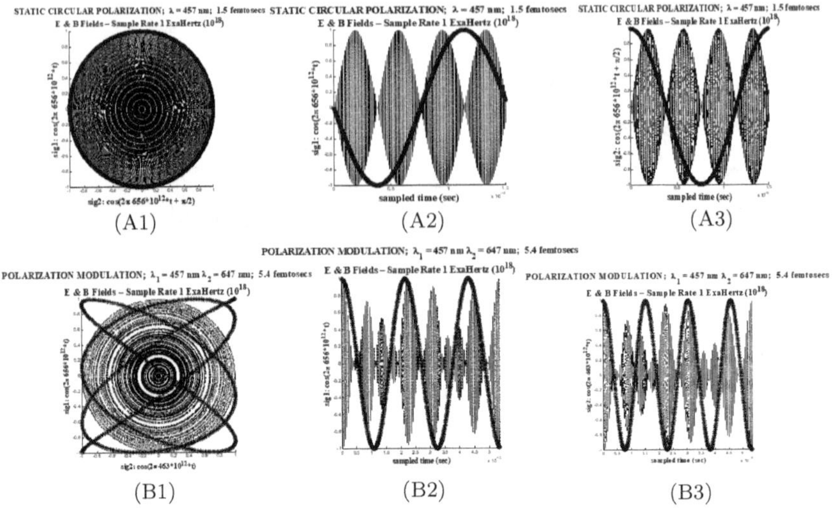

Fig. 6.10. Static circular polarization (top row) and polarization modulation (bottom row) — trivector representation. **E** and **B** fields are shown: head-on (A1) and (B1); side-on to beam 1 (A2) and (B2); side-on to beam 2 (A3) and (B3). The termination of the **E** vector is shown as hashed line. The extent of the area of the **B** bivector is shown. Sample rate is 1 ExaHertz (10^{18} Hz).

In (A) *(static circular polarization)* the representation is over 1.5 femtoseconds; in (B) *(polarization modulation)* the representation is over 5.4 femtoseconds.

In (A) beam 1 $= a_1 \cos\left(2\pi \times 29 \times 10^{12} \times t\right)$; beam 2 $= a_2 \cos(2\pi \times 29 \times 10^{12} \times t + \pi/2)$; $a_1 = a_2 = 1$.

In (B) beam 1 $= a_1 \cos\left(2\pi \times 32 \times 10^{12} \times t\right)$; beam 2 $= a_2 \cos(2\pi \times 29 \times 10^{12} \times t)$; $a_1 = a_2 = 1$.

the one based on quaternion numbers (S^3, the SU(2) group and geometric algebra) and the other based on complex numbers (S^2, the $U(1)$ group and tensor algebra). In this strong form of duality (vector-to-bivector) the fields **E** and **B** are dual.

But the concept of duality has multiple meanings besides this strong form. We show that it is possible to create knotted structures commencing with two orthogonal beams (as in POLMOD) that are dual in a weak sense — e.g., theorems and concepts applied to one beam, equally apply to the other beam. As an exercise, we proceed to create a trefoil knot on the extended Poincaré sphere — i.e.,

Table 6.1. Summary of trivector or directed volume representations of the **E** axial vector and the **B** bivector.

Figure	Modulation	Significance
6.2	None. Static unmodulated and unchanging circular polarization. Constituent orthogonally polarized beams are of the same frequency. Interbeam phase: $\pi/2$	A circular polarized **E** field shown as the termination of a polar vector and four expansions and contractions of the **B** field shown as a series of bivector areas
6.3	None. Static unmodulated and unchanging circular polarization As in Fig. 6.2	As in Fig. 6.2 but showing only one expansion and contraction of the **B** bivector area
6.4	Polarization modulation compared with static unmodulated and unchanging circular polarization	A heads-on (spatial domain) view, contrasting static (unchanging) circular polarization with polarization modulation
6.5	Polarization modulation compared with a static (unchanging) circular polarization	A side (time domain) view, contrasting static (unchanging) circular polarization with polarization modulation
6.6	Polarization modulation compared with a static (unchanging) circular polarization	Trivector representation contrasting static (unchanging) circular polarization with polarization modulation. Expansions and contractions of the **B** field shown as a series of bivector areas
6.7	None. Static unmodulated and unchanging circular polarization. As in Fig 6.2	Trivector representation of one cycle of the **E** field on the Poincaré sphere during which the **B** field progresses through the four quarter sections of a complete Lissajous circular polarization cycle
6.8 (A–C)	Polarization modulation compared with a static (unchanging) circular polarization at three different viewing angles	A circular polarized **E** field shown as the termination of a polar vector and four expansions and contractions of the **B** field shown as a series of bivector areas
6.9	Polarization modulation compared with a static (unchanging) circular polarization	Trivector representation **E** and **B** fields of Fig. 6.8 shown head-on as Lissajous patterns
6.10	Polarization modulation compared with a static (unchanging) circular polarization	Trivector representation **E** and **B** fields of Fig. 6.8 shown side-on

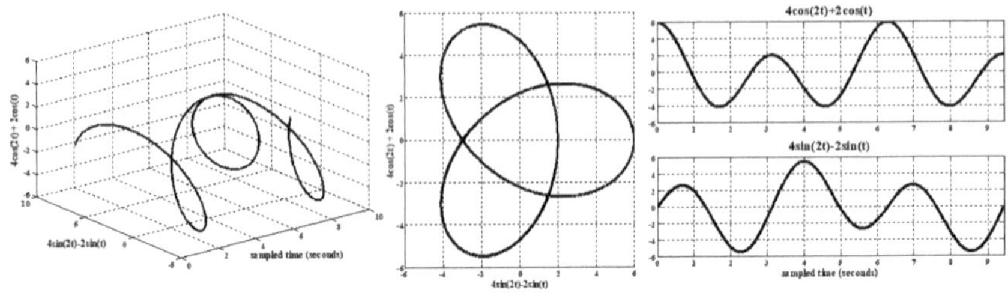

(a) (b) (c)

Fig. 6.11. (a) A trefoil knot created from 2 orthogonal signals:

$$4\cos(2t) + 2\cos(t),$$
$$4\sin(2t) - 2\sin(t),$$

(b) The trefoil knot of A viewed head-on (azimuth and elevation = (90,90), i.e., a Lissajous pattern with time suppressed).
(c) The two orthogonal signals:

$$4\cos(2t) + 2\cos(t), \text{ upper}$$
$$4\sin(2t) - 2\sin(t), \text{ lower.}$$

the S^2 sphere extended to a S^3 sphere by an indicated movement, demonstrating that knotted structures can be created from *two fields* which are dual in the weak sense, but a knotted theory does not necessarily provide a relation of $\mathbf{E}$ to $\mathbf{B}$, which are dual in the strong sense.

Figure 6.11(a) shows a continuous beam in 3D constructed from the two orthogonal signals:

$$\begin{aligned} &4\cos(2t) + 2\cos(t) \\ &4\sin(2t) - 2\sin(t) \end{aligned} \tag{6.7}$$

Viewed head-on (Fig. 6.11(b)), the trefoil knot is apparent. The two constituent signals are shown in Fig. 6.11(c).

Represented on an S^3 Poincaré sphere (i.e., movements are indicated across the S^3 sphere), Figs. 6.12(a)–(c), the trefoil knot is also apparent. Knotted structures can therefore be created from fields that are dual in the weak sense. One might state therefore, that the linking number is conserved (conservation of helicity) for fields

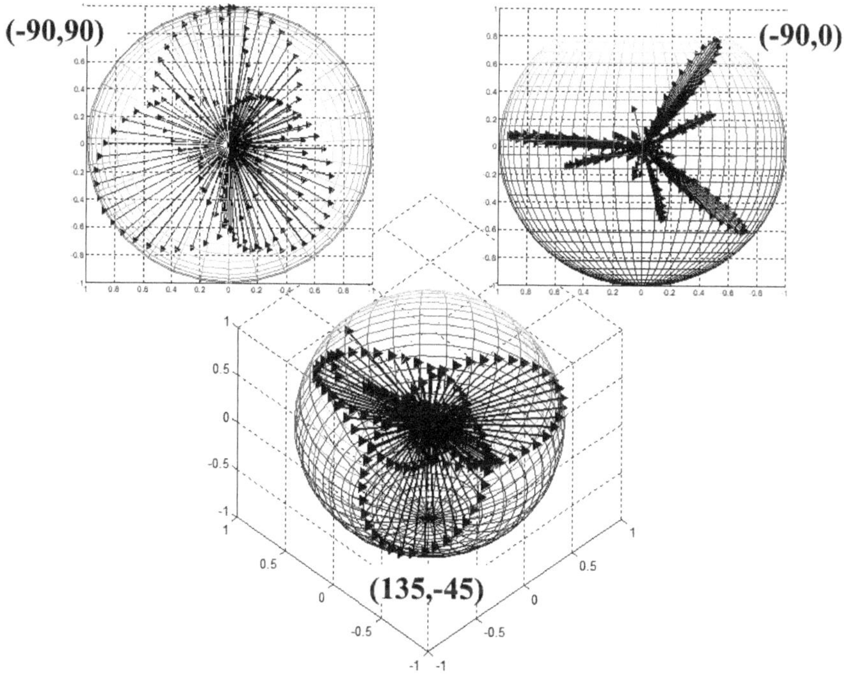

Fig. 6.12. Poincaré sphere representations of signal trefoil knots. The composite modulated signal of Fig. 6.11(a) projected on the Poincaré sphere and viewed (azimuth and elevation): Upper left: (–90,90). Notice the trefoil knot. Upper right: (–90, 0). Below: (–135, 45). Again, notice the trefoil knot.

defined by

$$\int \mathbf{E}_1 \cdot \mathbf{E}_2 = 0, \tag{6.8}$$

where $\mathbf{E}_1$ and $\mathbf{E}_2$ are two orthogonal **E** (polar vector) fields. The bivector **B** field is derived from these **E** fields.

Finally, Fig. 6.13 shows the Poincaré sphere, **E** field and Lissajous pattern representations of a three wavelengths — two beam configuration.

Knotted beams and other such complex beam combinations have not been used in either spectroscopy, imaging or sensing and their use awaits empirical testing.

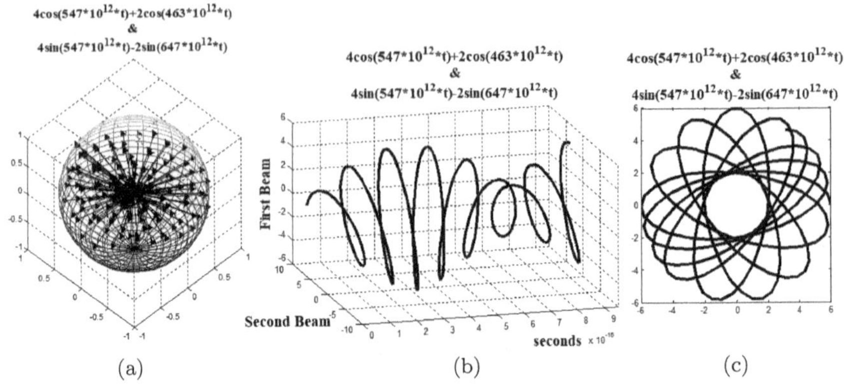

Fig. 6.13. Three wavelengths — two beam configuration.

The three beams are:

$$\text{beam } 1 = 4\cos(463 \times 10^{12} \times t);$$
$$\text{beam } 2 = 4\cos(547 \times 10^{12} \times t);$$
$$\text{beam } 3 = 4\cos(647 \times 10^{12} \times t).$$

Additives are as follows:

$$\text{additive } 1 = \text{beam } 2 + \text{beam } 1; \text{ same polarization}$$
$$\text{additive } 2 = \text{beam } 2 + \text{beam } 3; \text{ same polarization}.$$

Two beam combination: additive 1 + additive 2; polarized orthogonally.

(a) Poincaré sphere representation of POLMOD.

(b) **E** field (combined beam) vs. time.

(c) Head-on view of E Field (combined beam) — Lissajous pattern.

6.4. Anisotropy — A Geometric Algebra Definition of Unequal Physical Properties Along Different Axes in a Disturbed Medium

The goal of tests described in the previous chapter was to demonstrate optimum propagation through disturbed atmospheres/ ionospheres by means of polarization modulation (POLMOD) and axial or azimuthal modulation (AXMOD) techniques. The tests used two polarized continuous beam lasers of different wavelength, $\Delta\lambda = \lambda_2 - \lambda_1$, to give the combined beam either a polarization (combined beams are orthogonal) or rotational (combined beams are of the same polarization) *limited compatibility period* with induced polarizations/rotations in media with a duration $\tau = 1/\Delta f$, $\Delta f = (f_1 - f_2)/2$, $f_1 = c/\lambda_1$, $f_2 = c/\lambda_2$, which is less than the electronic, vibrational or rotational relaxation times of such media.

As described above, such continuous beams rapidly changing in polarization or rotation and forming wave packets, find a description in geometric and Clifford algebras[12] that are coordinate-free. In such algebras a distinction is made between the electric field, **E**, represented as a vector, or directed line segment, and the magnetic field, **B**, represented as a bivector or 2-vector, or oriented line segment.

Here we address and compare two aspects of an optimum propagation goal:

(1) to develop a *steady-state* geometric algebra description of the electromagnetic field-medium interaction.
(2) to commence an extension of that description to the *transient state* by addressing a model system — a liquid crystal medium — the transient state (onset and offset times) for which a considerable literature exists.

We show a relation of the angle, θ, between the electric field, **E**, and the electric displacement, **D**, adequately described in the geometric algebra description of electromagnetism to the angle, θ, defined with respect to a unit vector angle of incident radiation and to media with induced dipoles. The latter is related to the optical response times (T_{rise} and T_{decay}) that are linearly proportional to media director reorientation times (t_{rise} and t_{decay}).

A first observation is that there is a direct geometrical interpretation of anisotropy (the unequal physical properties of a media along different axes) without the intervention of any coordinate system. In Art. 794 (p. 443) of his 1873 Treatise, James Clerk Maxwell stated

[12]Hestenes and Sobczyk (1984); Chisholm and Common (1986); Baylis (1999); Doran and Lasenby (2003); Arthur (2011); Chappel *et al.* (2014).

that:

> "in certain media the specific capacity for electrostatic induction is different in different directions, or in other words, the electric displacement, instead of being in the same direction as the electromotive intensity, and proportional to it, is related to it by a system of linear equations."

With this definition of an electrically anisotropic medium, there have been three approaches to the description of anisotropic media: (1) the tensor method; (2) differential forms; and (3) Clifford geometric algebra method. Here, as in the previous section, we implement the geometric algebra method, adopting the recent application of this method by Matos *et al.* (2007), which, in turn, builds on the work of Hestenes and others, as well as the founders: Grassman,[13] Hamilton[14] and Clifford.

If a medium is electrically anisotropic, an angle between the electric field vector, **E**, (in the present instance **E** being a laser beam), and the electric displacement vector, **D**, depends on the direction of the Euclidean space along which **E** is directed. This observation amounts to the impossibility of writing $\mathbf{D} = \varepsilon_0\varepsilon\mathbf{E}$, where ε_0 is the permittivity of the vacuum, and ε is the relative permittivity of the medium. Whereas this problem is addressed by tensor algebra by introducing a 3×3 permittivity tensor, in geometric algebra, the anisotropy is written as:

$$\mathbf{D} = \varepsilon_0(\varepsilon\mathbf{E}), \tag{6.9}$$

where $(\varepsilon\mathbf{E})$ is a linear function that maps vectors to vectors and ε is a dielectric function that is defined *along a particular direction* for a particular medium.[15] The following definitions are required to develop the algebra.

The geometric product is defined as the graded sum:

$$u = \mathbf{ED} = \mathbf{E} \cdot \mathbf{D} + \mathbf{E} \wedge \mathbf{D} = \alpha + \mathbf{F}, \tag{6.10}$$

---

[13]Hermann Günther Grassman (1809–1877).

[14]William Rowan Hamilton (1805–1865).

[15]A dielectric tensor coordinate system can be recaptured by defining: $\varepsilon_{ij} = \mathbf{e}_j \cdot \varepsilon(\mathbf{e}_k)$.

where: $\boldsymbol{\alpha} = \mathbf{E} \cdot \mathbf{D}$ is the dot or inner product and symmetric, $\mathbf{F} = \mathbf{E} \wedge \mathbf{D}$ is the outer exterior product, is antisymmetric, associative, is a *bivector* and a directed plane segment.[16]

As noted above, the outer product of a vector and a bivector produces a *trivector*, $\mathbf{V}$, which is an oriented volume element; and now we introduce a *multivector*, which is a sum of a scalar, a vector, a bivector and a trivector, e.g.:

$$u = \alpha + \boldsymbol{a} + \mathbf{F} + \mathbf{V}. \tag{6.11}$$

The *reverse* of u is defined as $\tilde{u}$, so that:

$$\begin{aligned}
\tilde{u} &= \mathbf{DE} = \mathbf{D} \cdot \mathbf{E} - \mathbf{D} \wedge \mathbf{E} = \alpha + \mathbf{F}, \\
\alpha &= \mathbf{E} \cdot \mathbf{D} = (u + \tilde{u})/2, \\
\mathbf{F} &= \mathbf{E} \wedge \mathbf{D} = (u - \tilde{u})/2, \\
|u|^2 &= u\tilde{u} = \mathbf{EDDE} = \mathbf{E}^2\mathbf{D}^2 = (\alpha + \mathbf{F})(\alpha - \mathbf{F}) = \alpha^2 - \mathbf{F}^2 \\
&= \alpha^2 + \beta^2 = \rho^2,
\end{aligned}$$

If the unit bivector is such that $\hat{\mathbf{F}}^2 = -1$, then:

$$\begin{aligned}
u &= \alpha + \beta\hat{\mathbf{F}} = \rho\cos(\theta) + \hat{\mathbf{F}}\rho\sin(\theta), \\
\rho &= |\mathbf{E}|\,|\mathbf{D}| = \sqrt{\alpha^2 + \beta^2}, \text{and} \\
\mathbf{F} &= \beta\hat{F}.
\end{aligned}$$

Thus any bivector is the *dual* of a vector.

In geometric algebra, $\mathcal{C}\ell_3$, a multivector is an operation that projects onto a chosen grade k, where $k = 0, 1, 2, 3$, giving the structure:

$$\mathcal{C}\ell_3 = \mathbb{R} \oplus \mathbb{R}^3 \oplus \wedge^2\mathbb{R}^3 \oplus \wedge^3\mathbb{R}^3. \tag{6.12}$$

A *k-blade* of $\mathcal{C}\ell_3$ is an element u_k such that $u_k = \langle u_k \rangle_k$, where $\langle u_k \rangle_k$ is a homogeneous multivector of grade k.

We can write any trivector as $\mathbf{V} = \beta\mathbf{e}_{123}$, where $\beta \in \mathbb{R}$ and $\mathbf{e}_{123} = \hat{V}$ is the unit trivector such that $\mathbf{e}_{123}^2 = -1$. With these

[16]As previously noted, and in contrast, the Gibbs cross product is: $\mathbf{E} \times \mathbf{D}$, which is a vector, not associative, and is a directed line segment.

definitions, two vectors, $\boldsymbol{a}$ and $\boldsymbol{b}$ ($\boldsymbol{a}, \boldsymbol{b} \in \mathbb{R}^3$) are related by outer and cross products:

$$\begin{aligned} \boldsymbol{a} \wedge \mathbf{B} &= (\boldsymbol{a} \times \mathbf{B})\mathbf{e}_{123}, \\ \boldsymbol{a} \times \mathbf{b} &= -(\boldsymbol{a} \wedge \mathbf{b})\mathbf{e}_{123}. \end{aligned} \tag{6.13}$$

Geometric algebra $\mathcal{C}\ell_3$ is a linear space of dimensions $1+3+3+1 = 2^3 = 8$, with $\{\mathbf{e}_1, \mathbf{e}_2, \mathbf{e}_3\}$ as an orthonormal basis for vector space $\mathbb{R}^3$. The $\mathcal{C}\ell_3$ space is characterized by

$$\left\{ \underbrace{1}_{\text{scalars}}, \underbrace{\mathbf{e}_1, \mathbf{e}_2, \mathbf{e}_3}_{\text{vectors}}, \underbrace{\mathbf{e}_{12}, \mathbf{e}_{31}, \mathbf{e}_{23}}_{\text{bivectors}}, \underbrace{\mathbf{e}_{123}}_{\text{trivectors}} \right\} \tag{6.14}$$

where $\mathbf{e}_{12} = \mathbf{e}_1 \wedge \mathbf{e}_2 = \mathbf{e}_1\mathbf{e}_2, \mathbf{e}_{31} = \mathbf{e}_3 \wedge \mathbf{e}_1 = \mathbf{e}_3\mathbf{e}_1$, and $\mathbf{e}_{23} = \mathbf{e}_2 \wedge \mathbf{e}_3 = \mathbf{e}_2\mathbf{e}_3$, is the basis for the subspace $\wedge^2\mathbb{R}^3$ of *bivectors* or *2-blades*.

A medium is *anisotropic* if the angle, θ, between the electric field, $\mathbf{E}$, and the electric displacement, $\mathbf{D}$, is different for different directions of $\mathbf{E}$. Therefore, with the above definitions:

$$\begin{aligned} \beta &= \beta(\theta) = \rho \sin(\theta), \text{ or} \\ \beta &= |\mathbf{F}| = |\langle \mathbf{ED} \rangle_2| = |\mathbf{E} \wedge \mathbf{D}|, \end{aligned} \tag{6.15}$$

there is a dependency on the direction in which $\mathbf{E}$ is applied — $\mathbf{E}$ being, in the present circumstances, the incident laser beam.

With:

$$\begin{gathered} \mathbf{E} = |\mathbf{E}|\, \boldsymbol{s}, \\ \mathbf{D} = |\mathbf{D}|\, \boldsymbol{t} \text{ and} \\ \boldsymbol{s}^2 + \boldsymbol{t}^2 = 1, \text{ then} \\ \hat{\boldsymbol{F}} = \boldsymbol{s} \wedge \boldsymbol{t} / \sin(\theta) = \boldsymbol{sr} \\ \boldsymbol{r}^2 = 1. \end{gathered} \tag{6.16}$$

It follows that:

$$\begin{aligned} \mathbf{D} &= \mathbf{D}_{||} + \mathbf{D}_{\perp}, \\ \mathbf{D}_{||} &= \mathbf{s} \cdot \mathbf{D}_{||} = |\mathbf{D}| \cos(\theta), \\ \mathbf{D}_{\perp} &= \mathbf{r} \cdot \mathbf{D}_{\perp} = |\mathbf{D}| \sin(\theta). \end{aligned} \tag{6.17}$$

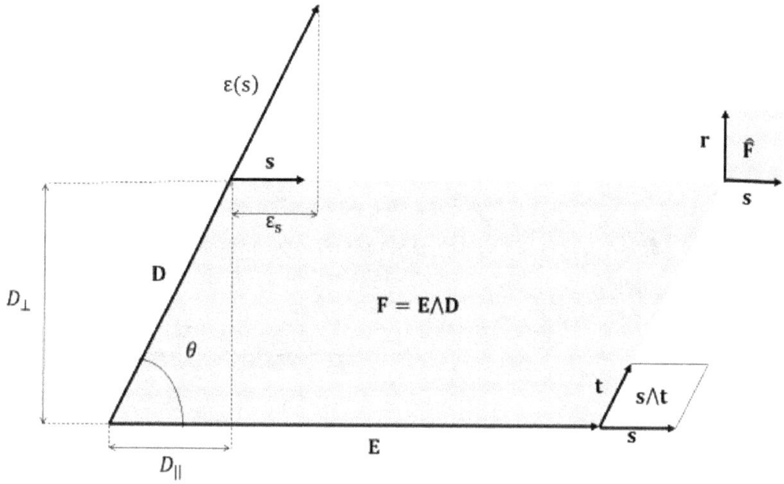

Fig. 6.14. Relation of E and D fields in geometric algebra description. After Matos, S.A., Ribeiro, M.A. & Paiva, C.R., Anisotropy without tensors: a novel approach using geometric algebra. *Optics Express*, 15, 15175–15186, 2007.

These relations are shown in Fig. 6.14 (Matos *et al.* 2007). Which permits a definition of relative permittivity:

$$\varepsilon_s = \mathbf{s} \cdot \varepsilon(\mathbf{s}), \tag{6.18}$$

with $\varepsilon_s = 0$ if $\mathbf{E} \perp \mathbf{D}$.

In the case of a lossless nonmagnetic medium,

$$\varepsilon(\boldsymbol{a}) = \lambda \boldsymbol{a}, \tag{6.19}$$

which provides three positive eigenvalues, $\varepsilon_1, \varepsilon_2$ and ε_3 corresponding to the three unit eigenvectors $\mathbf{e}_1$, $\mathbf{e}_2$ and $\mathbf{e}_3$.

6.5. Medium Transparency and Temporal Dependence of Anisotropy

As stated in Chapter 5, in order to defeat absorption/scattering we wish to obtain the temporal duration of the polarization/rotation dwell time of incident radiation to be less than the duration required of a molecular system for a TX → RX reception/absorption to occur.

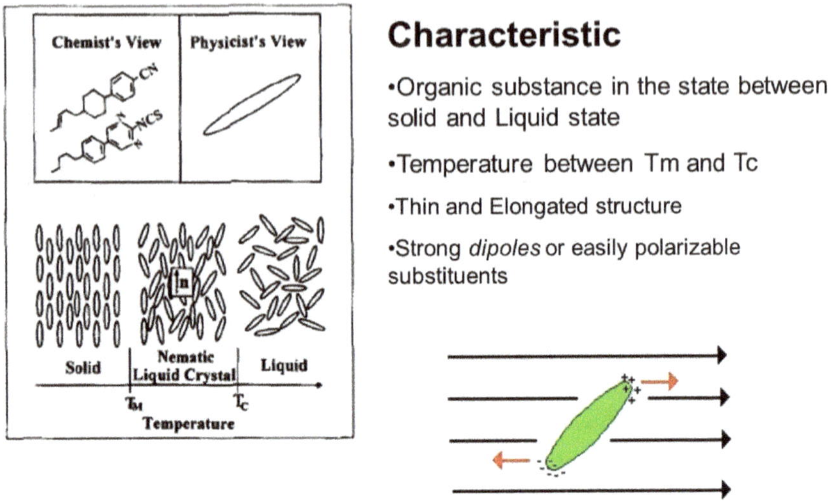

Fig. 6.15. In the case of positive dielectric anisotropy, the liquid crystal will aligned parallel to the electric field. Upon application of an electric field, the positive charge is displaced to one end of the molecule and the negative charge to the other end, thus creating an induced dipole moment. This results in the alignment of the longitudinal axis of liquid crystal molecules mutually parallel to the electric field direction. In the case of negative dielectric anisotropy, the liquid crystal will aligned perpendicular to the electric field. This means the induced index of refraction is larger along the long axis of the molecules, than perpendicular to it. After Wong (2005).

For molecular systems to function as "polarization/rotation selective receivers" — in other words, that molecular systems receive, or interact, optimally with radiation of a certain polarization — there is a finite time that such systems orient to the incident radiation, i.e., are induced to orient. Therefore we turn to the temporal dependence of the angle θ — see Figs. 6.15 and 6.16 — using liquid crystals (LCs) as a model system displaying a temporal dependence of dielectric anisotropy. As Fig. 6.14 shows, an LC will align perpendicular or parallel to incident radiation, and the angle θ is defined as the angle between the unit vector of the incident radiation and the orientation of the LC (Fig. 6.15). From the molecular point of view, the origin of the dielectric anisotropy is the anisotropic distribution of the molecular dipoles in the liquid crystal phases.

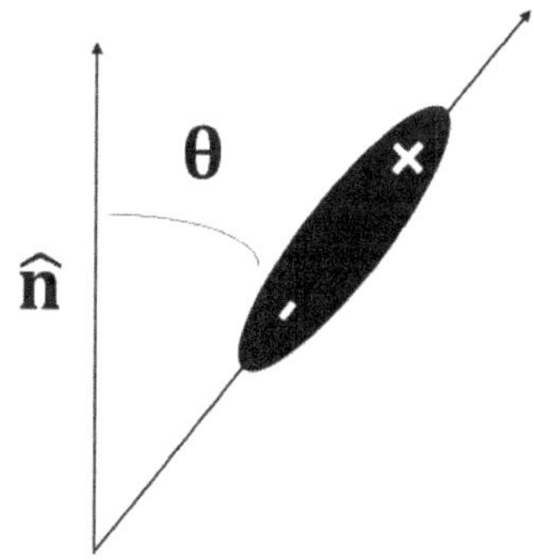

Fig. 6.16. The angle, θ, defined with respect to the unit vector angle of the incident of radiation and the LC. The unit vector is referred to as "the director".

Maier and Meier (1961) extended Onsager theory to nematic LCs where the molecules are oriented in parallel but not arranged in well-defined planes. In their theory, a molecule is represented by an anisotropic polarizability α with principal elements a_1 and a in a spherical cavity of radius a. Denoting the dipole moment with a_1 at an angle θ, the LC dielectric components $\varepsilon_{||}$, $\varepsilon_{\perp}$ and $\Delta\varepsilon$, can be expressed as:

$$\begin{aligned}
\varepsilon_{||} &= NhF\left\{\langle\alpha_{||}\rangle + (F\mu^2/3kT)[1-(1-3\cos^2\theta)S/2]\right\},\\
\varepsilon_{\perp} &= NhF\left\{\langle\alpha_{\perp}\rangle + (F\mu^2/3kT)[1+(1-3\cos^2\theta)S/2]\right\},\\
\Delta\varepsilon &= NhF\left\{\langle\alpha_1-\alpha_t\rangle + (F\mu^2/2kT)(1-3\cos^2\theta)\right\}.
\end{aligned} \tag{6.20}$$

where N is the molecular packing density, μ is the dipole moment, F is the Onsager reaction field, S is an order parameter described by: $S = (1-T/T_c)^{\beta}$, β is a coefficient dependent on molecular structure, T_c is the clearing temperature, $\langle n\rangle$ is the average refractive index, $\langle\varepsilon\rangle$ is the average dielectric constant, and h and F are defined as:

$$\begin{aligned}
h &= \frac{3\langle\varepsilon\rangle}{2\langle\varepsilon\rangle+1},\\
F &= \frac{(2\langle\varepsilon\rangle+1)(\langle n\rangle^2+2)}{3(2\langle\varepsilon\rangle+\langle n\rangle^2)},
\end{aligned} \tag{6.21}$$

Indicating that the dielectric anisotropy of an LC is a function of three factors: molecular structure, temperature and frequency.

A figure-of-merit (FoM) characterizes the overall LC performance (Khoo and Wu, 1993):

$$\text{FoM} = \frac{\Delta n^2}{\gamma_1 / K_{11}}, \tag{6.22}$$

where K_{11} is the splay elastic constant, Δn is the birefringence, γ_1 is the rotational viscosity. All three parameters are temperature dependent and the viscosity and elastic constants are dependent on the order parameter S.

When the applied voltage exceeds the Fréedericksz transition[17] voltage V_{th}, the LC molecules will rotate and be reoriented by the electric field, causing a change of the permittivity of the substrate. The electric energy under the applied voltage is given by (Wang, 2005):

$$f_{\text{Electric}} = \iiint \left(\frac{1}{2} D \cdot E\right) d\nu = \iiint \left(\frac{1}{2} \varepsilon \nabla V \cdot \nabla V\right) d\nu, \tag{6.23}$$

where V is the voltage distribution and ε is the dielectric tensor of the LC medium.

The relative dielectric tensor is

$$\varepsilon_r = \begin{bmatrix} \varepsilon_{xx} & \varepsilon_{xy} & \varepsilon_{xz} \\ \varepsilon_{yx} & \varepsilon_{yy} & \varepsilon_{yz} \\ \varepsilon_{zx} & \varepsilon_{zy} & \varepsilon_{zz} \end{bmatrix}, \tag{6.24}$$

and, referring to Fig. 6.17:

$$\begin{aligned} \varepsilon_{xx} &= n_o^2 + \left(n_e^2 - n_o^2\right) \cos^2\theta \cos^2\phi, \\ \varepsilon_{yy} &= n_o^2 + \left(n_e^2 - n_o^2\right) \cos^2\theta \sin^2\phi, \\ \varepsilon_{zz} &= n_o^2 + \left(n_e^2 - n_o^2\right) \sin^2\theta, \\ \varepsilon_{yz} &= \varepsilon_{zy} = \left(n_e^2 - n_o^2\right) \sin\theta \cos\phi \sin\phi, \\ \varepsilon_{xy} &= \varepsilon_{yx} = \left(n_e^2 - n_o^2\right) \cos^2\theta \sin\phi \cos\phi, \\ \varepsilon_{xz} &= \varepsilon_{zx} = \left(n_e^2 - n_o^2\right) \sin\theta \cos\theta \cos\phi, \end{aligned} \tag{6.25}$$

[17]The Fréedericksz transition is a phase transition in liquid crystals produced when a sufficiently strong electric or magnetic field is applied to a liquid crystal in an undistorted state.

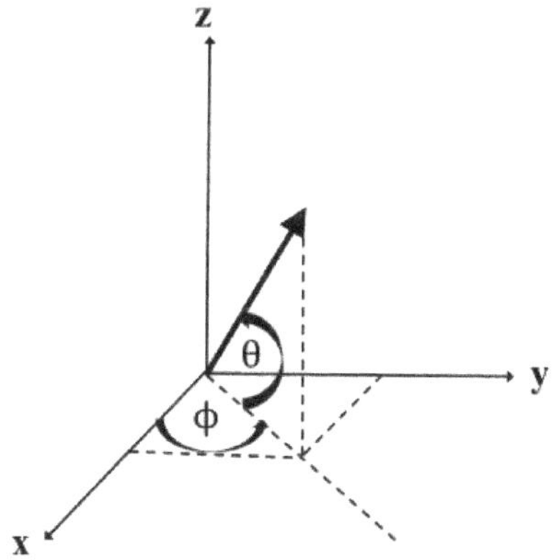

Fig. 6.17. The coordinate system of a liquid crystal (LC) director.

where n_o is the ordinary refractive index of the LC medium, n_e is the extraordinary refractive index of the LC medium, θ is the tilt angle of the LC director (Fig. 6.4.3), ϕ is the azimuthal angle between projection of the LC director on the x–y plane and the axis (Fig. 6.17).

The optical response time for amplitude modulation and phase response time for phase modulation is related to the LC director reorientation time. Erickson (1961) and Leslie (1968) described the dynamics of the LC director reorientation as:

$$\left(K_{11}\cos^2\phi + K_{33}\sin^2\phi\right)\frac{\partial^2\phi}{\partial z^2} + (K_{33} - K_{11})\sin\phi\cos\phi\left(\frac{\partial\phi}{\partial z}\right)^2 + \varepsilon_o\Delta\varepsilon E^2\sin\phi\cos\phi = \gamma_1\frac{\partial\phi}{\partial t}, \quad (6.26)$$

where γ_1 is the rotational viscosity, K_{11} and K_{33} are the splay and bend elastic constants, $\varepsilon_o\Delta\varepsilon E^2$ is the electric field energy density, $\Delta\varepsilon$ is the dielectric anisotropy, and ϕ is the tilt angle of the LC directors.

If the tilt angle is small ($\sin\phi \sim \phi$) and $K_{33} \sim K_{11}$ (small angle approximation), then a reduction is possible to (Wang, 2005):

$$K_{33}\frac{\partial^2\phi}{\partial z^2} + \varepsilon_o\Delta\varepsilon E^2\phi = \gamma_1\frac{\partial\phi}{\partial t}. \quad (6.27)$$

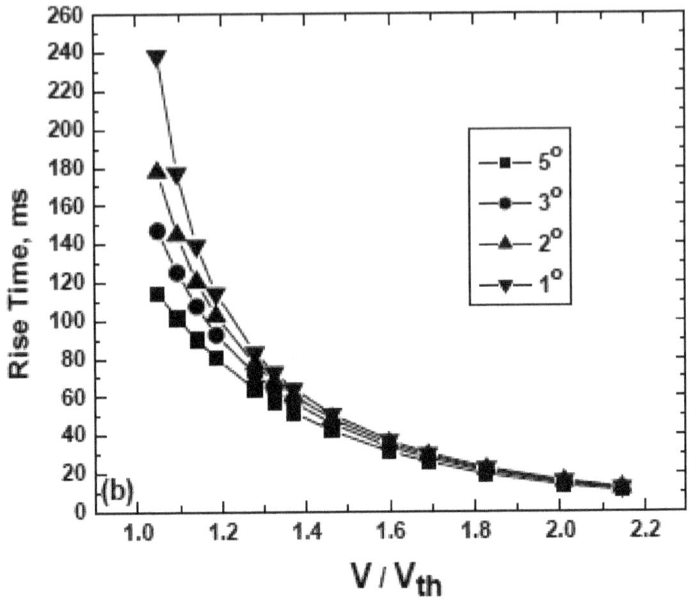

Fig. 6.18. Typical liquid crystal Fréedericksz transition optical rise time (10%–90%) as a function of V/V_{th} at four different pretilt angles, $\alpha = 1°$, $2°$, $3°$, and $5°$. After Wang (2005).

With α defined as the pretilt angle or the angle of the LC directors deviated from the LC normal, so that if $\alpha = 0$ the LC directors are aligned perpendicular to the incident beam, the optical rise time (10%–90%) as a function of V/V_{th} at four different pre-tilt angles, $\alpha = 1°$, $2°$, $3°$, and $5°$ can be calculated (Fig. 6.18). Optical response times (T_{rise} and T_{decay}) are linearly proportional to the LC director reorientation times (τ_{rise} and τ_{decay}).

Thus we have shown a relation of the angle, θ, between the electric field, **E**, and the electric displacement, **D**, adequately described in the geometric algebra description of electromagnetism to the angle, θ, defined with respect to a unit vector angle of incident radiation and to media with induced dipoles. The latter is related to the optical response times (T_{rise} and T_{decay}) that are linearly proportional to media director reorientation times (t_{rise} and t_{decay}).

In some special cases involving both the radiation and the medium, transparency to radiation is a transient phenomenon where the radiated medium can become opaque after a finite time delay.

We next consider three forms of absorption defeat:

(1) Self-induced transparency.
(2) Electromagnetically-induced transparency; and the subject of the previous and the present chapter.
(3) polarization/rotation modulation induced transparency.

All three methods to achieve transparency differ in specific requirements and parameters of (i) the incident radiation; and (ii) the medium. These methods are quite dissimilar. In other words, although the three forms are similar in requiring a specific "key" (the radiation) to fit a specific "lock" (the medium) in a "lock-and-key" arrangement, in each case the "locks" and "keys" are quite dissimilar.

6.5.1. *Self-induced transparency (SIT)*

McCall and Hahn[18] defined *self-induced transparency* as follows:

> "Above a critical power threshold for a given pulse width, a short pulse of coherent traveling-wave optical radiation is observed to propagate with anomalously low energy loss while at resonance with a two-quantum-level system of absorbers. The line shape of the resonant system is determined by inhomogeneous broadening, and the pulse width is short compared to dissipative relaxation times. A new mechanism of self-induced transparency, which accounts for the low energy loss, is analyzed in the ideal limit of a plane wave which excites a resonant medium with no damping present. The stable condition of transparency results after the traversal of the pulse through a few classical absorption lengths into the medium. This condition exists when the initial pulse has evolved into a symmetric hyperbolic-secant pulse function of time and distance, and has the area characteristic of a "2π pulse."[19] Ideal transparency then persists when coherent induced absorption of pulse energy

[18] McCall and Hahn (1967) were the first to describe and experimentally observe the self-induced transparency phenomenon.

[19] A 2π pulse or a self-induced transparency (SIT) soliton, has a group velocity that is smaller than the phase velocity of light in a specific medium which depends on the pulse duration. The shorter the pulse, the greater is its propagation velocity. The shape and group velocities of 2π pulses, as is typical for solitons, do not change.

> during the first half of the pulse is followed by coherent induced emission of the same amount of energy back into the beam direction during the second half of the pulse. The effects of dissipative relaxation times upon pulse energy, pulse area, and pulse delay time are ... to first order in the ratio of short pulse width to long damping time. The analysis shows that the 2π pulse condition can be maintained if losses caused by damping are compensated by beam focusing." (McCall & Hahn, 1969)

Self-induced transparency (SIT) of propagating coherent light pulses in absorbing media[20] is well proven. SIT has been demonstrated in gases, e.g., in gaseous SF_6 and SF_6-He (Patel and Slusher, 1967). As to be expected, the molecular dephasing relaxation time is longer than the pulse temporal length.

6.5.2. *Electromagnetically-induced transparency (EIT)*

Another induced transparency effect is *electromagnetically-induced transparency* (EIT), which is a state in which a coherent optical nonlinearity that renders a medium transparent over a narrow spectral range within an absorption line.[21] Extreme dispersion is also created within this transparency "window" which leads to "slow light".

Observation of EIT involves two optical fields (highly coherent light sources, such as lasers) which are tuned to interact with three quantum states of a material. Tuning the "probe" field near resonance between two of the states will measure the medium absorption transition spectrum. A much stronger "coupling" field is tuned near resonance at a different transition. If the states are selected properly, the presence of the coupling field will create a spectral "window" of transparency which will be detected by the probe. The coupling laser is sometimes referred to as the "control" or "pump", the latter in analogy to incoherent optical nonlinearities such as spectral hole burning or saturation.

EIT is based on the destructive interference of the transition probability between atomic states. Closely related to EIT are

[20] McCall and Hahn (1965, 1967, 1969); Lamb (1971).

[21] Kocharovskaya and Khanin (1986); Kasapi *et al.* (1995).

coherent population trapping (CPT) phenomena. EIT requires a coherently prepared media which has long quantum memory. The following protocol describes the underlying EIT mechanism:

> "... one applies two laser wavelengths whose frequencies differ by a Raman (non-allowed) transition of the medium... the electrons must be stopped from moving at the frequencies of the applied fields. If the electrons do not move, then they do not contribute to the dielectric constant. Non-movement will occur if, at each applied frequency, the electron is driven by two sinusoidal forces of opposite phase ... In quantum mechanical terms ... what happens is that the probability amplitude of ... a state is driven by two terms of equal magnitude and opposite sign. One driving term is proportional to the probability amplitude of the ground state. The other terms is oppositely phased and proportional to the probability amplitude of a third state and the expected value of the amplitude of the sinusoidal motion at each of the applied frequencies is zero." (Harris, 1997)

A major difference between EIT and SIT is that in the case of SIT only a single pulse is required whose area is 2π (see the Bloch sphere in the previous chapter) and only a ground and an excited state are involved (Fig. 6.19). In the case of EIT, there must be a coherent optical nonlinearity resulting in a medium transparency within a narrow spectral range around an absorption line.

6.5.3. *POLMOD & AXMOD-Induced transparency (PIT and AIT)*

A continuous polarization modulated (POLMOD) or axially modulated (AXMOD) wave can induce a transparency (PIT or AIT), which mitigates/defeats jitter, absorption and scattering. This occurs when the modulation has enough rapidity to produce a polarization/azimuthal dipole matching compatibility of the radiation to the existence of a molecular polarization/azimuthal state that is less than the dephasing relaxation time. In such a case, with a polarization/azimuthal-selective receiver (i.e., the molecular system), the radiation can only be sampled as a series of wave packets. This opens new possibilities for point-to-point communications and remote sensing in previously impossible cases.

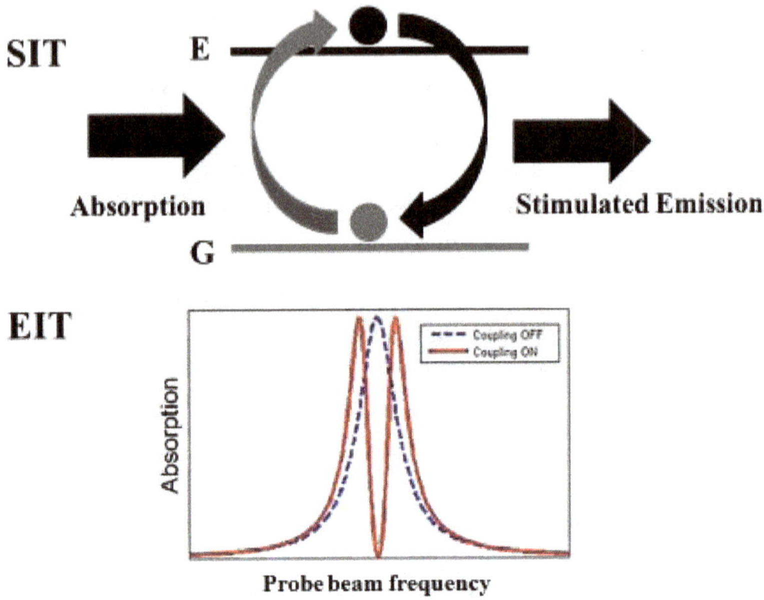

Fig. 6.19. Schematic of SIT and EIT mechanisms.

We can achieve polarization modulation (POLMOD & AXMOD) induced transparency (PIT and AIT), or the mitigation/defeat of jitter and absorption/scattering with a continuous rapidly modulated wave. The incident radiation must produce a transient polarization/axial dipole matching, or polarization/axial compatibility, such that the molecular polarization/axial state is less than the dephasing relaxation time. To a polarization-azimuthal-selective receiver (i.e., the molecular system), the radiation under these conditions will only be sampled as a series of wave packets. The aim is to match the incident wave modulation and frequency(ies), polarization and beat timing (the key) to similar properties in the material's molecular structure (the lock).

Both SIT and PIT/AIT achieve excitation durations that are less than the relaxation time of a detector or medium:

- SIT in the sense of the exciting pulse being shorter in duration than the relaxation time of a medium. A disadvantage of SIT is that

SIT is restricted to the generation of ultrashort, area 2π, pulses which carry much less energy than continuous waves, as well being more difficult to generate.

- PIT/AIT in the sense of the polarization compatibility of the exciting "pulses" (produced from continuous waves) with induced dipoles being of duration, again, less than the relaxation time of the medium. An advantage of PIT/AIT is that continuous beams of any duration can be used, but, as stated above, SIT is restricted to the generation of ultrashort, area 2π, pulses.

What evidence is there that molecular systems are polarization-selective receivers? In Sec. 5.3.4 of Chapter 5, we described the theoretical basis for Raman spectroscopy, or inelastic light scattering, which detects and requires molecular polarization preferences as evidenced by depolarization ratios. As further evidence, the selectivity of induced dipoles to linear vs. circular polarization in solid-state media has been demonstrated (Barrett *et al.* 1983; Barrett, 1983; Barrett, 1987a–c). The present analysis extends the PIT/AIT approach to dielectric media with the objective of defeating atmospheric/ionospheric propagation losses.

6.6. Conclusions and Future Directions

Geometric (Clifford) algebra provides new descriptive insights into the representation of POLMOD and AXMOD signals formed by beam combining for penetration of disturbing media at the molecular level and future communications as well as future remote sensing at the classical level. This algebraic description provides a way to model how radiation beams interact with media at both levels. Progress in advanced sensing systems will require analysis methods beyond the simple steady-state models, particularly when exploiting transient phenomena such as induced transparency. The present work addressed media in the "channel" medium between a transmitter and a receiver (as in communications). Similar approaches can be used to address the two-way channel between a transceiver and a target (as in radar and remote sensing) or the one-way multiple channels between a transmitter and multiple receivers (as in medical imaging).

Another application is to address the targets themselves, as selective receivers, to obtain optimum responses from those targets.

Previous work demonstrates the value of understanding the interaction of signals, targets and channel media (Barrett, 2012). We now know that signals and targets generate complex transient responses and yield exploitable *data* that can provide detailed information about the target. This chapter and the previous shows why we must also examine channel media at the molecular level.

The use of complex beam combinations, such as knotted beams, in transmission through disturbing media, and for imaging and sensing, has not yet been empirically tested. There is the possibility that their use may offer new capabilities. For example, complex beam combinations may produce new remote sensing applications at frequencies in the THz (0.3 to 3×10^{12} Hz) RF spectrum. Perhaps Lord Kelvin was on a (partially) right path when he applied knot theory to connect electromagnetic waves and material interactions.

Classical radar theory treats targets as reflectors with specified cross sections. It treats the channel media as a mass with steady-state characteristics, which works well for most practical locating and ranging applications. Induced transparency concepts expand radar and remote sensing from simple echo locating to advanced applications involving optical and molecular level channel materials properties. Among other applications, e.g., medical imaging, these advances can offer new remote sensing improvements and capabilities. Such applications will also require multidisciplinary approaches.

Bibliography

Arthur, J.W., *Understanding Geometric Algebra for Electromagnetic Theory*, Wiley IEEE Press, 2011.

Barrett, T.W., Wohltjen, H. and Snow, A., A new method of modulating electrical conductivity in phthalocyanines using light polarization. *Nature*, 301, 694–695, 1983a.

Barrett, T.W., Modulation of electrical conductance of phthalocyanine films by circularly polarized light, *Thin Solid Films*, 102, 231–244, 1983b.

Barrett, T.W., Wohltjen, H. and Snow, A., A new method of modulating electrical conductivity in phthalocyanines using light polarization, Chapter 24, in *Molecular Electronic Devices II*, F.L. Carter (ed), Marcel Dekker, New York, NY, 475–505, 1987a.

Barrett, T.W., Physical properties of ordered ultrathin organic films. *NSF Workshop "The Molecular Engineering of Ultrathin Polymeric Films"*, University of California at Davis, *Thin Solid Films*, 152, 67–98, 1987b.

Barrett, T.W., Physical properties of ordered ultrathin organic films. Chapter 5, in P. Stroeve and E. Ranses (eds) *Molecular Engineering of Ultrathin Polymeric Films*, Elsevier, New York, pp. 67–98, 1987c.

Barrett, T.W., *Resonance and Aspect Matched Adaptive Radar (RAMAR)*, World Scientific, Singapore, 2012.

Baylis, W.E., *Electrodynamics: A Modern Geometric Approach*, Birkhäuser, Boston, 1999.

Berry, M.V. and Dennis, M.R., Knotted and linked phase singularities in monochromatic waves. *Proc. R. Soc. A*, 457 2251–2263, 2001.

Chappell, J.M., Drake, S.P., Seidel, C.L., Gunn, L.J., Iqbal, A., Allison, A. and Abbott, D., Geometric algebra for electrical and electronic engineers. *Proc. IEEE*, 102, 1340–1363, 2014.

Chisholm, J.S.R. and Common, A.K., *Clifford Algebras and Their Applications in Mathematical Physics*, NATO Advanced Science Institutes Series, Reidel, 1986.

Dennis, M., Tying light in knots. *Phys. Org*, Jan. 17, 2010

Dennis, M.R., King, R.P., Jack, B., O'Holleran, K. and Padgett, M.J., Isolated optical vortex knots. *Nature Phys.*, 6, 118–121, 2010.

Doran, C. and Lasenby, A., *Geometric Algebra for Physicists*, Cambridge Univrsity Press, 2003.

Harris, S.E., Electromagnetically induced transparency. *Phys. Today*, 50(7) 36–42, 1997.

Hestenes, D. and Sobczyk, G., *Clifford Algebra to Geometric Calculus: A Unified Language for Mathematics and Physics*, Reidel, 1984.

Irvine, W.T.M, Linked and knotted beams of light, conservation of helicity and the flow of null electromagnetic fields. *J. Phys. A: Math. Theor.* 43(385203), 1–9, 2010.

Irvine, W.T.M. and Bouwmeister, D., Linked and knotted beams of light. *Nature Phys.*, 4, 716–720, 2008.

Kasapi, A., Jain, M., Yin, G.Y. and Harris, S.E., Electromagnetically induced transparency: propagation dynamics. *Phys. Rev. Lett.*, 74, 2447–2450, 1995.

Kedia, H., Bialynicki-Birula, I., Peralta-Salas, D. and Irvine, W.T.M., Tying knots in light fields. *Phys. Rev. Lett.*, 111, 150404, 2013.

Khoo, I.C. and Wu, S.T., *Optics and Nonlinear Optics of Liquid Crystals*, World Scientific, Singapore, 1993.

Kleckner, D. and Irvine, W.T.M., Creation and dynamics of knotted vortices. *Nature Phys.*, 9, 253–258, 2013.

Kocharovskaya, O.A. and Khanin, Ya. I., Population trapping and coherent bleaching of a three-level medium by a periodic train of ultrashort pulses. *Sov. Phys. JETP*, 63, 945–950, 1986.

Lamb, G.L., Jr., Analytic descriptions of ultrashort optical pulse propagation in a resonant medium. *Rev. Mod. Phys.*, 43, 99, 1971.

Maier, W. and Meier, G., A simple theory of the dielectric characteristics of homogeneous oriented crystalline-liquid phases of the nematic type. *Z. Naturforsch. Teil* A 16, 262 (1961).

Matos, S.A., Ribeiro, M.A. and Paiva, C.R., Anisotropy without tensors: a novel approach using geometric algebra. *Optics Express*, 15, 15175–15186, 2007.

Maxwell, J.C., *A Treatise on Electricity and Magnetism*, Vol. 2, 1873 (Dover, New York, 1954).

McCall, S.L. and Hahn, E.L., Self-induced transparency by pulsed coherent light. *Bull. Am. Phys. Soc.*, 10, 1189, 1965.

McCall, S.L. and Hahn, E.L., Self-induced transparency by pulsed coherent light. *Phys. Rev. Lett.*, 18, 908–911, 1967.

McCall, S.L. and Hahn, E.L., Self-induced transparency. *Phys. Rev.*, 183, 457–485, 1969.

Patel, C.K.N. and Slusher, R.E., Self-induced transparency in gases. *Phys. Rev. Lett.*, 19, 1019–1022, 1967.

Rañada, A.F., A topological theory of the electromagnetic field. *Lett. Math. Phys.*, 18, 97–106. 1989.

Rañada, A.F., Knotted solutions to the Maxwell equations in vacuum. *J. Phys. A: Math. Gen.* 23, L815–L820, 1990.

Rañada, A.F., On the magnetic helicity. *Eur. J. Phys.*, 13, 70–76, 1992a.

Rañada, A.F., Topological electromagnetism. *J. Phys. A: Math. Gen.* 25, 1621–1641, 1992b.

Rañada, A.F., Interplay of topology and quantization: topological energy quantization in a cavity. *Phys. Lett. A*, 310, 434–444, 2003.

Rañada, A.F. and Trueba, J.L., Electromagnetic knots. *Phys. Lett. A*, 202, 337–342, 1995.

Rañada, A.F. and Trueba, J.L., Ball lightning an electromagnetic knot? *Nature*, 383, 32, 1996.

Rañada, A.F. and Trueba, J.L., Two properties of electromagnetic knots. *Phys. Lett.*, 232, 25–33, 1997.

Rañada, A.F. and Trueba, J.L., A topological mechanism of discretization for the electric charge. *Phys. Lett. B*, 422, 196–200, 1998.

Rañada, A.F. and Trueba, J.L., Topological electromagnetism with hidden nonlinearity, in *Modern Nonlinear Optics, Part 3, Second Edition, Advances in Chemical Physics, Vol. 119*, Myron W. Evans (ed), Wiley, 2001.

Thomson, Sir William (Lord Kelvin). "On Vortex Atoms. *Proc. Roy. Soc. Edinburgh* VI, 94–105, 1867.

Wang, H., *Studies of Liquid Crystal Response Time*. Ph.D. dissertation, Department of Electrical and Computer Engineering, University of Central Florida, Orlando, FL, 2005.

Wong, K.C., *Liquid Crystal Overview*, ECE–E443, 2005. Slide Presentation,

www.ingramcontent.com/pod-product-compliance
Lightning Source LLC
LaVergne TN
LVHW052030200926
848481LV00005B/1629